Wavelet Analysis

Wavelet Analysis
Basic Concepts and Applications

Sabrine Arfaoui
University of Monastir, Tunisia
University of Tabuk, Saudi Arabia

Anouar Ben Mabrouk
University of Kairouan, Tunisia
University of Monastir, Tunisia
University of Tabuk, Saudi Arabia

Carlo Cattani
University of Tuscia, Italy

CRC Press
Taylor & Francis Group
Boca Raton London New York

CRC Press is an imprint of the
Taylor & Francis Group, an **informa** business
A CHAPMAN & HALL BOOK

First edition published 2021
by CRC Press
6000 Broken Sound Parkway NW, Suite 300, Boca Raton, FL 33487-2742

and by CRC Press
2 Park Square, Milton Park, Abingdon, Oxon, OX14 4RN

Library of Congress Cataloging-in-Publication Data

Names: Arfaoui, Sabrine, author. | Ben Mabrouk, Anouar, author. | Cattani, Carlo, 1954- author.
Title: Wavelet analysis : basic concepts and applications / Sabrine Arfaoui, University of Monastir, Anouar Ben Mabrouk, University of Kairouan, Carlo Cattani, University of Tuscia.
Description: Boca Raton : Chapman & Hall/CRC Press, 2021. | Includes bibliographical references and index.
Identifiers: LCCN 2020050757 (print) | LCCN 2020050758 (ebook) | ISBN 9780367562182 (hardback) | ISBN 9781003096924 (ebook)
Subjects: LCSH: Wavelets (Mathematics)
Classification: LCC QA403.3 .A74 2021 (print) | LCC QA403.3 (ebook) | DDC 515/.2433--dc23
LC record available at https://lccn.loc.gov/2020050757
LC ebook record available at https://lccn.loc.gov/2020050758

ISBN: 978-0-367-56218-2 (hbk)
ISBN: 978-1-003-09692-4 (ebk)

Typeset in Latin Modern font
by KnowledgeWorks Global Ltd.

Contents

List of Figures

Preface

Nowadays, wavelets are applied almost everywhere in science. Both pure fields, such as mathematics and theoretical physics, and applied ones, such as signal/image processing, finance and engineering, apply wavelets. Although the references and/or the documentation about wavelets and their applications are wide, it seems that with the advancement of technology and the appearance of many phenomena in nature and in life there still exist some places for more efforts and developments to understand the new problems, as the existing wavelet methods do not provide good understanding of them. The new COVID-19 pandemic may be one of the challenges that should be understood.

On the other hand, especially for young researchers, existing references such as books in wavelet theory are somehow very restricted. The majority are written for specific communities. This is, in fact, not surprising and may be due to the necessity of developing such references to overcome the concerned problems in that time.

Next, with the inclusion of wavelet theory in academic studies such as in master's and PhD programs, the scientific and academic communities have had a great need to develop references in other forms. Students and generally researchers need sometimes self-containing references responding to their need, to avoid losing time in redeveloping existing results, which is a necessary step for both the generalization and the experiments.

The present volume is composed of eight chapters. In the first introductory chapter, a literal introduction is developed discussing generally the topic. Chapter 2 is concerned with the presentation of the original developments of wavelet theory on the real Euclidean space. This is also a preliminary chapter that will be of great help for young researchers. Chapter 3 is more specialized and constitutes a continuation of the previous one, in which some extending cases of wavelet theory and applications have been provided. Chapter 4 is a very specialized part that is developed for the first time to our knowledge. It is concerned with the presentation of wavelet theory in a general functional framework based on Clifford algebras. This is very important as these algebras contain all the Euclidean structures and gather them in one structure to facilitate calculus. Readers will notice clearly that Clifford wavelet theory induces naturally the Euclidean ones such as real and complex numbers, circles and spheres. Chapter 5 is a continuation of the development of the theory in specialized fields such as quantum theory. Next, in Chapter 6, statistical application of wavelets has been reviewed. Topics such as density estimation, thresholding concepts, variance and covariance have been detailed. Chapter 7 is devoted to wavelets applied in solving partial differential equations. Recall that this field needs many assumptions on the functional bases applied, especially the explicit form of the basis elements and their regularities. The last chapter is devoted to the link and/or the use of wavelet theory in characterizing fractal and multifractal functions and their application. Each chapter contains a series of exercises and experimentations to help understand the theory and also to show the utility of wavelets.

The present book stems, in fact, from lectures and papers on the topics developed, which have been gathered, re-developed, improved and sometimes completed with necessary missing developments. However, naturally it is not exhaustive and should be always criticized, sometimes corrected and improved by readers. So, we accept and wait for any comments and suggestions.

We also want to stress the fact that we have provided in some chapters, especially those on preliminary concepts that may be useful to young researchers, some exercises and applications that are simple to handle with the aim to help the readers understand the theory. We apologize if there are simpler applications and details that may be more helpful to the readers but that have been left out from inclusion in this book. This, in fact, needs more time and may induce delays in the publication of the book. We hope that with the present form the readers become acquainted with the topics presented.

The aim of this book is to provide a basic and self-contained introduction to the ideas underpinning wavelet theory and its diversified applications. Readers of our proposed book would include master's degree students, PhD students and senior researchers. It may also serve scientists and research workers from industrial settings, where modeling real-world phenomena and data needs wavelets such as finance, medicine, engineering, transport, images and signals. Henceforth, the book will interest practitioners and theorists alike. For theorists, rigorous mathematical developments will be presented with necessary prerequisites that make the book self-containing. For the practitioner, often interested in model building and analysis, we provide the cornerstone ideas.

As with any scientific production and reference, the present volume could not have been realized without the help of many persons. We thus owe thanks to many persons who have helped us in any direction such as encouragements, scientific discussions and documentation. We thank the Taylor & Francis Publishing Group for giving us the opportunity to write and publish the present work. We also would like to express our gratitude to our professors, teachers, colleagues, and universities. Without their help and efforts, no such work might be realized. We would also like to thank all the members of the publishing house, especially the editorial staff for the present volume, Callum Fraser and Mansi Kabra, for their hospitality, cooperation, collaboration and for the time they have spent on our project.

CHAPTER 1

Introduction

Wavelets were discovered in the eighteenth century, essentially in pertroleum extraction. They have induced a new type of analysis extending the Fourier one. Recall that Fourier analysis has been for long a time the essential mathematical tool, especially in harmonic analysis and related applications such as physics, engineering, signal/image processing and PDEs. Next, wavelets were introduced as new extending mathematical tools to generalize the Fourier one and to overcome, in some ways, the disadvantages of Fourier analysis. For a large community, especially non-mathematicians and non-physicists, a wavelet may be defined in the most simple sense as a wave function that decays rapidly and has a zero mean.

Compared to the Fourier theory, wavelets are mathematical functions permitting themselves to cut up data into different components relative to the frequency spectrum and next focus on these components somehow independently, extracting their characteristics and lifting to the original data. One main advantage of wavelets is the fact that they are more able than Fourier modes to analyze discontinuities and/or singularities efficiently (see [20], [177], [218], [248], [279], [295], [305]).

Wavelets have been also developed independently in the fields of mathematics, quantum physics, electrical engineering and seismic geology. Next, interchanges between these fields have yielded more understanding of the theory and more and more bases as well as applications such as image compression, turbulence, human vision, radar and earthquake.

Nowadays, wavelets have become reputable and successful tools in quasi all domains. The particularity in a wavelet basis is that all its elements are deduced from one source function known as the mother wavelet. Next, such a mother gives rise to all the elements necessary to analyze objects by simple actions of translation, dilatation and rotation. The last parameter was introduced by Antoine and his collaborators ([11], [13]) to obtain some directional selectivity of the wavelet transform in higher dimensions. Indeed, unlike Fourier analysis, there are different ways to define multidimensional wavelets. Directional wavelets based on the rotation parameter just evoked. Another class is based on tensor products of one-dimensional wavelets. Also, some wavelets are related to manifolds, essentially spheres, where the idea is based on the geometric structure of the surface where the data lies. This gives rise to the so-called isotropic and anisotropic wavelets (see [11], [12], [20], [205], [218], [219], [220], [229], [230], [248], [279], [295], [305], [357], [358], [363]).

Wavelet theory provides for functional spaces and time series good bases, allowing their decomposition into spaces associated with different horizons known as the levels of decomposition. A wavelet basis is a family of functions obtained from one function known as the mother wavelet, by translations and dilations. Due to the power of their theory, wavelets have many applications in different domains such as mathematics, physics, electrical engineering and seismic geology. This tool permits the representation of L^2-functions in a basis well localized in time and in frequency.

Wavelets are also associated with many special functions such as orthogonal polynomials and hypergeometric series. The most well known may be the Bessel functions that have been developed in both classic theory of Bessel functional analysis and the modified versions in fractional and quantum calculus. As its name indicates, Bessel wavelets are related to Bessel special function. Historically, special functions differ from elementary ones such as powers, roots, trigonometric and their inverses, mainly with the limitations that these latter classes are known for. Many fundamental problems such as orbital motion, simultaneous oscillatory chains and spherical body gravitational potential were not best described using elementary functions. This makes it necessary to extend elementary functions' classes to more general ones that may describe well unresolved problems.

Wavelets are also developed and applied in financial time series such as market indices and exchange rates. In [42], for example, a study of the largest transaction financial market was carried out. The exchange market gave some high-frequency data. Compared to other markets, such data can be available at long periods and with high frequency. The data were detected for very small periods, which means that the market is also liquid. Until 1990, economists were interested in intra-daily data because of which the detection of some behaviors did not appear in the daily analysis of data such as homogeneity.

A well-known hypothesis in finance is the homogeneity of markets where all investigators have almost the same behavior. The idea of nonhomogeneous markets is more recent, and it suggests that investigators have different perceptions and different laws. For the exchange market, for example, investigators can differ in profiles, geographic localizations and also in institutional constraints. Another natural suggestion can be done about traders. Naturally, traders investigating at short time intervals allow some high-frequency behaviors in the change market. Long-time traders are interested in the general tendency and the volatility of the market along a microscopic greed. Short-time traders, however, are interested in fractional perceptions and so in macroscopic greed. This leads to the wavelet analysis of financial time series.

Recently, other models have been introduced in modeling financial time series by means of fractals, which are in turn strongly related to wavelets. For example, in Olsen & Associates, operating the largest financial database, has noticed that the tick frequency has strongly increased in one decade, causing problems in studying the time series extracted from such a database. Such problems can be due to transmission delays, input errors and machine damages. So, some filtering procedure has to be done before using the data. The first point that one must take into account in filtering time series is their scaling behavior. Scaling laws were empirically observed by Olsen et al until 1990 (see [306]). A time series $X(t)$ has a scaling law if its so-called partition function has the form

$$S_q(t) = \frac{t}{N} \sum_{j=1}^{N/t} |X(jt)|^q \sim t^{\tau_q+1}$$

where N stands for the size of the series for some appropriate function τ. This estimation is well understood when merging wavelet tools into fractal models. (See Chapter 8.) In such models, Fourier analysis could not produce good results in estimating such behavior. Indeed, Fourier transform of time series is generally limited because a single analysis window cannot detect features in the signals that are either much longer or shorter than the window size. Moving-window Fourier transform (MWFT) slides a fixed-size analysis window along the time axis and is able to detect non-stationarities. The fixed-size window algorithm of MWFT limits the detection of cycles at wavelengths that are longer than the analysis windows, and non-stationarities in short wavelengths (i.e., high frequencies) are smoothed. Use of the wavelet transform solves this problem, because it uses narrow windows at high frequencies and wide windows at low frequencies.

This book is devoted to developing the basic concepts of wavelet analysis necessary for young researchers doing their Master's level in science and researchers doing doctoral studies in pure mathematical/physical sciences, as well as applied and interacted ones by providing the basic tools required, with simple and rigorous methods. It also aims to serve researchers at advanced levels by providing them the necessary tools that will allow them to understand and adapt wavelet theory to their needs such as supervision and development of research projects.

The book provides some highly flexible methods and ideas that can be manipulated easily by undergraduate students, and thus may be of interest for Bachelors in science by providing them a clear idea on what wavelets are, and thus permitting them to decide in their scientific future.

Organization of the book

Chapter 2 presents the notion of wavelets as analyzing functions and as mathematical tools for analyzing square integrable functions known in signal theory as finite variance and/or finite energy signals. The analysis passes through two essential types of transforms: the continuous wavelet transforms and the discrete wavelet transforms. Such transforms are applied to represent the analyzed signals by means of wavelet series in the time–frequency domain and thus use the modes generating such series to localize singularities. Furthermore, we review multi-resolution analysis as a basic construction tool related to wavelets. It allows to split the whole space of analyzed signals into sub-spaces known as approximation spaces and detail ones. Multi-resolution analysis is based on nested sub-spaces that are related to each other by means of specific algorithms such as the decomposition low-pass filter-based algorithm and the inverse high-pass one. These algorithms are efficient for understanding the behavior of a series and eventual predictions.

Chapter 3 is a mixture of different concepts. We aim precisely to develop it as a continuation of the previous one in order to show some extensions of wavelet theory and also some other point of view to its theoretical introduction, essentially affine group method, and also to show to the readers that although manifolds are parts of Euclidean spaces for many cases, the concept of introducing wavelets on them may differ. We considered the special case of spheres.

Chapter 4 is a new developed academic reference devoted to Clifford wavelet analysis. It offers a general context of Euclidean wavelet analysis by a higher-dimensional analogue. The notion of monogenic function theory is reviewed: monogenic polynomials and their application to yield Clifford wavelets. Mathematical formulations of harmonic analysis such as Fourier–Plancherel and Parseval are established in the new context. Applications in image processing are also developed. Readers will notice clearly that Clifford wavelet theory induces naturally the Euclidean ones such as real and complex circles and spheres.

Chapter 5 is a continuation of the development of the theory in specialized fields such as quantum theory. We precisely present in detail quantum and fractional Bessel functions and associated wavelet theory. Plancherel/Parseval as well as reconstruction formula has been investigated. Bessel wavelets are applied in various domains, especially partial differential equations, wave motion, diffusion, etc.

Next, in Chapter 6, statistical application of wavelets is reviewed. Density estimation, thresholding concepts, and variance and covariance topics are discussed in detail. Recall that statistical and time series constitute a very delicate area of study due to specific characteristics. Most of the time-varying series are nonlinear, in particular, the financial and economic series, which present an intellectual challenge. Their behavior seems to change dramatically, and uncertainty is always present. To understand and to discover hidden characteristics and behavior of these series, wavelet theory has been proved to be useful and necessary compared to existing previous tools in statistics.

Chapter 7 is devoted to wavelets applied in solving partial differential equations. Recall that this field needs many assumptions on the functional bases applied, especially the explicit form of the basis elements and their regularities. We propose to show the contribution of wavelet theory in PDEs solving.

The idea is generally not complicated, and it consists essentially of developing the unknown solution of the PDE into its eventual wavelet series decomposition (and its derivatives included in the PDE) and then using the concept of wavelet basis to obtain algebraic and/or matrices/vector equations on the wavelet coefficients that should be resolved. The crucial point in this theory is the so-called connection coefficients of wavelets. These have been investigated by many authors. In the present work, we develop a somehow new procedure to compute them. Some applications are also provided at the end of the chapter.

The last chapter is devoted to the link and/or the use of wavelet theory in characterizing fractal and multifractal functions and their applications. Concepts such as Hölder regularity, spectrum of singularity, multifractal formalism for function and self-similar-type functions based on wavelets are discussed with necessary developments.

CHAPTER 2

Wavelets on Euclidean Spaces

2.1 INTRODUCTION

Wavelets were discovered in the eighteenth century in petroleum exploration, and since their discovery, they have proven to be powerful tools in many fields from pure mathematics to physics to applied ones such as images, signals, medicine, finance and statistics. The study of their constructions and their properties, especially in functional/signals decompositions on functional wavelet bases, has indeed grown considerably.

In pure mathematics, wavelets constitute a refinement of Fourier analysis as they compensate and/or resolve some anomalies in Fourier series. The first wavelet basis has been, in fact, used before the pure mathematical discovery of wavelet bases since the introduction of Haar system, which by the next has been proved indeed to be a possible wavelet basis reminiscent of the regularity. Such a system dates back to the beginning of the 20th century and was precisely discovered in the year 1909. It was introduced in order to construct a functional basis permitting the representation of all continuous functions by means of a uniformly convergent series. Recall that in Fourier series, there are, as usual, many kinds of convergence that may be investigated such as the point-wise convergence subject of the well-known Dirichlet theorem, the uniform convergence which needs more assumptions on the series and the function, the convergence in norm and Carleson's or almost everywhere convergence. Each kind of convergence requires special assumptions on the function. Although a Fourier series converges in sense of Dirichlet, it does not imply that the graph of the partial Fourier series converges to that of the function. This phenomenon is known as the Gibbs phenomenon and is related to the presence of oscillations in the Fourier series near the discontinuity points of the function. This means that the uniform convergence is not sufficient. One of the challenging concepts in wavelet analysis is its ability to describe well the behavior of the analyzed function near its singularities and join or more precisely extend the notion of Littlewood-Palay decomposition.

To overcome some drawbacks of Fourier analysis, mathematicians have introduced a bit of modification called the windowed Fourier transform by computing the original Fourier transform of the analyzed signals on a special localized extra function called the window. However, some situations remain non resolved especially with the emergence of irregular signals or high-frequency variations. The major problem in the windowed Fourier extension is due to the use of fixed window, which may not be well adapted to other problems such as high fluctuations of non stationary signals. This led researchers to think about a stronger tool taking into account nonlinear algorithms, nonstationary signals, nonperiodical, volatile and/or fluctuated ones. It holds that wavelets, since their discovery, have permitted to overcome these obstacles.

These powers are related simultaneously to many properties of wavelets. Indeed, wavelet decomposition of functions joins Littlewood-Paley decomposition in many cases. Wavelets provide simultaneous local analyses related to time–frequency. They can be adapted to study-specific operators, especially differential and stochastic ones. From the numerical and/or applied point of view, wavelets provide fast and accurate algorithms, multi-resolution analyses as well as recursive schemes. These are very important especially in big data analysis, image processing and also in numerical resolution of partial differential equations.

In this chapter, we propose to review the basic concepts of wavelets as well as their basic properties.

2.2 WAVELETS ON $\mathbb{R}$

The first wavelet bases and thus analyses have been constructed on the real line $\mathbb{R}$ and have been next extended to the general cases of the real/complex Euclidean spaces $\mathbb{R}^m$ -$\mathbb{C}^m$ using different methods such as the natural tensor product.

Mathematically speaking, a wavelet or an analyzing wavelet on the Euclidean space $\mathbb{R}^m$ may be defined in a large way as a function with specific properties that may or may not be required necessarily as simultaneous assumptions. More precisely, wavelet analysis is based primarily on the following points:

- An effective representation for standard functions,

- Robustness to the specification models,

- A reduction in the computation time,

- Simplicity of the analysis,

- An easy generalization and efficient, depending on the dimension,

- A location in time and frequency.

A wavelet is a function $\psi \in L^2(\mathbb{R})$ that satisfies the following conditions:

- Admissibility,

$$\int_{\mathbb{R}+} |\hat{\psi}(\omega)|^2 \frac{d\omega}{|\omega|} = C_\psi < \infty. \tag{2.1}$$

- Zero mean,

$$\hat{\psi}(0) = \int_{-\infty}^{+\infty} \psi(u)du = 0. \tag{2.2}$$

- Localization in time/frequency domains

$$\widehat{\psi}(0) = \int_{-\infty}^{+\infty} |\psi(u)|^2 du = 1. \tag{2.3}$$

- Enough vanishing moments,

$$p = 0, ..., m - 1, \quad \int_{\mathbb{R}} \psi(t) t^p dt = 0. \tag{2.4}$$

To analyze a signal by wavelets, one passes as in Fourier analysis by the wavelet transform of the signal. A wavelet transform (WT) is a re representation of a time–frequency signal. It replaces the Fourier sine by a wavelet. Generally, there are two types of processing: the continuous wavelet transform (CWT) and the discrete wavelet transform.

2.2.1 Continuous wavelet transform

The CWT is based firstly on the introduction of a translation parameter $u \in \mathbb{R}$ and another parameter $s > 0$ known as the scale to the analyzing wavelet ψ, which plays the role of Fourier sine and cosine and will be subsequently called mother wavelet. The translation parameter determines the position or the time around which we want to assess the behavior of the signal, while the scale factor is used to assess the signal behavior around the position. That is, it allows us to estimate the frequency of the signal at that point. Let

$$\psi_{s,u}(x) = \frac{1}{\sqrt{s}} \psi\left(\frac{x-u}{s}\right). \tag{2.5}$$

The CWT at the position u and the scale s is defined by

$$d_{u,s}(f) = \int_{-\infty}^{\infty} \psi_{u,s}(t) f(t) dt, \quad \forall\, u, s. \tag{2.6}$$

By varying the parameters s and u, we may cover completely all the time–frequency plane. This gives a full and redundant representation of the whole signal to be analyzed (see [295]). This transform is called continuous because of the nature of the parameters s and u that may operate at all levels and positions. The original signal S can be reproduced knowing its CWT by the following relationship:

$$S(t) = \frac{1}{C_\psi} \int \int_{\mathbb{R}} d_{u,s}(S) \psi\left(\frac{x-u}{s}\right) \frac{dsdu}{s^2}. \tag{2.7}$$

It remains to notice that CWT is suitable for continuous time signals and those representing varying singularities.

The function $\psi \in L^2(\mathbb{R})$ satisfies some conditions such as the admissibility condition and somehow describes Fourier-Plancherel identity and says that

$$\int_{\mathbb{R}^+} |\widehat{\psi}(\omega)|^2 \frac{d\omega}{|\omega|} = C_\psi < \infty. \tag{2.8}$$

The function ψ has to also satisfy a number of vanishing moments, which is related in wavelet theory to its regularity order. It states that

$$p = 0, ..., m - 1, \quad \int_{\mathbb{R}} \psi(t) t^p dt = 0. \tag{2.9}$$

Sometimes, we say that ψ is C^m on $\mathbb{R}$. The time localization chart is a normalization form that is resumed in the identity

$$\int_{-\infty}^{+\infty} |\psi(u)|^2 du = 1. \tag{2.10}$$

To analyze a signal by wavelets, one passes via the so-called wavelet transforms. A wavelet transform is a representation of the signal by means of an integral form similar to Fourier one in which the Fourier sine and/or cosine is replaced by the analyzing wavelet ψ. In Fourier transform, the complex exponential source function yields the copies $e^{is\cdot}$ indexed by $s \in \mathbb{R}$, which somehow represent frequencies. This transform is continuous in the sense that it is indexed on the whole line of indices $s \in \mathbb{R}$.

In wavelet theory, the situation is more unified. A CWT is also well known. Firstly, a frequency, scale or a dilation or compression parameter $s > 0$ and a second one related to time or position $u \in \mathbb{R}$ have to be fixed. The source function ψ known as the analyzing wavelet is next transformed to yield some copies (replacing the $e^{is\cdot}$):

$$\psi_{s,u}(x) = \frac{1}{\sqrt{s}} \psi \left(\frac{x - u}{s} \right). \tag{2.11}$$

The CWT of a real-valued function f defined on the real line at the position u and the scale s is defined by

$$d_{s,u}(f) = \int_{-\infty}^{\infty} f(t) \psi_{s,u}(t) dt, \quad \forall u, s. \tag{2.12}$$

By varying the parameters s and u, we cover completely all the time–frequency plane. This gives a full and redundant representation of the whole signal to be analyzed (see [295]). This transform is called continuous because of the nature of the parameters s and u that may operate at all levels and positions.

So, wavelets operate according two parameters: the parameter u, which permits translation of the graph of the source mother wavelet ψ, and the parameter s, which permits compression or dilation of the graph of ψ. Computing or evaluating the coefficients $d_{u,s}$ means analyzing the function f with wavelets.

Theorem 2.1 *The wavelet transform $d_{s,u}(f)$ possesses some properties such as*

1.   The linearity, in the sense that

$$d_{s,u}(\alpha f + \beta g) = \alpha d_{s,u}(f) + \beta d_{s,u}(g), \quad \forall f, g.$$

2. *The translation-invariance, in the sense that*

$$d_{s,u}(\tau_t f) = d_{s,u-t}(f), \forall f; \text{ and } \forall u, s, t,$$

and where

$$(\tau_t f)(x) = f(x - t).$$

3. *The dilation-invariance, in the sense that*

$$d_{s,u}(f_a) = \frac{1}{\sqrt{a}} d_{as,au}(f), \forall f; \text{ and } \forall u, s, a,$$

and where for $a > 0$,

$$(f_a)(x) = f(ax).$$

The proof of these properties is easy and the readers may refer to [163] for a review.

In wavelet theory, as in Fourier analysis theory, the original function f can be reproduced via its CWT by an L^2-identity.

Theorem 2.2 *For all $f \in L^2(\mathbb{R})$, we have the L^2-equality*

$$f(x) = \frac{1}{C_\psi} \int \int d_{s,u}(f) \psi(\frac{x-u}{s}) \frac{dsdu}{s^2}.$$

The proof of this result is based on the following lemma:

Lemma 2.3 *Under the hypothesis of Theorem (2.2), we have*

$$\int \int d_{s,u}(f) \overline{d_{s,u}(g)} \frac{dsdu}{s} = C_\psi \int f(x) \overline{g(x)} dx, \ \forall \ f, g \in L^2(\mathbb{R}).$$

Proof. We have

$$d_{s,u}(f) = f * \psi_s(u) = \frac{1}{\sqrt{s}} \int f(x) \psi(\frac{x-u}{s}) dx = \frac{1}{2\pi} \{ (\hat{f}(y) \overline{\hat{\psi}(sy)} e^{-iuy}).$$

Consequently

$$\int_u d_{s,u}(f) \overline{d_{s,u}(g)} du = \frac{1}{2\pi} \int_y \hat{f}(y) \overline{\hat{g}(y)} |\hat{\psi}(sy)|^2 dy.$$

By application of Fubini's rule, we get

$$\int_{s>0} \int_u d_{s,u}(f) \overline{d_{s,u}(g)} \frac{dsdu}{s} = \frac{1}{2\pi} \int_{s>0} \int_y \hat{f}(y) \overline{\hat{g}(y)} |\hat{\psi}(sy)|^2 \frac{dsdy}{s}$$

$$= \frac{1}{2\pi} C_\psi \int_y \hat{f}(y) \overline{\hat{g}(y)} dy$$

$$= C_\psi \int_y f(y) \overline{g(y)} dy.$$

Proof of Theorem 2.2. By applying the Riesz rule, we get

$$||F(x) - \frac{1}{C_\psi} \int_{1/A \le a \le A} \int_{|b| \le B} C_{a,b}(F) \psi(\frac{x-b}{a}) \frac{dadb}{a^2}||_{L^2}$$

$$= \sup_{||G||=1} \left(\int F(x) - \frac{1}{C_\psi} \int_{1/A \le a \le A} \int_{|b| \le B} C_{a,b}(F) \psi(\frac{x-b}{a}) \frac{dadb}{a^2} \right) \overline{G(x)} dx.$$

Next, using Fubini's rule, we observe that the last line is equal to

$$\sup_{||G||=1} \left(\int F(x)\overline{G(x)} dx - \frac{1}{C_\psi} \int_{1/A \le a \le A} \int_{|b| \le B} C_{a,b}(F) \overline{C_{a,b}(G)} \frac{dadb}{a} \right)$$

$$= \sup_{||G||=1} \frac{1}{C_\psi} \int_{(a,b) \notin [1/A,A] \times [-B,B]} C_{a,b}(F) \overline{C_{a,b}(G)} \frac{dadb}{a}$$

which by Cauchy–Schwarz inequality is bounded by

$$\frac{1}{C_\psi} \left[\int_{(a,b) \notin [1/A,A] \times [-B,B]} |C_{a,b}(F)|^2 \frac{dadb}{a} \right]^{1/2}$$

$$\left[\sup_{||G||=1} \int_{(a,b) \notin [1/A,A] \times [-B,B]} |C_{a,b}(G)|^2 \frac{dadb}{a} \right]^{1/2}.$$

Now, Lemma 2.3 shows that the last quantity goes to 0 as A, B tends to $+\infty$.

2.2.2 Discrete wavelet transform

To analyze statistical series or discrete time signals and avoid redundancy problems and integrals calculations appearing in the CWT, one makes use of the discrete wavelet transform. It is to restrict to discrete calculations grids for scale parameters and position instead of browsing the entire domain. The most used method is dyadic grid based on taking $s = 2^{-j}$ and $u = k2^{-j}$. In this case, the wavelet copy $\psi_{u,s}$ will be denoted by $\psi_{j,k}$ and defined by

$$\psi_{j,k}(x) = 2^{-j/2} \psi(2^j x - k), \quad j, k \in \mathbb{Z}.$$

The discrete wavelet transform will be defined by

$$d_{j,k} = \int_{-\infty}^{\infty} \psi_{j,k}(t) S(t) dt. \tag{2.13}$$

These are often called wavelet coefficients or detail coefficients of the signal S.

It holds that the set $(\psi_{j,k})_{j,k \in \mathbb{Z}}$ constitutes an orthonormal basis of $L^2(\mathbb{R})$ and it is called wavelet basis. A signal S with finite energy (in $L^2(\mathbb{R})$) is then decomposed according to this basis into a series

$$S(t) = \sum_{j=0}^{\infty} \sum_k d_{j,k} \psi_{j,k}(t) \tag{2.14}$$

called the wavelet series of S which replaces the reconstruction formula for the CWT.

2.3   MULTI-RESOLUTION ANALYSIS

Multi-resolution analysis (MRA) is a functional framework for representing a series of approximations to different levels called resolutions. MRA is a family of vector spaces $(V_j)_{j\in\mathbb{Z}}$ of the whole space of finite variance (energy) signals $L^2(\mathbb{R})$ nested in the sense of a scaling law. Recall that

$$L^2(\mathbb{R}) = \{S : \mathbb{R} \longrightarrow \mathbb{R}; \int_R |S(t)|^2 dt < \infty\}. \tag{2.15}$$

For each $j \in \mathbb{Z}$, V_j is called the approximation space at the scale or the level j. More precisely, we have the following definition ([295]).

Definition 2.4 *A multiresolution analysis is a sequence of closed subsets $(V_j)_{j\in\mathbb{Z}}$ of $L^2(\mathbb{R})$ that satisfies the following points.*

a) $\forall j \in \mathbb{Z}; V_0 \subset V_1 \subset .... \subset V_j \subset V_{j+1}.$

b) $\forall j \in \mathbb{Z}; f \in V_j \Leftrightarrow f(2.) \in V_{j+1}$

c) *There exists $\varphi \in V_0$ such that $\{\varphi_{0,k} = \varphi(. - k); k \in \mathbb{Z}\}$ is a Riesz basis of V_0.*

d) $\bigcap_{j\in\mathbb{R}} V_j = \{0\}.$

e) $\overline{\bigcup_{j\in\mathbb{R}} V_j} = L^2(\mathbb{R}).$

f) $\forall j \in \mathbb{Z}; f \in V_j \Longleftrightarrow f(x - k) \in V_j$

The property (a) reflects that the approximation of a signal at the resolution's level $j + 1$ contains the necessary information to yield the approximation at the level j. Assertion (b) called dilatation's property permits passing from a level of resolution to another. The space V_{j+1} contains signals that are coarser than V_j. Assertion (c) means that a scaling function exists and permits the decomposition of the signal at the starting level 0. The property (d) means that at very low resolution level ($2^{-j} \longrightarrow 0$ as $j \longrightarrow +\infty$) we lose all the details of the signal. At a minimal resolution, we lose all the information about the signal. Assertion (e) implies that the signal may be approximated with elements in V_j. At a maximal resolution, we reconstruct all the whole signal. Finally, the property (f) of translation means that the space V_j is invariant under integer translation.

Definition 2.5 *The source function φ is called the scaling function of the MRA or also the father wavelet.*

It holds that this function generates all the subspaces V_j's of the MRA by acting as dilation/translation parameters. Indeed, the property (d) combined with (f) in Definition 2.4 implies that, for all $j, \in \mathbb{Z}$, the set $\left(\varphi_{j,k}(x) = 2^{j/2}\varphi(2^j x - k)\right)_k$ is an orthogonal basis of V_j. From assertion (a), it holds that we may complete V_j in V_{j+1}

in the sense of the direct sum. Let W_j be the orthogonal supplementary of V_j in V_{j+1}, that is

$$V_{j+1} = V_j \oplus^\perp W_j. \tag{2.16}$$

We will see that W_j plays a primordial role in representing the details of the analyzed signal. This is why it is called the detail space at the level j. Iterating the relation (2.16), we obtain an orthogonal decomposition for all $J \in \mathbb{Z}$,

$$V_J = V_0 \oplus^\perp \bigoplus_{j=0}^{J,\perp} W_j. \tag{2.17}$$

By exploiting (c), this leads to

$$L^2(\mathbb{R}) = \bigoplus_{j \in Z} W_j. \tag{2.18}$$

$L^2(\mathbb{R})$ is decomposed into subspaces that are mutually orthogonal.

Definition 2.6 *The space W_j, $j \in \mathbb{Z}$, is called detail space at the scale or the level j.*

In wavelet theory, the following result is proved.

Theorem and Definition 2.7 *There exists a function $\psi \in W_0$ that satisfies*

- $(\psi(t - k))_k$ *is an orthogonal basis of W_0.*

- $\left(\psi_{j,k}(t) = 2^{j/2}\psi(2^j t - k)\right)_k$ *is an orthogonal basis of W_j.*

The function ψ is called the wavelet function or the mother wavelet associated with the scaling function φ of the MRA.

We will now explain the relationship between φ and ψ. It holds from (2.18) that any $S \in L^2(\mathbb{R})$ is decomposed into components according to the subspaces W_j's in the sense that

$$S = \sum_{j=-\infty}^{+\infty} S_{W_j},$$

where the S_{W_j} for $j \in \mathbb{Z}$ designates the orthogonal projection of S on W_j. Recall next that W_j is generated by the orthogonal basis $\left(\psi_{j,k}(t) = 2^{j/2}\psi(2^j t - k)\right)_k$ for any $j \in \mathbb{Z}$. Consequently,

$$S_{W_j} = \sum_k <S, \psi_{j,k}> \psi_{j,k}.$$

Which leads to

$$S = \sum_{j=-\infty}^{+\infty} \sum_k <S, \psi_{j,k}> \psi_{j,k}.$$

Observing now that the wavelet coefficients are already defined by (2.12) or (2.13) as

$$d_{j,k}(S) = \int_{-\infty}^{+\infty} S(t)\psi_{j,k}(t)dt = <S, \psi_{j,k}>$$

we obtain the wavelet series of S defined already by (2.14). Returning next to (2.16) or also (2.17), we may write that

$$S = \sum_{j=-\infty}^{-1} \sum_k \, < S, \psi_{j,k} > \psi_{j,k} + \sum_{j=0}^{+\infty} \sum_k \, < S, \psi_{j,k} > \psi_{j,k}. \tag{2.19}$$

It results from the nesting property of the V_j's and the fact that $W_j \subset V_{j+1}$ that the first part in (2.19) is an element of V_0 which is generated by the basis $(\varphi_k(t) = \varphi(t-k))_k$. Consequently, we may also write that

$$\sum_{j=-\infty}^{-1} \sum_k \, < S, \psi_{j,k} > \psi_{j,k} = \sum_k C_k \varphi_k.$$

Hence, the wavelet series decomposition of S becomes

$$S(x) = \sum_k C_k \varphi_k(x) + \sum_{j=0}^{\infty} \sum_k d_{j,k} \psi_{j,k}(x). \tag{2.20}$$

In fact, it may truncate at any level J and apply the $\varphi_{J,k}$ instead of φ_k to obtain

$$S(x) = \sum_k C_{J,k} \varphi_{J,k}(x) + \sum_{j=J}^{\infty} \sum_k d_{j,k} \psi_{j,k}(x) \tag{2.21}$$

which is known as the wavelet series decomposition of S at the level J. In fact, the coefficients that appears in this decomposition are evaluated via the relation

$$C_{J,k} = \int_{-\infty}^{+\infty} S(t) \varphi_{J,k}(t) dt$$

and are called the scaling or approximation coefficients of S. The first component in V_J reflects the global behavior or the tendency of the whole signal S and the second component relative to the $d_{j,k}$'s represents the details of S and thus reflects the dynamic behavior of the signal.

2.4   WAVELET ALGORITHMS

The strongest point in MRA and wavelet theory is that the scaling function and the analyzing wavelet lead each one to the other. Indeed, recall that φ belongs to $V_0 \subset V_1$ and the latter is generated by the basis $(\varphi_{1,k})_k$. Hence, φ is expressed by means of $(\varphi_{1,k})_k$. More precisely, we have the following result.

Theorem and Definition 2.8   • *The scaling function satisfies the so-called two-scale relation*

$$\varphi(x) = \sqrt{2} \sum_k h_k \varphi(2x - k) \tag{2.22}$$

where the coefficients h_k are

$$h_k = \int_{\mathbb{R}} \varphi(x) \overline{\varphi(2x - k)} dx.$$

- *The mother wavelet ψ is expressed as*

$$\psi(x) = \sqrt{2} \sum_k g_k \varphi(2x - k)$$

where the g_k's are evaluated by

$$g_k = (-1)^k h_{1-k}.$$

For more details, we refer to [177], [236], [295]. We will now prove that these relations allow obtaining all the decomposition of a signal from each other through specific algorithms. Indeed, consider a signal S and its approximation coefficients $C_{j,k}$ and details $d_{j,k}$. For $j \in \mathbb{Z}$, we have

$$C_{j,k} = \langle S, \varphi_{j,k} \rangle.$$

It follows from the two-scale relation (2.22) that

$$\varphi_{j,k} = \sum_l h_l \varphi_{j+1,l+2k}.$$

Hence,

$$C_{j,k} = \sum_l h_l C_{j+1,2k+l}.$$

This means that the approximation at level j is obtained from the level $j+1$ by the intermediate of a filter. We have, in fact, the following definition.

Definition 2.9 *The sequence $H = (h_k)_k$ is called the discrete low-pass filter.*

Analogously, we have for j, k fixed,

$$\psi_{j,k}(x) = \sum_l g_l \varphi_{j+1,l+2k}.$$

Hence,

$$d_{j,k} = \langle S, \psi_{j,k} \rangle = \sum_l g_l C_{j+1,l+2k}.$$

This means that the approximation at level j is obtained from the level $j+1$ by the intermediate of a filter. We have here also the following definition.

Definition 2.10 *The sequence $G = (g_k)_k$ is called the discrete high-pass filter.*

Figure 2.1 presents the decomposition algorithm due to Mallat [295]. It consists of a cascade algorithm permitting to obtain all the levels of resolution.

We will now explain the inverse algorithm which permits to obtain the level $j+1$ from the level j. Recall that the orthogonal projection of S on the approximation space V_j is given by (Hardle et al (1997) [236])

$$S_{V_j} = \sum_k C_{j,k} \varphi_{j,k} + \sum_k d_{j,k} \psi_{j,k}$$

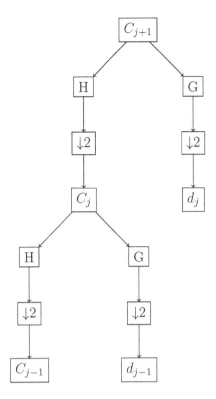

Figure 2.1 Decomposition algorithm.

and that

$$
\begin{aligned}
C_{j+1,k} &= \ <S_{V_j}, \varphi_{j+1,k}> \\
&= \ \sum_k C_{j,k} <\varphi_{j,k}, \varphi_{j+1,k}> + \sum_k d_{j,k} <\psi_{j,k}, \varphi_{j+1,k}>
\end{aligned}
$$

or else

$$
C_{j+1,n} = \sum_k h_{n-2k} C_{j,m} + \sum_k g_{n-2k} d_{j,m}. \tag{2.23}
$$

This reconstruction is illustrated by means of Figure 2.2.

Finally, we recall some properties related to the filters H and G ([295]).

Proposition 2.11 *The filters* H *and* G *satisfy*

1. $\sum_n h_n h_{n+2j} = 0; \qquad \forall j \neq 0.$

2. $\sum_n h_n^2 = 1.$

3. $g_n = (-1)^n h_{l-n}; \ \forall n.$

4. $\sum_n h_n g_{n+2j} = 0, \ \forall j \in \mathbb{N}.$ *(called mutual orthogonality).*

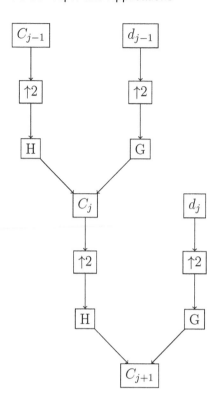

Figure 2.2 Inverse algorithm.

2.5 WAVELET BASIS

Providing functional spaces with bases is the most important task in functional analysis. It permits in some sense to reduce the problem of proving properties and/or demanding characteristics from a function—usually unknown—to doing it and/or demanding it from the elements of the basis. In finite dimensional spaces, this is somehow equivalent. In general and especially for almost all functional spaces where the dimension is usually infinite, we intend that the basis satisfy more properties to permit the reduction.

In wavelet bases, for example, one of the important properties required is the localization and regularity of the mother wavelet ψ, which is obviously inherited by all the elements $\psi_{j,k}$.

Wavelet bases may also be related to the concept of multiresolution analysis where an important property is also required and looks like the localization and consists of the existence of Riesz basis to define a multiresolution analysis. Recall that in a separable Hilbert space, a countable collection $(e_k)_{k \in \mathbb{Z}}$ is said to be a Riesz basis if all its finite linear combinations are dense in the whole Hilbert space and in addition if it satisfies an equivalence inequality of the form

$$K_1 \sum_k |\lambda_k|^2 \leq \| \sum_k \lambda_k e_k \|^2 \leq K_1 \sum_k |\lambda_k|^2$$

for some positive constants $K_1 \leq K_2$.

The following result provides a characterization of the scaling function—father wavelet—of a multiresolution analysis to be effectively a good candidate.

Proposition 2.12 *Let $\varphi \in L^2(\mathbb{R})$ and denote $\Gamma_\varphi(\xi) = \sum_{k \in \mathbb{Z}} |\widehat{\varphi}(\xi + 2k\pi)|^2$. Then, the collection $(\varphi(x - k))_{k \in \mathbb{Z}}$ is a Riesz basis for V_0 if and only if*

$$0 < \inf \Gamma_\varphi \leq \Gamma_\varphi(\xi) \leq \sup \Gamma_\varphi < \infty, \ \forall \xi.$$

Proof 2.1 *Assume that the collection $(\varphi(x - k))_{k \in \mathbb{Z}}$ is a Riesz basis for V_0 and let $(\lambda_k)_{k \in \mathbb{Z}} \in \ell^2(\mathbb{R})$. Denote $F_\varphi(x) = \sum_{k \in \mathbb{Z}} \lambda_k \varphi(x - k)$. Its Fourier transform is $\widehat{F_\varphi}(\xi) = \sum_{k \in \mathbb{Z}} \lambda_k e^{ik\xi} \widehat{\varphi}(\xi)$. We next have*

$$
\begin{aligned}
\|\widehat{F_\varphi}\|_{L^2(\mathbb{R})}^2 &= \int_{\mathbb{R}} |\sum_k \lambda_k e^{ik\xi} \widehat{\varphi}(\xi)|^2 d\xi \\
&= \sum_{l \in \mathbb{Z}} \int_{2\pi l}^{2\pi(l+1)} |\sum_k \lambda_k e^{ik\xi} \widehat{\varphi}(\xi)|^2 d\xi \\
&= \sum_{l \in \mathbb{Z}} \int_0^{2\pi} |\sum_k \lambda_k e^{ik\xi} \widehat{\varphi}(\xi + 2\pi l)|^2 d\xi \\
&= \int_0^{2\pi} |\sum_k \lambda_k e^{ik\xi}|^2 (\sum_{l \in \mathbb{Z}} |\widehat{\varphi}(\xi + 2\pi l)|^2) d\xi.
\end{aligned}
$$

As a result,

$$\|F_\varphi\|_{L^2(\mathbb{R})}^2 = \frac{1}{2\pi} \|\widehat{F_\varphi}\|_{L^2(\mathbb{R})}^2 = \frac{1}{2\pi} \int_0^{2\pi} |\sum_k \lambda_k e^{ik\xi}|^2 \Gamma_\varphi(\xi) d\xi.$$

Now, observing that the map $\Phi : \lambda = (\lambda_k) \to \Phi(\lambda) = \frac{1}{\sqrt{2\pi}} \sum \lambda_k e^{ik\xi}$ is an isomorphism from $\ell^2(\mathbb{R})$ to $L^2([0, 2\pi])$, to get the proposition, the map $\mathcal{K}$ defined on $L^2([0, 2\pi])$ by $\mathcal{K}(f) = f \sqrt{\Gamma_\varphi}$ should also be an isomorphism. This is the case if and only if Γ_φ and $\frac{1}{\Gamma_\varphi}$ are both bounded (in $L^\infty([0, 2\pi])$).

Sometimes we need more regularity on the wavelet basis especially in the case of partial differential equations. Besides, we sometimes need compactly supported wavelets and/or wavelets with fast decay which are more adaptable and suitable for boundary and/or limit conditions due to partial differential equations.

It is sometimes necessary to assume that the scaling function φ and its derivatives to an order r (usually the order of the MRA) are of fast decay. This permits to avoid the assumption on the overlap function Γ_φ of φ to be upper bounded, as this characterization becomes already holding. Indeed, whenever φ is of fast decay, its Fourier transform $\widehat{\varphi}$ is therefore infinitely continuously differentiable and is also of fast decay. It satisfies, in particular,

$$|\widehat{\varphi}(\xi)| \leq \frac{K}{1 + |\xi|}$$

for some constant K. This leads to an upper-bounded overlap function Γ_φ. As a consequence of these facts, we may assume that $\Gamma_\varphi \equiv 1$. This results in a correspondence between $\ell^2(\mathbb{R})$, which are r-order fast decay, and the elements of V_0, which are fast decay, as well as their derivatives to the order r.

In some other applications, one may need more assumptions on the wavelet basis, especially in numerical analysis where the exponential decay behavior is more suitable than just fast decay. It holds, in fact, that whenever the sequence $(\lambda_k)_k$ is exponentially decaying in $\ell^2(\mathbb{R})$, the associated function F_φ is of exponential decay in V_0. More generally, we may prove the following result.

Lemma 2.13 Let $m(\omega) = \sum_k \lambda_k e^{ik\omega}$. Then, m is analytic in some annulus $|Im\,\omega| < R$ (for some positive constant R) if and only if $(|\lambda_k|)_K$ has an exponential decay.

Proof 2.2 *Assume that the trigonometric series m is analytic on the annulus $|Im\,\omega| < R$. Observe next that m is also periodic with period 2π. As a consequence, the function $\omega \longmapsto m(\omega + 2\pi) - m(\omega)$ is vanishing on all the annulus. Now applying the complex integral, we get*

$$\lambda_k = \int_0^{2\pi} m(\omega) e^{-ik\omega} d\omega = \int_{-iR}^{-iR+2\pi} m(z) e^{-ikz} dz$$

Observe next that

$$2\pi \sup_{z \in [-iR, iR+2\pi]} |m(z)| e^{-kImz} \leq C e^{-kR}.$$

Consequently,

$$|\lambda_k| \leq C e^{-kR}.$$

We will now show that in Definition 2.4 some properties and/or assumptions may be deduced from others, which thus reduces the number of assumptions and gives more flexibility in the introduction of multiresolution analyses. We have precisely the following result.

Properties 2.14 *Assume that assertions (a), (b) and (c) in Definition 2.4 hold and denote Φ, the function defined by its Fourier transform*

$$\widehat{\Phi}(\omega) = \widehat{\varphi}(\omega) \sum_k |\widehat{\varphi}(\omega + 2k\pi)|^2)^{-\frac{1}{2}}.$$

It holds that

$$\int_{-\infty}^{+\infty} \varphi(x) dx = 1 \text{ and } \sum_k \varphi(x - k) = 1.$$

Moreover, assertions (d), (e), and (f) in Definition 2.4 hold.

Proof 2.3 *Observe from assertion (a) that*

$$\varphi(x) = \sum_k h_{j,k} 2^{\frac{j}{2}} \varphi(2^j x - k),$$

where the coefficients $h_{j,k}$ are the analogues of the filter coefficients h_k in the two-scale relation. More precisely,

$$h_{j,k} = \int_{-\infty}^{+\infty} \varphi(x) 2^{\frac{1}{2}} \varphi(2^j x - k) dx.$$

Denote next $I_\varphi = \int_{+\infty}^{+\infty} \varphi(x) dx$. We may write

$$
\begin{aligned}
\left| h_{j,k} - I_\varphi 2^{-\frac{j}{2}} \varphi(\frac{k}{2^j}) \right| &= \left| h_{j,k} - \int_{+\infty}^{+\infty} \varphi(\frac{k}{2^j}) 2^{\frac{j}{2}} \varphi(2^j x - k) dx \right| \\
&\leq C \int_{+\infty}^{+\infty} |x - \frac{k}{2^j}| 2^{\frac{j}{2}} |\varphi(2^j x - k)| dx \\
&\leq C 2^{-3\frac{j}{2}} \int_{+\infty}^{+\infty} |x| |\varphi(x)| dx \\
&\leq C 2^{-3\frac{j}{2}}.
\end{aligned}
$$

As a result,

$$h_{j,k} = I_\varphi 2^{-\frac{j}{2}} \varphi(\frac{k}{2^j}) + O(2^{-3\frac{j}{2}}).$$

Consequently,

$$\varphi(x) = I_\varphi \sum_k \varphi(\frac{k}{2^j}) \varphi(2^j x - k) + O(2^{-3\frac{j}{2}}).$$

We now claim that $I_\varphi = 1$. Indeed, let x_0 be such that $\varphi(x_0) \neq 0$. Let $\delta, \eta > 0$ be such that $|\varphi(x) - \varphi(x_0)| \leq \epsilon$ whenever $|x - x_0| \leq \eta$. Therefore,

$$
\begin{aligned}
\left| \varphi(x_0) - I_\varphi \sum_k \varphi(x_0) \varphi(2^j x - k) \right| &\leq I_\varphi \sum_{|\frac{k}{2^j} - x_0| \leq \eta} \epsilon |\varphi(2^j x - k)| \\
&+ \sum_{|\frac{k}{2^j} - x_0| > \eta} \left[|\varphi(\frac{k}{2^j})| + |\varphi(x_0)| \right] |\varphi(2^j x - k)| \\
&+ C\epsilon + \bigcirc(2^{-3\frac{j}{2}}).
\end{aligned}
$$

Observe next whenever $|x - x_0| \leq \dfrac{\eta}{2}$ then we necessarily get $|\dfrac{k}{2^j} - x_0| > \eta$, which implies that $|\dfrac{k}{2^j} - x| \geq \dfrac{\eta}{2}$. Hence,

$$|\varphi(2^j x - k)| \leq \frac{C}{(1 + |2^j x - k|)^4} \leq \frac{C}{(1 + |2^j x - k|)^2 (2^j \frac{\eta}{2})^2}.$$

As a consequence,

$$
\begin{aligned}
\left| \varphi(x_0) - I_\varphi \sum_k \varphi(x_0) \varphi(2^j x - k) \right| &\leq C \sum_k \frac{1}{(1 + |2^j x - k|)^2 (2^j \frac{\eta}{2})^2} + C\epsilon + O(2^{-3\frac{j}{2}}) \\
&\leq C 2^{-2j} \eta^{-2} + C\epsilon + O(2^{-3\frac{j}{2}}).
\end{aligned}
$$

It follows consequently that for j large enough

$$\left| \varphi(x_0) - I_\varphi \sum_k \varphi(x_0)\varphi(2^j x - k) \right| \leq C\epsilon.$$

This yields that $M \neq 0$, and whenever $|y - 2^j x_0| \leq 2^j \frac{\eta}{2}$, we get

$$\left| \sum_k \varphi(y - k) - \frac{1}{I_\varphi} \right| \leq C\epsilon.$$

Now observe that the function $y \longmapsto \sum_k \varphi(y - k)$ is 1-periodic. Hence, for j large enough, we get that such a function remains constant on the whole line $\mathbb{R}$, which means that

$$\sum_k \varphi(y - k) = \frac{1}{M}.$$

This yields

$$M = \int_\infty^{+\infty} \varphi(y)dy = \sum_{k\in\mathbb{Z}} \int_k^{k+1} \varphi(y)dy = \int_0^1 (\sum_{k\in\mathbb{Z}} \varphi(y - k))dy = \frac{1}{M}.$$

Hence, $M^2 = 1$ and thus may be chosen to be equal to 1.

We now prove that $\bigcap_j V_j = \{0\}$. Let F be an element in the intersection. Hence, for all j, we may write

$$F = \sum_k a_{j,k}\varphi_{j,k},$$

where the $a_{j,k}$ are uniformly bounded. Moreover,

$$\|F\|^2 = \sum_k |a_{j,k}|^2.$$

Therefore, for all j we get

$$|F| \leq C \sum_k |\varphi_{j,k}| \leq C 2^{\frac{j}{2}}$$

which means that $F = 0$.

It remains now to show assertion (e) in Definition 2.4. To do this, we prove that any piecewise constant function may be approximated by elements of the union of the V_j's. So, let $[a, b]$ be an interval of $\mathbb{R}$ and let for $j \in \mathbb{Z}$ the function

$$F_j = \sum_{\frac{k}{2^j}\in[a,b]} 2^{j/2}\varphi_{j,k}.$$

We immediately observe that for all x we have

$$|F_j(x)| \leq C \sum_{\frac{k}{2^j}\in[a,b]} \frac{1}{(1 + |2^j x - k|^2)} \leq \frac{C}{2^j dist(x, [a, b])}.$$

On the other hand,

$$|1 - F_j(x)| = |\sum_{\frac{k}{2^j} \notin [a,b]} \varphi(2^j x - k)| \leq \frac{c}{2^j dist(x, [a,b]^c)}.$$

Now observe that

$$\int |\chi_{[a,b]}(x) - F_j(x)|^2 dx = \int_{dist(x,[a,b]^c) \geq 2^{-\frac{j}{2}}} |\chi_{[a,b]}(x) - F_j(x)|^2 dx$$
$$+ \int_{dist(x,[a,b]) \geq 2^{-\frac{j}{2}}} |\chi_{[a,b]}(x) - F_j(x)|^2 dx$$
$$+ \int_{dist(x,\partial[a,b]) \leq 2^{-\frac{j}{2}}} |\chi_{[a,b]}(x) - F_j(x)|^2 dx.$$

Next, observing the estimations above we obtain

$$\int |\chi_{[a,b]}(x) - F_j(x)|^2 dx \leq C2^{-2j}2^j + C2^{-2j}2^j + C2^{-\frac{j}{2}}.$$

Consequently,

$$\|\chi_{[a,b]} - F_j\|_2^2 \leq C2^{-\frac{j}{2}}.$$

Letting $j \to \infty$ we observe F_j goes to $\chi_{[a,b]}$ in L^2.

2.6 MULTIDIMENSIONAL REAL WAVELETS

Many ideas have been exploited to introduce multidimensional wavelet analysis. Some are based on the adoption of multiresolution analysis on $\mathbb{R}$ to the multidimensional case. The first constructed bases were separable ones. Their construction focus on an analogy with the Haar one. Recall that in one-dimensional case, this basis is defined by

$$\begin{cases} \psi_{j,k}(x) = 2^{j/2}\psi(2^j x - k) \ ; \ j,k \in \mathbb{Z} \\ \psi = \xi_{[0,1/2[} - \xi_{[1/2,1[} \end{cases}$$

Generally, let (V_j^1) be a multiresolution analysis of $L^2(\mathbb{R})$ with a scaling function φ and a wavelet ψ, and let the orthogonal projection on V_j be denoted by P_j. The main idea in the multiresolution analysis is the ability to construct an orthonormal wavelet basis $\{\psi_{j,k} \ ; \ j,k \in Z\}$, $\psi_{j,k}(x) = 2^{j/2}\psi(2^j x - k)$, such that

$$P_{j+1}f = P_j f + \sum_k <f, \psi_{j,k}> \psi_{j,k} \ \forall f \in, L^2(\mathbb{R}).$$

The adoption of these one-dimensional MRA and wavelets will be the starting point to construct the multidimensional case. Indeed, let (V_j^1) be a multiresolution analysis of $L^2(\mathbb{R})$ with a scaling function φ, a wavelet ψ and an orthogonal projections P_j^1. Consider then orthogonal projection P_j^d in $L^2(\mathbb{R}^d)$ defined as the tensor product of d copies of P_j^1

$$P_j^d = P_j^1 \otimes P_j^1 \otimes ... \otimes P_j^1.$$

Denote $V_j^d = P_j^d(L^2(\mathbb{R}^d))$. We have $V_j^d = V_j^1 \otimes V_j^1 \otimes \ldots \otimes V_j^1$. The closure in $L^2(\mathbb{R}^d)$ of V_j^d has an orthonormal basis

$$\begin{cases} \varphi_{j,k}^d = 2^{jd/2}\varphi^d(2^j x - k)\,;\, j \in \mathbb{R}\,,\, k \in \mathbb{R}^d \\ \varphi^d = \varphi \otimes \varphi \otimes \ldots \otimes \varphi \end{cases}$$

The detail spaces W_j^d will be defined by

$$W_j^d = \oplus_{\varepsilon \neq (0,\ldots,0)} V_j^{\varepsilon_1} \otimes V_j^{\varepsilon_2} \otimes \ldots \otimes V_j^{\varepsilon_d}.$$

This yields an orthonormal basis of $L^2(\mathbb{R}^d)$ associated with P_j^d

$$\begin{cases} \psi_{j,k}^\varepsilon(x) = 2^{jd/2}\psi^{\varepsilon_1}(2^j x_1 - k_1)\ldots\psi^{\varepsilon_d}(2^j x_d - k_d)\,;\, j, k_i \in \mathbb{Z}, \\ \psi^1 = \psi\,,\, \psi^0 = \varphi = \xi_{[0,1[}. \end{cases}$$

This last formula looks better than the one defined by tensor products and yields simple separable wavelets. In contrast, nonseparable wavelets remain difficult to be used and to construct. However, in analysis and in nature, one can speak about propagations in privileged directions. One plans to study their behavior by means of well-adapted wavelets. An important example of directional phenomena is supplied by spirals, such as the domain between the two curves of equations (in polar coordinates)

$$r = \theta^{-\alpha} \quad \text{and} \quad r = (\theta + \pi)^{-\alpha}.$$

Another example that bears similarities with spirals is the set

$$\mathcal{C}_\alpha = \bigcup_n \left[\frac{1}{(2n+1)^\alpha}, \frac{1}{(2n)^\alpha}\right].$$

If the aim is a pointwise analysis, without particular emphasis on directions, the topic will be more economical. However, if the signal to be analyzed has a preferred direction, then one needs a wavelet with good angular selectivity.

2.7 EXAMPLES OF WAVELET FUNCTIONS AND MRA

In this section, we propose to present some examples of wavelets and associated MRA. The readers can be referred to [163], [248], [289], [295] for more details and examples of original wavelet analysis on the real line and Euclidean spaces in general.

2.7.1 Haar wavelet

The example of Haar is the simplest example in the theory of wavelets and MRA. It is based on the Haar scaling function explicitly given by

$$\varphi = \chi_{[0,1[}$$

and characterized by the possibility of explicit computations of the transforms and coefficients. The basic approximation space is given by

$$V_0 = \left\{ f \in L^2(\mathbb{R})\,;\, f = \sum_{k \in \mathbb{Z}} a_k \varphi_k\,;\, (a_k)_k \subset \mathbb{R} \text{ such that } \sum_{k \in \mathbb{Z}} a_k^2 < \infty \right\}.$$

For $k \in \mathbb{Z}$, we denote φ_k by the function

$$\varphi_k(t) = \varphi_{0,k}(t) = \chi_{[0,1[}(t - k) = \chi_{[k,k+1[}(t).$$

Consequently, observing that the φ_k's have disjoints supports, any element $S \in V_0$ may be written in the form

$$f(t) = a_k \, ; \;\; t \in [k, k+1[.$$

hence, V_0 is the subspace of signals that are constant on intervals of the form $[k, k+1[$, $k \in \mathbb{Z}$. By exploiting the assertion (d) in Definition 2.4, we obtain for any $j \in \mathbb{Z}$,

$$V_j = \left\{ f \in L^2(\mathbb{R}) \, ; \, f = \sum_{k \in \mathbb{Z}} a_{j,k} \varphi_{j,k} \, ; \, (a_{j,k})_k \subset \mathbb{R} \text{ such that } \sum_{k \in \mathbb{Z}} a_{j,k}^2 < \infty \right\}$$

where we recall that

$$\varphi_{j,k}(t) = 2^{j/2} \varphi(2^j t - k) = 2^{j/2} \chi_{[0,1[}(2^j t - k) = 2^{j/2} \chi_{[k/2^j,(k+1)/2^j[}(t).$$

V_j is by analogy the set of functions that are constant on intervals of the form $[\frac{k}{2^j}, \frac{k+1}{2^j}[, \, k \in \mathbb{Z}$.

Next, we propose to compute the coefficients of the filters H and G, and consequently the expression of the associated Haar mother wavelet. The starting point is the two-scale relation (2.22), which we recall hereafter

$$\varphi(x) = \sqrt{2} \sum_k h_k \varphi(2x - k).$$

Having in hand the explicit expression of Haar scaling function $\varphi = \chi_{[0,1[}$, the latter becomes

$$\chi_{[0,1[}(x) = \sqrt{2} \sum_k h_k \chi_{[k/2,(k+1)/2[}(x).$$

Next, because of the supports of the different functions appearing in the relation, we obtain

$$h_k = 0 \, , \quad \text{for all } k \neq 0, 1$$

and thus,

$$\chi_{[0,1[}(x) = \sqrt{2} h_0 \chi_{[0,1/2[}(x) + \sqrt{2} h_1 \chi_{[1/2,1[}(x)$$

which yields immediately that

$$h_0 = h_1 = \frac{1}{\sqrt{2}}.$$

It results that the filter G is given by

$$g_0 = -g_1 = \frac{1}{\sqrt{2}} \quad \text{and} \quad g_k = 0 \, , \quad \text{for all } k \neq 0, 1.$$

The associated Haar mother wavelet is then expressed by

$$\psi = \chi_{[0,1/2[} - \chi_{[1/2,1[}.$$

We hereafter give the graphic illustrations of the Haar scaling function φ and the Haar mother wavelet ψ (see Figure 2.3).

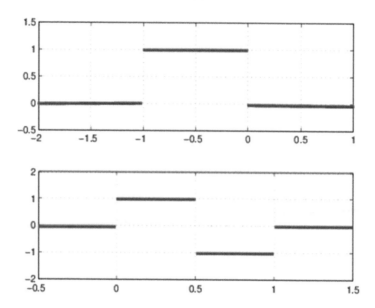

Figure 2.3 Haar scaling and wavelet functions.

2.7.2 Faber–Schauder wavelet

It is based on the explicit Schauder scaling function

$$\varphi(x) = (1 - |x|)\chi_{[-1,1[}(x)$$

which is also characterized by the possibility of explicit computations of the transforms and coefficients. The basic approximation space V_0 is composed of all functions $f \in L^2(\mathbb{R})$, which may be written in the form

$$f = \sum_{k \in \mathbb{Z}} a_k \varphi_k$$

where, as usual, $(a_k)_k$ satisfies

$$\sum_{k \in \mathbb{Z}} a_k^2 < \infty.$$

For $k \in \mathbb{Z}$, φ_k is given by

$$\varphi_k(x) = \varphi_{0,k}(x) = (1 - |x - k|)\chi_{[-1,1[}(x - k) = (1 - |x - k|)\chi_{[k-1,k+1[}(x).$$

Here, also, using the fact that the supports are disjoint, we obtain

$$f(x) = a_k(1 - |x - k|) \; x \in [k - 1, k + 1[.$$

V_0 is then the subspace of square integrable functions that are affine on the intervals $[k - 1, k + 1[, \; k \in \mathbb{Z}$. More precisely,

$$f(x) = \begin{cases} a_k(1 - k + x) & , \quad x \in [k - 1, k[, \\ a_k(1 + k - x) & , \quad x \in [k, k + 1[, \\ 0 & , \quad \text{others.} \end{cases}$$

The subspace V_j is composed of square integrable functions that are of the form

$$f(x) = \begin{cases} a_{j,k}2^{j/2}(1 - k + 2^j x) & , \quad x \in [(k-1)/2^j, k/2^j[, \\ a_{j,k}2^{j/2}(1 + k - 2^j x) & , \quad x \in [k/2^j, (k+1)/2^j[, \\ 0 & , \quad \text{others.} \end{cases}$$

where for all $j \in \mathbb{Z}$, $(a_{j,k})_k$ satisfies

$$\sum_{k \in \mathbb{Z}} a_{j,k}^2 < \infty.$$

We now evaluate the filter coefficients. The two-scale relation (2.22) is written in the present case as

$$(1 - |x|)\chi_{[-1,1[}(x) = \sqrt{2} \sum_{k=-2}^{2} h_k(1 - |2x - k|)\chi_{[(k-1)/2,(k+1)/2[}(x).$$

The coefficients h_k are vanishing for $|k| \geq 3$. Next by choosing for x the values -1, $\dfrac{-1}{2}$, 0, $\dfrac{1}{2}$ and 1, we obtain

$$h_{-2} = h_2 = 0, \, h_{-1} = h_1 = \frac{1}{2\sqrt{2}} \quad \text{and } h_0 = \frac{1}{\sqrt{2}}.$$

The G filter will be

$$g_0 = g_2 = \frac{1}{2\sqrt{2}} \quad \text{and} \quad g_1 = \frac{-1}{\sqrt{2}}.$$

The Schauder mother wavelet will be expressed by

$$\psi(x) = \frac{1}{2}(1 - |2x|)\chi_{[-1/2,1/2[}(x) - (1 - |2x - 1|)\chi_{[0,1[}(x) + \frac{1}{2}(1 - |2x - 2|)\chi_{[1/2,3/2[}(x).$$

The following graph (Figure 2.4) represents Schauder φ and ψ.

2.7.3 Daubechies wavelets

In 1988, Daubechies built with the concept of multiresolution analysis a compact support wavelet family. These wavelets are interesting as they are orthogonal and well localized in time or space. The specificity of this wavelet is the absence of an explicit function. In addition, the wavelet provides the ability to choose the degree of regularity by imposing the number of vanishing moments. We will describe the construction of Daubechies wavelets and show some basic properties. The two basic relationships to build a multiresolution analysis are the following:

$$\forall x, \quad \Phi(x) = \Sigma h_k \Phi(2x - k)$$

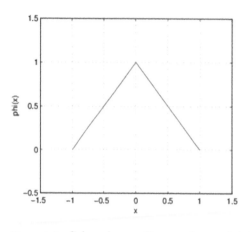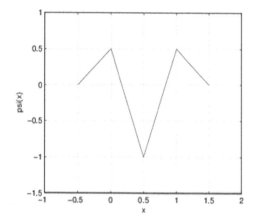

Figure 2.4 Schauder scaling and wavelet functions.

and

$$\forall\, x, \quad \psi(x) = \sqrt{2} \sum g(n)\phi(2x - k)$$

where

$$\sum_n h(n - 2k)h(n - 2l) = \delta_{kl},$$

$$\sum_n h(n) = \sqrt{2},$$

and

$$g(n) = (-1)^n h(1 - n)$$

If the filter $H = (h(n))_{n \in \mathbb{Z}}$ is finite, the support of the scaling function is also finite. Let

$$supp(\phi) \subset [N_{min}, N_{max}].$$

Hence,

$$supp(\psi) \subset [1 \div 2(1 - N_{min} - N_{max}), 1 \div 2(1 + N_{min} - N_{max})]$$

For example for $N = 2$, we obtain the so-called *Db4* wavelet, whose low-pass filter coefficients h_k are

$$h_0 = \frac{1 + \sqrt{3}}{4\sqrt{2}}, \quad h_1 = \frac{3 + \sqrt{3}}{4\sqrt{2}}$$

$$h_2 = \frac{3 - \sqrt{3}}{4\sqrt{2}}, \quad h_3 = \frac{1 - \sqrt{3}}{4\sqrt{2}}$$

The next graph illustrates some wavelet function psi due to Daubechies. The representation illustrates clearly the strong relation between the regularity, the number of vanishing moments and the support length N.

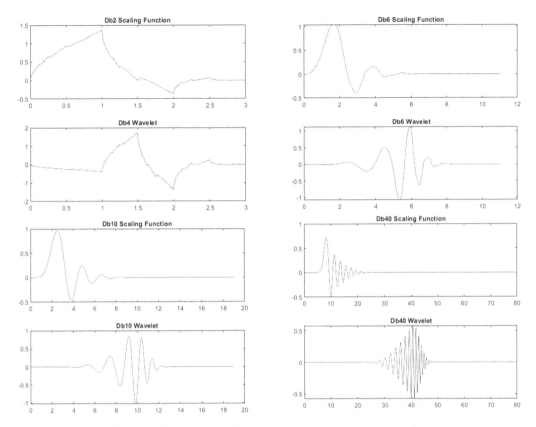

Figure 2.5 Some Daubechies scaling functions and associated wavelets.

2.7.4 Symlet wavelets

The symlets are nearly symmetrical wavelets proposed by Daubechies as modifications to the db family. The properties of the two wavelet families are similar. The symlet (symN) wavelets are also known as Daubechies' least-asymmetric wavelets. In symN, N is the number of vanishing moments. These filters are also referred to in the literature by the number of filter taps, which is 2N. Here are the wavelet functions psi.

2.7.5 Spline wavelets

A fundamental remark that we can notice from the previous constructions of multiresolution analysis (and its associated scaling and wavelet functions) is the lack of regularity. However, this fact is very important in many cases of applications such as the resolution of partial differential equations where higher order regularities are required. Moreover, the regularity is sometimes related to the supports of wavelets and scaling functions as in Daubechies cases. This is also important in approximating time series where the size is large.

In the present section, we aim to redevelop a case of explicit construction of wavelets and thus multiresolution analysis by taking into account the characteristics of regularity.

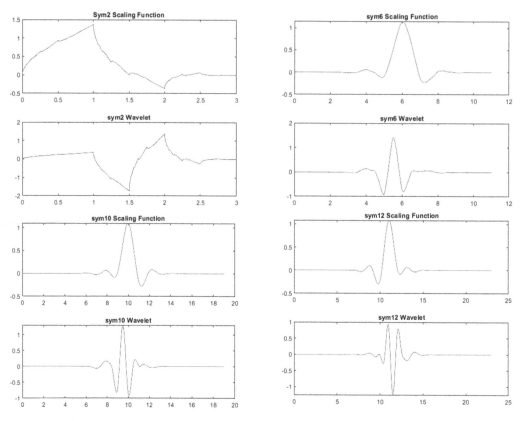

Figure 2.6 Some symlet scaling functions and associated wavelets.

We will start by the most simple example of wavelets and multiresolution analysis due to the Haar case and show that, even though this is a nonregular case, it may serve to construct higher order multiresolutions. These are the so-called spline wavelets.

So, let $\varphi = \chi_{[0,1]}$ be the Haar scaling function, which is equivalently the characteristic function of the interval $[0,1]$. Let $r > 0$ be in $\mathbb{N}$. The spline wavelets are defined inductively by

$$\varphi_0 = \varphi \quad \text{and} \quad \varphi_r = \varphi_{r-1} * \varphi, \; \forall r \geq 1.$$

Definition 2.15 *The function φ_r is said to be the spline wavelet of order r.*

Obviously, the starting multiresolution space V_0 will be the closure in $L^2(\mathbb{R})$ of $spann(\varphi_r(. - k))_{k \in \mathbb{Z}}$ and for $j \in \mathbb{W}$ the j-level approximation space V_j may be obtained by scaling the elements of V_0 by 2^j. In other words, $F(.) \in V_j$ whenever $F(\frac{.}{2^j}) \in V_0$ and vice versa.

We will show in the next part that spline wavelets may be seen as piecewise polynomial functions on intervals of the form $[k, k+1[$, $k \in \mathbb{Z}$ with degrees at most the order r of the spline. More precisely, we have the following result.

Proposition 2.16 *Let P_r be the space of all $(r-1)$ continuously differentiable functions with their restriction on each interval $[k, k+1[$, $k \in \mathbb{Z}$, a polynomial of degree at most r. Then, $\varphi_r \in P_r$.*

Proof 2.4 *We proceed by recurrence on r. The function φ_0 is constant. Therefore, it may be seen as a polynomial of degree 0. Assume next that on each interval $[k, k+1[$, $k \in \mathbb{Z}$, the function φ_r is a polynomial of degree at most r. Observing next that $\varphi_{r+1} = \varphi_r * \varphi$, it suffices to show that $x^r * \varphi \in P_r$. Indeed, for all $x \in [0,1]$ we explicitly have*

$$(x^r * \varphi)(t) = \int_0^1 (y-t)^r dy = \sum_{l=0}^r C_r^l \frac{(-1)^{r-l}}{r+1-l} x^l$$

which is q polynomial of degree r.

In fact, we may prove more properties of these wavelets such as the minimality of supports. We now investigate briefly the important assumption in the construction of multiresolution analysis stating that the collection $\varphi_r(x-k)$, $k \in \mathbb{Z}$, constitutes a Riesz basis of the space that they span. We thus have to show that there exist constants K_1, K_2 with $0 < K_1 < K_2 < \infty$ such that almost everywhere we have

$$K_1 \leq \sum |\widehat{\varphi_r}(\omega + 2k\pi)|^2 \leq K_2.$$

Indeed, it is straightforward that

$$\widehat{\varphi_r}(\omega) = \left(\frac{e^{i\omega}-1}{i\omega}\right)^{r+1}.$$

Notice now that the overlap function $\Gamma_\varphi = \sum |\widehat{\varphi_r}(\omega + 2k\pi)|^2$ is periodic with period 2π and it is upper bounded on $[-\pi, \pi]$. On the other hand, the function $\Theta(\omega) = |\frac{\exp i\omega - 1}{i\omega}|$ is lower bounded away from 0 on $[-\pi, \pi]$. So is the function Γ_φ. As a consequence, the function φ_r yields a multiresolution on $\mathbb{R}$.

2.7.6 Anisotropic wavelets

The first anisotropic wavelet is due to Morlet ([279]). Its basic function is

$$\psi(X) = e^{iK.X}e^{-\|X\|^2/2}$$

which is in fact Gaussian modulated according to the direction of K.

It is not, in any rigor, a wavelet; additional terms must be added to it for reason of oscillation. Note that it is not isotropic but has a privileged direction given by K. It is used to analyze images having anisotropic characteristics.

A second example of anisotropic wavelets is the so-called Mexican hat ([279]). It is based on the function

$$\psi(X) = (2 - XAX)e^{-XAX/2}$$

where A is an unspecified symmetric matrix of $M_2(\mathbb{R})$.

It differs from that of Morlet due the fact that the anisotropy is in its module. If $A = \lambda I$, the Mexican hat is of a radial symmetry. However, if the spectrum of the matrix A is not a singleton, the wavelet will be anisotropic.

2.7.7 Cauchy wavelets

Cauchy wavelets are one step in the direction of introducing spherical wavelets as they aim to take into account the angular behavior of the analyzed functions. In one-dimensional case, Cauchy wavelets are defined via their Fourier transform

$$\widehat{\psi}_m(\omega) = \begin{cases} 0 & \text{for} \quad \omega < 0 \\ \omega^m \, e^{-\omega} & \text{for} \quad \omega \geq 0. \end{cases}$$

with $m > 0$. In one dimension, the positive half-line is a convex cone. Thus, a natural generalization to two dimensions will be a wavelet whose support in spatial frequency space is contained in a convex cone with apex at the origin. Let $C \equiv C(\alpha, \beta)$ be the convex cone determined by the unit vectors e_α, e_β, where $\alpha < \beta$, $\beta - \alpha < \pi$ and for all θ, $e_\theta \equiv (\cos\theta, \sin\theta)$. The axis of the cone is $\xi_{\alpha\beta} = e_{\frac{\alpha+\beta}{2}}$. In other words,

$$C(\alpha, \beta) = \left\{ k \in \mathbb{R}^2, \ \alpha \leq arg(k) \leq \beta \right\}$$

$$= \left\{ k \in \mathbb{R}^2, \ k.\xi_{\alpha\beta} \geq e_\alpha.\xi_{\alpha\beta} = e_\beta.> 0 \right\}.$$

The dual cone to $C(\alpha, \beta)$ is

$$\widetilde{C}(\alpha, \beta) = \left\{ k \in \mathbb{R}^2, \ k.k' > 0, \ \forall k' \in C(\alpha, \beta) \right\}.$$

Remark that $\widetilde{C}(\alpha, \beta)$ may be also seen as

$$\widetilde{C}(\alpha, \beta) = C(\widehat{\alpha}, \widehat{\beta}).$$

where

$$\widehat{\alpha} = \beta - \frac{\pi}{2}, \quad \widehat{\beta} = \alpha + \frac{\pi}{2}$$

and

$$e_\alpha.e_{\widehat{\alpha}} = e_\beta.e_{\widehat{\beta}} = 0.$$

Thus, the axis of $\widetilde{C}$ is $\xi_{\alpha\beta}$.

The two-dimensional Cauchy wavelet is defined via its Fourier transform

$$\widehat{\psi}_{lm}^{C,\eta}(k) = \begin{cases} (k.e_{\widetilde{\alpha}})^l \, (k.e_{\widetilde{\beta}})^m \, e^{-k.\eta} & , \quad k \in C(\alpha, \beta), \\ 0 & , \quad \text{otherwise.} \end{cases}$$

where $\eta \in \widetilde{C}$ and $l, m \in \mathbb{N}^*$. Note that such a wavelet is also supported by the cone C. It satisfies the admissibility condition

$$A_{\psi_{lm}^{C,\eta}} \equiv (2\pi)^2 \int |\widehat{\psi}_{lm}^{C,\eta}(k)|^2 \frac{d^2k}{|k|^2} < \infty.$$

The following result of Antoine et al is proved in [13] and yields an explicit form for the two-dimensional Cauchy wavelet.

Proposition 2.17 *For even* $\eta \in \tilde{C}$ *and* $l, m \in \mathbb{N}^*$, *the two-dimensional Cauchy wavelet* $\psi_{lm}^{C,\eta}(x)$ *with support in* C *belongs to* $L^2(\mathbb{R}^2, dx)$ *and is given by*

$$\psi_{lm}^{C,\eta}(x) = \frac{i^{l+m+2}}{2\pi} l! m! \frac{[\sin(\beta - \alpha)]^{l+m+1}}{[(x+i\eta).e_\alpha]^{l+1} [(x+i\eta).e_\beta]^{m+l}}.$$

We can, with analogous techniques, define multidimensional Cauchy wavelets. See [13] and the references therein for more details.

2.8 EXERCISES

Exercise 1.

Denote $\phi = \chi_{[0,1[}$ and $\phi_k^j(t) = \sqrt{2^j}\phi(2^j t - k)$, $t \in \mathbb{R}$. Denote further $E = L^2([0,1[)$ the vector space of square-integrable functions on $[0,1[$. Next for j fix V^j, the vector space of constant functions, on $\{\frac{k}{2^j}; \frac{k+1}{2^j}\}$ for $k \in \{0; 2^j - 1\}$.

a. Show that $(\phi_k^j)_{k \in \{0, 2^j-1\}}$ is an orthonormal basis of V^j. $< (\phi_k^j); (\phi_h^j) >= 0 \forall (k, h) \in \{0, 2^j - 1\} \times \{0, 2^j - 1\}$. and that the elements ϕ_k^j are unitary.

b. Sketch the graphs of $(\phi_k^1)_{k \in \{0,1\}}$ and $(\phi_k^2)_{k \in \{0,3\}}$.

Exercise 2.

Consider the same assumptions as in Exercise 1 above.

a. Find two functions $(\psi_k^1)_{k \in \{0,1\}}$ constituting an orthonormal basis of a vector space W^1 such that $V^2 = V^1 \oplus W^1$.

b. Express the functions $(\psi_k^1)_{k \in \{0,1\}}$ and $(\phi_k^1)_{k \in \{0,1\}}$ by means of $(\phi_k^2)_{k \in \{0,3\}}$.

c. Express conversely the functions $(\phi_k^2)_{k \in \{0,3\}}$ in terms of $(\phi_k^1)_{k \in \{0,1\}}$ and $(\psi_k^1)_{k \in \{0,1\}}$.

d. Express the function $\psi_k^1 = \psi(2t - k)$.

Exercise 3.

Consider the same assumptions as in Exercises 1 and 2 above.

a. Provide the 2^j Haar wavelets $(\psi_k^1)_{k \in \{0,1\}}$ which constitute an orthonormal basis of $W^j = V^{j+1} \ominus V^j$.

b. Write the (ϕ_k^j) and (ψ_k^j) by means of (ψ_k^{j+1}).

c. Write the (ϕ_k^{j+1}) by means of (ϕ_k^j) and (ψ_k^j).

d. Write the approximation coefficients a_k^j and the detail coefficients (d_k^j) by means of the (a_k^{j+1}).

e. Write the coefficients (a_k^{j+1}) by means of (a_k^j) and d_k^j.

f. Compare the relation between the coefficients in $V^j \oplus W^j$ to the one in V^{j+1}.

Exercise 4.

Consider the discrete signal $S = [2 \ 4 \ 8 \ 12 \ 14 \ 0 \ 2 \ 1]$ and the Haar multiresolution analysis on $\mathbb{R}$.

a. Decomposer S in $V^0 \oplus_{j=0}^2 W^j$

b. Sketch the graphs of S relatively to V^2, V^1 and V^0.

b. Sketch the graphs of S relatively to the details W^2, W^l and W^0

Exercise 5.

Let $\psi : \mathbb{R} \to \mathbb{C}$ be wavelet in the Schwarz class such that $\widehat{\psi}(0) = 0$). Denote

$$C = \int_{\mathbb{R}} |x\psi(x)| dx.$$

Denote also for any function $f : \mathbb{R} \to \mathbb{C}$, $s > 0$ and $u \in \mathbb{R}$,

$$W_f(s, u) = \frac{1}{\sqrt{s}} \int_{\mathbb{R}} \psi\left(\frac{x - u}{s}\right) \overline{f(x)} dx.$$

1. Let $f : \mathbb{R} \to \mathbb{C}$ be 1-Lipschitzian. Prove that for all s and u we have

$$|W_f(s, u)| \leq C s^{\frac{3}{2}}.$$

2. a) Show that there exists $\phi : \mathbb{R} \to \mathbb{C}$ such that $\psi = \phi'$ and $\psi(x) \to 0$ as $|x| \to \pm\infty$.
b) Let $a \in \mathbb{R}$ be fixed and consider $f = 1_{[a,+\infty]}$. Compute W_f by means of ϕ.
3. Let $f : \mathbb{R} \to \mathbb{C}$ be piecewise continuous with finite number of discontinuities $x_1 < \ldots < x_n$ and denote for $k \leq n$ $\alpha_k = f(x_k^+) - f(x_k^-)$. Assume further that f is 1-Lipschitzian on any interval not containing any of the x_k's. Sketch the graph of $W_f(s, .)$ for s small enough.

Exercise 6.

Let $f, \psi : \mathbb{R} \to \mathbb{C}$ be in the Schwarz class such that $\psi(0) = 0$) and denote

$$F(t) = \int_0^{+\infty} \int_{\mathbb{R}} \psi_s(t - u) W_f(u, s) du \frac{dt}{s^2}.$$

Show that $F = C_\psi f$, where C_ψ is a constant depending on ψ and not on f.

Exercise 7.

Let ψ be a wavelet such that $\widehat{\psi}(0) = 0$, where

$$\widehat{\psi}(w) = \int_{\mathbb{R}} \psi(t) e^{-izt} dt, \ \forall w \in \mathbb{R}.$$

Show the following Heisenberg uncertainty inequality:

$$\|\psi\|_2^2 \leq \frac{2}{\sqrt{2\pi}}\|t\psi(t)\|_2\|\xi\widehat{\psi}(\xi)\|_2.$$

(Hint. We may use the derivative $(t\psi(t))'$.)

Exercise 8.

Let $f \in L^2(\mathbb{R})$ be such that $\widehat{f}(\xi) = 0$ for $\xi < 0$. For $z \in \mathbb{C}$ such that $I,(z) \geq 0$, let

$$f_p(z) = \frac{1}{\pi} \int_0^{+\infty} (i\xi)^p \widehat{f}(\xi)e^{iz\xi}d\xi.$$

a) Prove that f_p is the derivative of f at the order p whenever f is $\mathcal{C}^p$.
b) Prove that f_p is analytic on the upper half complex plane $Im(z) > 0$.
c) Prove that for $z = x + iy$, $y > 0$, we have

$$f_p(z) = y^{-p-1/2}d_{y,x}(f),$$

calculated with an analytic wavelet ψ that you will specify.

Exercise 9.

Let for $N \in \mathbb{N}$, $f : [0, N] \to \mathbb{R}$ be continuous and $(V_m)_{m\geq0}$ be a multiresolution analysis on $[0, N]$ with a scaling function φ.
1) Verify that for all $\varepsilon > 0$ there exists $m \geq 0$ such that for all elements $t_1, t_2 \in [0, N]$ with $|t_1 - t_2| < 2^{-m}$ we have $|f(t_1) - f(t_2)| < \varepsilon$.
2) Denote next $t_{m,n+1/2} = (n + \frac{1}{2})2^{-m}$ and

$$f_\varepsilon(t) = \sum_{n=0}^{2^N-1} f(t_{m,n+1/2})\varphi_{m,n}(t).$$

Prove that
$$|f(t) - f_\varepsilon(t)| < \varepsilon, \ \forall t \in [n2^{-m}, (n + 1)2^{-m}].$$

3) Show that
$$\|f - f_\varepsilon\|_2 \leq \varepsilon\sqrt{N}.$$

Exercise 10.

Part A. Consider the Haar system on $\mathbb{R}$ with its scaling function $\varphi = \chi_{[0,1[}$ and wavelet function $\psi(x) = \chi_{[0,1/2[} - \chi_{[1/2,1[}$. Let next $f(t) = (t^2 + 1)\chi_{[-1,1[}$ and its projections on V_j and W_j be, respectively,

$$f_j = \sum_k a_{j,k}\varphi_{j,k} \quad \text{and} \quad d_j = \sum_k d_{j,k}\psi_{j,k}.$$

a) Evaluate $f_0(t)$, $f_1(t)$, $f_2(t)$, $d_0(t)$ and $d_1(t)$.
b) Draw the graphs of $f(t)$, $f_0(t)$, $f_1(t)$, $f_2(t)$, $d_0(t)$ and $d_1(t)$.

Part B. Denote for $(x, y) \in \mathbb{R}^2$

$$\phi(x, y) = \varphi(x)\varphi(y), \quad \Psi^0(x, y) = \varphi(x)\psi(y),$$

$$\Psi^1(x, y) = \psi(x)\varphi(y), \quad \Psi^2(x, y) = \psi(x)\psi(y).$$

a) Prove that the system $(\Phi, \Psi^i, i = 0, 1, 2)$ constitutes a scaling function and a wavelet on $\mathbb{R}^2$.

b) Express the approximation and detail spaces V_j and W_j associated.

c) Let $f(x, y) = \cos(x) \sin(x+y)$ on the cube $C = [-\pi, \pi] \times [-\pi, \pi]$ and its projections on V_j and W_j be, respectively,

$$f_j = \sum_k a_{j,k} \Phi_{j,k} \quad \text{and} \quad d_j^i = \sum_k d_{j,k} \Psi_{j,k}^i, \quad i = 0, 1, 2.$$

a) Evaluate $f_0(x, y)$, $f_1(x, y)$, $f_2(x, y)$, $d_0(x, y)$ and $d_1(x, y)$.

b) Draw the graphs of $f(x, y), f_0(x, y), f_1(x, y), f_2(x, y), d_0(x, y)$ and $d_1(x, y)$.

CHAPTER 3

Wavelets extended

3.1 AFFINE GROUP WAVELETS

Wavelet transform of functions may be understood as a group action on the space of square integrable functions $L^2(\mathbb{R})$ for suitable group. In this chapter, we aim to present the eventual link between wavelet transforms and the real affine group. To make the section self-contained and easy for readers from different fields, we firstly recall the basic definition of groups.

Definition 3.1 *Let G be a nonempty set. G is said to be a group if there exists an operation $*$ defined on G for which*

1. *$x*y \in G$ for all $x, y \in G$ (known as closure property).*

2. *$(z*y)*z = z*(y*2)$ for all $x, y, z \in G$ (known as associative property).*

3. *There is an element $e \in G$ such that $e*x = x*e = x$ for all $x \in G$ (e is known as the identity element).*

4. *For each $x \in G$, there is an element $\widetilde{x} \in G$ such that $x*\widetilde{x} = \widetilde{x}*x = e$ ($\widetilde{x}$ is called the inverse of x).*

If, in addition, $x*y = y*x$ for all $x, y \in G$, the group G is said to be commutative.

A subset H of a group G is a subgroup of G if it is a group under the same operation as G.

The group G may have additional structure. For instance, if G is a topological space and the operation $*$ is continuous on $G \times G$ with respect to this topology, G will be a topological group. Furthermore, if G has a differentiable structure, G will be a differentiable manifold, and is called a Lie group. A differentiable manifold, for our purpose, is a set that looks like Euclidean space. For example, $\mathbb{R}^n$ is one since it is Euclidean space. On Euclidean spaces, we can choose different parametrizations; however, the change of coordinates must be differentiable. Rectangular coordinates and polar coordinates are familiar examples of such parametrizations of $\mathbb{R}^n$. We recall that change of coordinates must also be invertible locally, that is, the Jacobian of the transformation should be invertible.

In our case, we will focus on the so-called linear or affine group on $\mathbb{R}^n$. These are groups that can be realized as groups of linear transformations of a vector space or, equivalently, as matrix groups. We will see that group representation theory may help explain the essential properties of wavelet transforms.

Consider the connected affine group G_+, also called the $ax + b$ group, consisting of transformations of $\mathbb{R}$ of the type $x \longmapsto gx = ax + b$, $x \in \mathbb{R}$, where $a > 0$, $b \in \mathbb{R}$ and where we denoted $g = (b, a) \in G_+$ with the multiplication law

$$g_1 g_2 = (b_1 + a_1 b_2, a_1 a_2).$$

It holds that an invariant measure on G is $d_\mu(a, b) = \dfrac{dadb}{a^2}$.

A nontrivial unitary irreducible representations of G on Hilbert space $L^2(\mathbb{R}, dx)$ will be

$$(U(a, b)f)(x) = \frac{1}{\sqrt{a}} f(\frac{x - b}{a}), \ f \in L^2(\mathbb{R}, dx), \ (a, b) \in G.$$

For the sequel, we define the following operators on the set of functions defined on the Euclidean space $\mathbb{R}$:

- The translation: $T_b f(x) = f(x - b)$, $a \in \mathbb{R}$.

- The dilation: $D_a f(x) = \sqrt{a} f(ax)$, $a > 0$.

- The modulation: $E_a f(x) = e^{2i\pi ax} f(x)$, $a \in \mathbb{R}$.

- The multiplication: $M_a f(x) = e^{2i\pi a} f(x)$, $a \in \mathbb{R}$.

These operators are easily seen to be unitary. Using these operators, we immediately deduce that

$$(U(a, b)f) = D_{a^{-1}} T_b f.$$

As a consequence, the wavelet transform of f at the scale $a > 0$ and the position $b \in \mathbb{R}$ may be expressed by means of the representation U as

$$C_f(a, b) = \langle f, U(a, b)\psi \rangle = \langle f, D_{a^{-1}} T_b \psi \rangle.$$

This inverse transform will be expressed as

$$f(\bullet) = \frac{1}{\mathcal{A}_\psi} \langle C_f(a, b), U(a, b)\psi(\bullet) \rangle_{d_\mu(a,b)},$$

where $\mathcal{A}_\psi$ is the admissibility constant expressed as

$$\mathcal{A}_\psi = \int_\mathbb{R} \frac{|\widehat{\psi}(\xi)|^2}{|\xi|}.$$

The results presented in this section may be extended in a usual way to multidimensional case.

3.2 MULTIRESOLUTION ANALYSIS ON THE INTERVAL

Different ways have been proposed to adapt the AMR of $L^2(\mathbb{R})$ to $L^2(]0,1[)$. [68, 69, 70, 71, 72]. Most of these constructions are based on the idea of considering the scaling functions $\varphi_{j,k}$ with supports in $[0,1]$ and adding suitable linear combinations of elements $\varphi_{j,k}$ whose support crosses the left (respect. the right) boundary of the interval.

3.2.1 Monasse–Perrier construction

Monasse and Perrier [304] have proceeded as follows: for all integer $j \geq j_0 = \log(4N)$, the space $V_j^{]0,1[}$ is spanned by the following functions.

- $2^{j/2} \tilde{\varphi}_k^0(2^j x)|_{]0,1[}, \quad k = 0, \cdots, M,$

- $\varphi_{j,k} \quad K = N, \cdots, 2^j - N,$

- $2^{j/2} \tilde{\varphi}_k^1(2^j(x-1))|_{]0,1[}, \quad k = 0, \cdots, M,$

where the functions $\tilde{\varphi}_k^0$ and $\tilde{\varphi}_k^1$ are defined as linear combinations of $\varphi_{j,k}$ with supports containing 0 or 1 and with suitable polynomial coefficients such as

- $\tilde{\varphi}_k^0(x) = \displaystyle\sum_{-N+1}^{N-1} P_k^0(l)\varphi(x-l),$

- $\tilde{\varphi}_k^1(x) = \displaystyle\sum_{-N+1}^{N-1} P_k^1(l)\varphi(x-l).$

$P_0^0, P_1^0, ..., P_M^0$ and $P_0^1, P_1^1, ..., P_M^1$ are two bases of the space of polynomials of degree at most M. The functions $\varphi_{j,k}, k = N, ..., 2^j - N$, whose supports are contained in $[0,1]$, are called interior scaling functions. The functions $\tilde{\varphi}_k^0$ and $\tilde{\varphi}_k^1, k = 0, ..., M$ are called edge scaling functions.

3.2.2 Bertoluzza–Faletta construction

Recently, Bertoluzza and Faletta [70] have proposed a new construction of AMR on the interval by considering the following spaces

$$V_j^{loc} = Span < \varphi_{j,k}, k \in \mathbb{Z} >_{L_{loc}^2(\mathbb{R})},$$

where φ is the scaling function of an AMR in $L^2(\mathbb{R})$ with compact support. Let V_j^* be the subspace of V_j^{loc} of functions with coefficients having polynomial behavior on some neighborhood of 0 or 1. The degrees of polynomials depend possibly on j. More precisely, they have proceeded as follows. For $j \geq 0$, let $N_j = 2^j - 2N + 2M + 3, M_j = min(M, N_j - 1)$ and

$$V_j^* = \left\{ \sum_{k \in \mathbb{Z}} f_k^j \varphi_{j,k}; \exists P_l, P_r \in P_{M_j}, f_k^j = P_l(k), k \leq N - 1, f_k^j = P_r(k), k \geq 2^j - N + 1 \right\}.$$

V_j^* is well defined for all $j \geq \hat{j}_0 = [\log_2(2(N-M)-1)]$. Compared to j_0 of Monasse–Perrier, the integer $\hat{j}_0$ depends directly on $N-M$. For example, if we consider the compactly supported Daubechies wavelets, then $N-M = 1$ and V_j^* are defined for any $j \geq 0$. Besides, the polynomials P_l and P_r of V_j^* are not necessarily independent. In fact, for all $j \leq [\log_2(2(N-1)-M)]$, the polynomials are the same. Finally, the parameters N_j and M_j are, respectively, the dimensions of V_j^* and the degree of polynomials exactly reproduced in V_j^*.

Proposition 3.2 *The sequence* $(V_j^*)_{j \geq j_0}$ *satisfies*
1/ $V_j^* \subset V_{j+1}^*$.
2/ $P_{M_j} \subset V_j^*$.

We will recall the important points in this construction. For $j \geq j_0$, let

$$I_j = \{N-M-1, N-M, ..., 2^j - N + M + 1\}.$$

For a given sequence $a = (a_k)_{k \in I_j}$, we consider

$$P_l(a, x) = \sum_{m=N-M-1}^{N-M-1+M_j} a_m \mathcal{L}_{M_{j,m}}(x),$$

$$P_r(a, x) = \sum_{m=2^j-N+M+1-M_j}^{2^j-N+M+1} a_m \mathcal{L}_{M_{j,m}}(x),$$

where

$$\mathcal{L}_{M_{j,m}}(x) = \prod_{m=N-M-1 \neq i}^{N-M-1+M_j} \frac{x-i}{m-i},$$

$$\mathcal{L}_{M_{j,m}}(x) = \prod_{m=2^j-N+M+1-M_{j \neq i}}^{2^j-N+M+1} \frac{x-i}{m-i}.$$

The linear extension operator $E_j : S(I_j) \longrightarrow S(\mathbb{Z})$ is defined for all $a = (a_k)_{k \in I_j}$ in $S(I_j)$ by $E_j(a) = \left(E_{j,k}(a)\right)_{k \in \mathbb{Z}}$, where

$$E_{j,k}(a) = \begin{cases} P_l(a, k) & , & k \leq N - M - 2 \\ a_k & , & N - M - 1 \leq k \leq 2^j - N + M + 1 \\ P_r(a, k) & , & k \geq 2^j - N + M + 2. \end{cases}$$

$S(A)$ for $A \subseteq \mathbb{Z}$ stands for the set of real sequences indexed by A. Let ε_j be the surjective operator which corresponds to $f = \sum_k f_{j,k} \varphi_{j,k}$ of V_j^{loc}, the element in V_j^* defined by

$$\varepsilon_j(f) = \sum_k E_{j,k}(f^j) \varphi_{j,k} |]0, 1[,$$

where $f^j = (f_{j,k})_{k \in \mathbb{Z}}$.

Proposition 3.3 *For all $j \geq j_0$, we have*

$$V_j^* |]0, 1[\equiv V_j^{]0,1[}.$$

For all $j \geq \hat{j}_0$, $V_j^{]0,1[}$ is defined as the restriction of V_j^* on $]0, 1[$.

Proof 3.1 *It follows immediately from a simple comparison of the dimensions of the two spaces.*

3.2.3   Daubechies wavelets versus Bertoluzza–Faletta

In this section, we propose to apply the method of Bertoluzza and Faletta to Daubechies wavelets in order to obtain a wavelet basis on $]0, 1[$. Let N be the larger integer such that $|supp(\varphi)| \geq 2N - 1$. Without loss of generality, we can suppose that this support coincides with the interval $[-N + 1, N]$. The two-scaler relation takes the form

$$\varphi = \sum_{k=-N+1}^{N} h_k \varphi_{1,k}$$

and the wavelet ψ has the form

$$\psi = \sum_{k=-N+1}^{N} h_k \varphi_{1,k}.$$

Recall that in this link we have

$$M = N - 1, \ N_j = 2^j + 1, \ M_j = N - 1, \ \hat{j}_0 = 0, \ I_j = \{1, 1, ..., 2^j\}.$$

The polynomial $P_l(c, .)$ of degree $M = N - 1$ interpolates (c) at the nodes $0, ..., N - 1$ and $P_r(c, .)$ of the same degree $M = N - 1$ interpolates (c) at the nodes $2^j - N + 1, ..., 2^j$. More precisely, we have

$$P_l(c, x) = \sum_{m=0}^{N-1} c_m L_{M,m}^0(x),$$

where $L_{M,m}^0$ is the Lagrange polynomial of degree M.

$$L_{M,m}^0(x) = \prod_{i=0, i \neq m}^{M} \frac{x - i}{m - i}.$$

$$P_r(c, x) = \sum_{m=2^j-N+1}^{2^j} c_m L_{M,m}^1(x),$$

where $L_{M,m}^1$ is the Lagrange polynomial of degree $M = N - 1$, taking the value 1 at $x = m$ and 0 at $x = i \neq m \in \{2^j - N + 1, ..., 2^j\}$

$$L_{M,m}^1(x) = \prod_{i=2^j-N+1 \neq m}^{2^j} \frac{x - i}{m - i}.$$

The operator $E_j : S(I_j) \to S(Z)$ becomes for all $c = (c_k)_k$ in $S(I_j)$

$$\left(E_j(c)\right)_k = \begin{cases} P_l(c,k) & , \quad k \leq -1 \\ c_k & , \quad 0 \leq k \leq 2^j \\ P_r(c,k) & , \quad k \geq 2^j + 1 \end{cases}$$

Via this operator, we define $\varepsilon_j : V_j^{loc} \to V_j^*$, which associates with each element

$$f = \sum_k f_k^j \varphi_{j,k} \in V_j^{loc}, \quad \text{the image} \quad \varepsilon_j(f) = \sum_{k \in I_j} f_k^j \varphi_{j,k}.$$

Thanks to the previous facts, the following definition will be consistent.

Definition 3.4 *For all $j \geq 0$, we define $V_j^{]0,1[}$ to be the restriction of V_j^* to the unit interval. That is,*

$$V_j^{]0,1[} = V_j^* |]0,1[.$$

To construct a wavelet basis for $V_j^{]0,1[}$, we simply apply the operator ε_j to the subset of $\{\varphi_{j,k}\}$ which cross the interval $]0,1[$. So that, we obtain for all $k \in I_j$,

$$\varphi_{j,k} = |]0,1[).$$

In [70], it has been proven that for all $f \in V_j^*$,

$$f_{|]0,1[} = \sum_{k \in I_j} \int_{\mathbb{R}} f \varphi_{j,k} \bar{\varphi}_{j,k}.$$

3.3 WAVELETS ON THE SPHERE

3.3.1 Introduction

Spherical wavelets are adapted for understanding complicated functions defined or supported by the sphere. The classical are essentially done by convolving the function against rotated and dilated versions of one fixed function ψ.

Gegenbauer polynomials are also applied to yield first examples of spherical wavelets ([205]). Let S^{n+1} be the unit sphere in $\mathbb{R}^{n+2}$ and σ the Lebesgue measure on S^{n+1}. Let $P_k^{n/2}$ be the kth Gegenbauer polynomial of order $n/2$ and define for $G \in L^1([-1,1])$

$$\widehat{G}(k) = \frac{(4\pi)^{n/2}\Gamma(k+1)\Gamma(n/2)}{\Gamma(k+n)} \int_{-1}^1 G(t) P_k^{n/2}(t)(1-t^2)^{(n-1)/2} dt.$$

Let $\psi_r \in L^1([-1,1])$, $r \geq 0$. The family $\{\,\psi_r\,\}_r$ is said to be spherical wavelet of order p if

$$\begin{cases} \hat{\psi}_r(0) = \hat{\psi}_r(1) = ... = \hat{\psi}_r(p) = 0 \\ \int_0^\infty (\hat{\psi}_r(k))^2 dr = 1 \text{ for } k \geq p+1. \end{cases}$$

The associated spherical wavelet transform is defined on $L^2(S^{n+1})$ by

$$C_\psi F(r, \eta) = \int_{S^{n+1}} F(\xi) \psi_r(\xi.\eta) d\sigma(\xi).$$

To introduce a special wavelet analysis on the sphere related to zonals, we recall firstly some useful topics. Let $F \in L^2[-1, 1]$ and L_n be the Legendre polynomial of degree n. The coefficients

$$\widehat{F}(n) = 2\pi < F, L_n > = 2\pi \int_{-1}^{1} F(x) L_n(x) \, dx, \quad n \in \mathbb{N}$$

are called the Legendre coefficients or the Legendre transforms of F. It is proved in harmonic Fourier analysis that F may be expressed via a series form

$$F = \sum_{n=0}^{\infty} \widehat{F}(n) \frac{2n + 1}{4\pi} L_n \tag{3.1}$$

called the Legendre series of F.

3.3.2 Existence of scaling functions

Definition 3.5 *A family $\{\phi_j\}_{j \in \mathbb{N}} \subset L^2[-1, 1]$ is called a spherical scaling function system if the following assertions hold.*

1. For all $n, j \in \mathbb{N}$, we have $\widehat{\phi}_j(n) \leq \widehat{\phi}_{j+1}(n)$.

2. $\lim_{j \to \infty} \widehat{\phi}_j(n) = 1$ for all $n \in \mathbb{N}$.

3. $\widehat{\phi}_j(n) \geq 0$ for all $n, j \in \mathbb{N}$.

Above, $\widehat{\phi}_j(n)$ is the Legendre transform of $\widehat{\phi}_j$.

We will now investigate a way of construction of a scaling function (see [218], [219], [220]).

Definition 3.6 *A continuous function $\gamma : \mathbb{R}^+ \longmapsto \mathbb{R}$ is said to be admissible if it satisfies the admissibility condition*

$$\sum_{n=0}^{\infty} \frac{2n + 1}{4\pi} \left(\sup_{x \in [n, n+1]} |\gamma(x)| \right)^2 < +\infty.$$

γ is called an admissible generator of the function ψ given by

$$\psi = \sum_{n=0}^{\infty} \frac{2n + 1}{4\pi} \gamma(n) L_n.$$

We immediately obtain the following characteristics [357].

Proposition 3.7 *The following assertions are true:*

1. If γ is an admissible generator, then its generated function $\psi \in L^2([-1,1])$.

2. For all $n \in \mathbb{N}$, $\widehat{\psi}(n) = \gamma(n)$.

Proof. 1. Since the Legendre polynomials form an orthogonal basis for $L^2[-1,1]$ with

$$< L_n, L_n >_{L^2[-1,1]} = \frac{4\pi}{2n+1},$$

the admissibility condition imposed on γ yields that

$$\begin{aligned}
\|\psi\|^2_{L^2[-1,1]} &= \sum_{n=0}^{\infty} \frac{2n+1}{4\pi}(\gamma_0(n))^2 \\
&\leq \sum_{n=0}^{\infty} \frac{2n+1}{4\pi}\left(\sup_{x \in [n,n+1]} |\gamma_0(x)|\right)^2 \\
&< +\infty.
\end{aligned}$$

2. It is an immediate result from (3.1).

We now investigate the idea to construct a whole family of admissible functions starting from one source admissible function.

Definition 3.8 *The dilation operator is defined for $\gamma : [0,\infty) \to \mathbb{R}$ and $a > 0$ by*

$$D_a\gamma(x) = \gamma(ax), \quad \forall x \in [0,\infty).$$

For $a = 2^{-j}$, $j \in \mathbb{Z}$, we denote $\gamma_j = D_j\gamma = D_{2^{-j}}\gamma$.

Definition 3.9 *An admissible function $\varphi : [0,\infty) \to \mathbb{R}$ is said to be a generator of a scaling function if it is decreasing, continuous at 0 and satisfies $\varphi(0) = 1$. The system $\{\phi_j\}_{j\in\mathbb{N}} \subset L^2[-1,1]$, defined by*

$$\phi_j = \sum_{n=0}^{\infty} \frac{2n+1}{4\pi}\varphi_j(n)L_n,$$

is said to be the corresponding spherical scaling function associated with φ.

It holds sometimes that for all j, the sequence $(\widehat{\phi_j}(n))_n$ is stationary with zero stationary value. In this case, the system $\{\phi_j\}_{j\in\mathbb{N}} \subset L^2[-1,1]$ is called bandlimited. It holds that for a bandlimited scaling functions, each ϕ_j is a one-dimensional polynomial, and for all $F \in L^2(S^2)$, $\phi_j * F$ is a polynomial on S^2. The following theorem affirms that scaling functions permit to approximate L^2-ones with polynomial approximates (see [357]).

Theorem 1 *Let $\{\phi_j\}_{j\in\mathbb{N}}$ be a scaling function and $F \in L^2(S^2)$. Then*

$$\lim_{j\to\infty} \|F - \phi_j^{(k)} * F\|_{L^2(S^2)} = 0, \quad \forall k \in \mathbb{N}.$$

Here, for a function $\Phi \in L^2$, we designate by $\Phi^{(k)}$ the k-times self-convolution of Φ with itself. The last approximation is called spherical approximate identity.

Proof. Observe firstly that

$$\phi_j^{(k)} * F = \sum_{n=0}^{+\infty} \sum_{j=1}^{2n+1} \widehat{\Phi}_J(n) \widehat{F}(n,j) Y_{n,j}.$$

Thus,

$$F - \phi_j^{(k)} * F = \sum_{n=0}^{+\infty} \sum_{j=1}^{2n+1} (1 - \widehat{\Phi}_J(n)) \widehat{F}(n,j) Y_{n,j},$$

which, by applying Parseval identity, yields that

$$\|F \phi_j^{(k)} * F\|_2^2 = \sum_{n=0}^{+\infty} \sum_{j=1}^{2n+1} (1 - \widehat{\Phi}_J(n))^2 (\widehat{F}(n,j))^2.$$

Now, observing that the last series is J-uniformly convergent and the fact that

$$\lim_{J \to +\infty} (1 - \widehat{\Phi}_J(n)) = 0, \quad \forall n,$$

it results that

$$\lim_{j \to \infty} \|F - \phi_j^{(k)} * F\|_{L^2(S^2)} = 0.$$

3.3.3 Multiresolution analysis on the sphere

In this section, we will show that such scaling functions are suitable candidates to approximate functions in L^2 as it is needed in wavelet theory in general and thus they are suitable sources to define multi-resolution analysis and/or a wavelet analysis on the sphere. The next theorem shows the role of spherical scaling functions in the construction of multi-resolution analysis on the sphere.

Theorem 2 *Let for $j \in \mathbb{Z}$,*

$$V_j = \left\{ \phi_j^{(2)} * F \,\middle|\, F \in L^2(S^2) \right\}$$

where $\{\phi_j\}_{j \in \mathbb{N}} \subset L^2[-1,1]$ is a scaling function. Then, the sequence $(V_j)_j$ defines a multi-resolution analysis on the sphere. That is,

1. $V_j \subset V_{j+1} \subset L^2(S^2)$, $\forall j \in \mathbb{N}$.

2. $\overline{\bigcup_{j=0}^{\infty} V_j} = L^2(S^2)$.

For $j \in \mathbb{Z}$, the spaces V_j represent the so-called scale or approximation space at the level j.

Proof. 1. As $\Phi \in L^2$ and also F, the convolution $\Phi * F$ is also L^2.

Consider next, for $J \in \mathbb{Z}$, the function

$$\gamma_J(n) = \left(\frac{\widehat{\Phi}_J(n)}{\widehat{\Phi}_{J+1}(n)} \right)^2 \widehat{F}(n,j) \quad \text{if} \quad \Phi_{J+1}(n) \neq 0$$

and 0 else, and define the function G by

$$G = \sum_{n=0}^{+\infty} \sum_{j=1}^{2n+1} \gamma_j(n) Y_{n,j}.$$

It is straightforward that $G \in L^2$ and that $\widehat{G}(n,j) = \gamma_j(n)$. Furthermore,

$$
\begin{aligned}
\phi_{J+1}^{(2)} * G &= \sum_{n=0}^{+\infty} \sum_{j=1}^{2n+1} \widehat{\Phi}_{J+1}(n) \widehat{G}(n,j) Y_{n,j} \\
&= \sum_{n=0}^{+\infty} \sum_{j=1}^{2n+1} \widehat{\Phi}_J(n) \widehat{F}(n,j) Y_{n,j} \\
&= \phi_J^{(2)} * F.
\end{aligned}
$$

Hence,

$$\phi_J^{(2)} * F = \phi_{J+1}^{(2)} * G \in V_{J+1}.$$

Consequently,

$$V_J \subset V_{J+1}.$$

2. The density property is an immediate consequence of the spherical approximate identity proved in Theorem 1.

3.3.4 Existence of the mother wavelet

Based on this multi-resolution analysis of $L^2(S^2)$, we will be able to introduce spherical wavelets.

Definition 3.10 *Let* $\Phi = \{\phi_j\}_{j\in\mathbb{N}}$ *in* $L^2([-1,1])$ *be a scaling function and* $\Psi = \{\psi_j\}_{j\in\mathbb{N}\cup\{-1\}}$ *and* $\widetilde{\Psi} = \{\widetilde{\psi}_j\}_{j\in\mathbb{N}\cup\{-1\}}$ *be in* $L^2([-1,1])$ *satisfying the so-called refinement equation*

$$\widehat{\psi_j}(n)\widehat{\widetilde{\psi}_j}(n) = (\widehat{\phi_{j+1}}(n))^2 - (\widehat{\phi_j}(n))^2, \quad \forall n, j \in [0, +\infty).$$

Then,

a. Ψ *and* $\widetilde{\Psi}$ *are called, respectively, (spherical) primal wavelet and (spherical) dual wavelet relatively to* Φ.

b. *The functions* ψ_0 *and* $\widetilde{\psi}_0$ *are called the primal mother wavelet and the dual mother wavelets, respectively.*

Here, we set $\psi_{-1} = \widetilde{\psi}_{-1} = \phi_0$.

The following result due to Volker in [357] shows the existence of primal and dual wavelets.

Theorem 3 *Let* φ_0 *be a generator of a scaling function and* ψ_0 *and* $\widetilde{\psi}_0$ *be admissible function such that*

$$\psi_0 \widetilde{\psi}_0(x) = (\varphi_0(\frac{x}{2}))^2 - (\varphi_0(x))^2, \quad \forall x \in \mathbb{R}^+.$$

Then, ψ_0 *and* $\widetilde{\psi}_0$ *are generators of primal and dual mother wavelets, respectively.*

Proof. We will prove precisely that the dilated copies $\{\psi_j\}_{j\in\mathbb{N}\cup\{-1\}}$ and $\{\tilde{\psi}_j\}_{j\in\mathbb{N}\cup\{-1\}}$ defined via their Legendre coefficients by dilating ψ_0 and $\tilde{\psi}_0(x)$ as

$$\widehat{\psi_j}(n) = \psi_j(n) = \psi_0(2^{-j}n), \quad \forall n, j \in \mathbb{N},$$

$$\widehat{\tilde{\psi}_j}(n) = \tilde{\psi}_j(n) = \tilde{\psi}_0(2^{-j}n), \quad \forall n, j \in \mathbb{N}$$

and

$$\widehat{\psi_{-1}}(n) = \widehat{\tilde{\psi}_{-1}}(n) = \varphi_0(n), \quad \forall n \in \mathbb{N}$$

are a primal and dual wavelets, respectively. Indeed, considering these dilated copies, we obtain for all $n, j \in \mathbb{N}$,

$$
\begin{aligned}
\widehat{\psi_j}(n)\widehat{\tilde{\psi}_j}(n) &= \psi_0(2^{-j}n)\tilde{\psi}_0(2^{-j}n) \\
&= (\varphi_0(2^{-j-1}n))^2 - (\varphi_0(2^{-j}n))^2 \\
&= (\widehat{\phi_{j+1}}(n))^2 - (\widehat{\phi_j}(n))^2.
\end{aligned}
$$

A fundamental property of spherical wavelets is the scale-step property proved hereafter and which prepares to introduce the detail spaces.

Theorem 4 *Let $\Psi = \{\psi_j\}_{j\in\mathbb{N}\cup\{-1\}}$ and $\tilde{\Psi} = \{\tilde{\psi}_j\}_{j\in\mathbb{N}\cup\{-1\}}$ be a primal and a dual wavelet corresponding to the scaling function $\{\phi_j\}_{j\in\mathbb{N}} \subset L^2[-1,1]$. The following assertions hold for all $F \in L^2(S^2)$.*

i. $\phi_{J_2}^{(2)} * F = \phi_{J_1}^{(2)} * F + \displaystyle\sum_{j=J_1}^{J_2-1} \tilde{\psi}_j * \psi_j * F, \quad \forall J_1 < J_2 \in \mathbb{N}.$

ii. $F = \phi_J^{(2)} * F + \displaystyle\sum_{j=J}^{\infty} \tilde{\psi}_j * \psi_j * F, \quad \forall J \in \mathbb{N}.$

Proof. i. We will evaluate the last right-hand series term in the assertion. Using the definition of primal and dual wavelets, we obtain

$$
\begin{aligned}
\tilde{\psi}_j * \psi_j * F &= \sum_{n=0}^{+\infty}\sum_{s=1}^{2n+1} \widehat{\psi_j}(n)\widehat{\tilde{\psi}_j}(n)\widehat{F}(n,s)Y_{n,s} \\
&= \sum_{n=0}^{+\infty}\sum_{j=1}^{2n+1} \left[(\widehat{\phi_{j+1}}(n))^2 - (\widehat{\phi_j}(n))^2\right]\widehat{F}(n,s)Y_{n,s} \\
&= \phi_{j+1}^{(2)} * F - \phi_j^{(2)} * F.
\end{aligned}
$$

As a result,

$$\sum_{j=J_1}^{J_2-1} \tilde{\psi}_j * \psi_j * F = \phi_{J_2}^{(2)} * F - \phi_{J_1}^{(2)} * F.$$

ii. It is an immediate consequence of assertion **i.**

Theorem 5 *Let, for $j \in \mathbb{Z}$,*

$$W_j = \{\tilde{\psi}_j * \psi_j * F / F \in L^2(S^2)\}.$$

Then, for all $J \in \mathbb{Z}$,

$$V_{J+1} = V_J + W_J.$$

Proof. The inclusion $V_J \subset V_{J-1} + W_{J-1}$ is somehow easy and it is a consequence of Theorem 4. We will prove the opposite inclusion. So, let $F_1 \in V_J$ and $F_2 \in W_J$. We seek a function $F \in L^2$ for which we have

$$\Phi_{J+1}^{(2)} * F = F_1 + F_2.$$

Since $F_1 \in V_J$ and $F_2 \in W_J$, there exist G_1 and G_2 in L^2 such that

$$F_1 = \Phi_J^{(2)} * G_1 \quad \text{and} \quad F_2 = \tilde{\Psi}_J * \Psi_J * G_2.$$

Now, consider the function γ defined by

$$\gamma(n, j) = \left(\frac{(\widehat{\Phi}_J(n))^2 \widehat{G}_1(n, j) + ((\widehat{\Phi}_{J+1}(n))^2 - (\widehat{\Phi}_J(n))^2)\widehat{G}_2(n, j)}{\widehat{\Phi}_{J+1}(n)} \right)^2$$

whenever $\Phi_{J+1}(n) \neq 0$ and 0 else, and define the function F by

$$F = \sum_{n=0}^{+\infty} \sum_{j=1}^{2n+1} \gamma(n, j) Y_{n,j}.$$

It is straightforward that $F \in L^2$ and that $\widehat{F}(n, j) = \gamma(n, j)$. Furthermore,

$$
\begin{aligned}
\Phi_{J+1}^{(2)} * F &= \sum_{n=0}^{+\infty,*} \sum_{j=1}^{2n+1} (\widehat{\Phi}_{J+1}(n))^2 \widehat{F}(n, j) Y_{n,j} \\
&= \sum_{n=0}^{+\infty} \sum_{j=1}^{2n+1} (\widehat{\Phi}_J(n))^2 \widehat{G}_1(n, j) Y_{n,j} \\
&\quad + \sum_{n=0}^{+\infty} \sum_{j=1}^{2n+1} ((\widehat{\Phi}_{J+1}(n))^2 - (\widehat{\Phi}_J(n))^2)\widehat{G}_2(n, j) Y_{n,j} \\
&= \sum_{n=0}^{+\infty} \sum_{j=1}^{2n+1} (\widehat{\Phi}_J(n))^2 \widehat{G}_1(n, j) Y_{n,j} \\
&\quad + \sum_{n=0}^{+\infty} \sum_{j=1}^{2n+1} \widehat{\tilde{\Psi}}_J(n)\widehat{\Psi}_J(n)\widehat{G}_2(n, j) Y_{n,j} \\
&= \phi_J^{(2)} * G_1 + \tilde{\Psi}_J * \Psi_J * G_2 \\
&= F_1 + F_2.
\end{aligned}
$$

Consequently,

$$F_1 + F_2 \in V_{J+1}.$$

Definition 3.11 *For $j \in \mathbb{Z}$, the space W_j is called the detail space at the level j and the mapping*

$$(SWT)_j : L^2(S^2) \to L^2(S^2)$$

$$F \longmapsto \psi_j * F$$

is called the spherical wavelet transform at the scale j.

Based on this definition and the results above, any function $F \in L^2(S^2)$ will be represented by means of an L^2-convergent series

$$F = \sum_{j=-1}^{\infty} \tilde{\psi}_j * (SWT)_j(F). \qquad (3.2)$$

3.4 EXERCISES

Exercise 1.

Let Φ^1 be a scaling function generating a multiresolution analysis on $L^2([0,\pi])$ and Γ be a scaling function generating a multiresolution analysis on $L^2(\mathbb{R})$.

1. Prove that the function $\Phi^2(x) = \sum_{k \geq 0} \Gamma(x-k)$ is a scaling function that generates a multiresolution analysis on $L^2([0,2\pi])$.

2. Prove, for $j, k_1, k_2 \in \mathbb{Z}$,

$$\Phi_{j,k_1,k_2}(\theta,\varphi) = \Phi_{j,k_1}(\theta)\Phi_{j,k_2}(\varphi)$$

generates a multiresolution analysis on $L^2(\mathbb{S}^2)$, where $\mathbb{S}^2$ is the unit sphere in $\mathbb{R}^3$.

Exercise 2.

Let Γ be a C^1-curve in $\mathbb{R}^2$ with a parametrization $\gamma(t) = (u(t), v(t))$, $t \in \mathbb{R}$, where u and v are C^1 on $\mathbb{R}$. Assume that γ is a homeomorphism between $\mathbb{R}$ and Γ.
1) Prove that the map V defined on the cylinder by $V(\cos\theta, \sin\theta, z) = (u(z)\cos\theta, u(z)\sin\theta, v(z))$ transforms the cylinder homeomorphically to a surface of revolution $\mathfrak{G}$ about the z-sxis.
2) Prove that the surface element $d_{\sigma_\mathfrak{G}}$ on the cylinder is transformed to

$$d_{\sigma_\mathfrak{G}} = w(z)d\theta dz,$$

where $w(z) = |u(z)|[u'(z)^2 + v'(z)^2]^{\frac{1}{2}}$.
3) Prove that V induces a unitary map $\hat{V} : L^2(\mathfrak{C}, \partial_3\partial\theta) \to L^2(\mathfrak{G}, d_{\sigma_\mathfrak{G}})$ by

$$(\hat{V}_f)(u(z)\cos\theta, u(z)\sin\theta, v(z))) = (w(z))^{-\frac{1}{2}} f((\cos\theta, \sin\theta, z)).$$

Exercise 3.

Consider on the unit sphere $\mathbb{S}^2$ of $\mathbb{R}^3$ the map $V : \mathfrak{E} \to S^2$ defined by

$$V(\cos\theta, \sin\theta, z) = (\cosh^{-1}(z)\cos\theta, \sinh^{-1}(z)\sin\theta, \tanh(z)).$$

Show that V induces a unitary map $\tilde{V} : L^2(\mathfrak{E}, dzd\theta) \to L^2(\mathbb{S}^2, \cosh^{-2}(z)dzd\theta)$ by

$$(\tilde{V}f)\left(\cosh^{-1}(z)\cos\theta, \sinh^{-1}(z)\sin\theta, \tanh(z)\right) = \cosh(z).f(\cos\theta, \sin\theta, z).$$

Let now $G = \mathbb{R} \times \mathbb{R}/\mathbb{Z}$ and $H_0 = (\mathbb{R} \times \mathbb{R}/\mathbb{Z}) \times (\hat{\mathbb{R}} \times \mathbb{R}/\mathbb{Z}) \times \mathbb{T}$, where $\pi = \mathbb{R}/\mathbb{Z}$ and $\mathbb{T}$ is the torus in $\mathbb{R}^3$. 3) Show that the map $\tilde{U}$ defined by

$$(\tilde{U}((x,\varphi);(w,k):\eta)f)\begin{pmatrix} \cosh^{-1}(z)\cos\theta \\ \cosh^{-1}(z)\sin\theta \\ \tanh(z) \end{pmatrix} = \Phi f \begin{pmatrix} \cosh^{-1}(z-x)\cos(\theta-\varphi) \\ \cosh^{-1}(z-x)\sin(\theta-\varphi) \\ \tanh(z-x) \end{pmatrix}$$

where

$$\Phi = \Phi((x,\varphi);(w,k):\eta;z)) = \eta[\frac{\cosh(z)}{\cosh^{-1}(z-x)}]e^{2\pi i(w.z+k\theta)},$$

is a representation of H_G on $L^2(S^2, \cosh^{-2}(z)dzd\theta)$.

Exercise 4.

For $\alpha, \gamma > 0$, let $\varepsilon_{\alpha,\gamma} = \left\{(x,y,z); \dfrac{x^2+y^2}{\alpha^2} + \dfrac{z^2}{\gamma^2} = 1\right\}$ be a two-dimensional ellipsoid of revolution and consider the map $V : \mathfrak{E} \to \varepsilon_{\alpha,\gamma}$ defined by

$$V : (\cos\theta, \sin\theta, z) \mapsto \left(\alpha\cosh^{-1}(z)\cos(\theta), \alpha\cosh^{-1}(z)\sin(\theta), \gamma\tanh(z)\right).$$

Show that V induces a unitary map

$$\tilde{V} : L^2(\mathfrak{E}, dzd\theta) \to L^2(\varepsilon_{\alpha,\gamma}, \rho_{\varepsilon_{\alpha,\gamma}}(z)dzd\theta)$$

by

$$(\tilde{V}f)\begin{pmatrix} \alpha\cosh^{-1}(z)\cos\theta \\ \alpha\cosh^{-1}(z)\sin\theta \\ \gamma\tanh(z) \end{pmatrix} = \frac{1}{\sqrt{\rho_{\varepsilon_{\alpha,\gamma}}(z)}}f\begin{pmatrix} \cos\theta \\ \sin\theta \\ z \end{pmatrix}$$

where we have

$$\rho_{\varepsilon_{\alpha,\gamma}}(z) = \frac{\alpha}{\cosh^3 z}\sqrt{\alpha^2\sinh^2 z + \gamma^2}$$

.

Exercise 5.

Let, for $(x, y) \in \mathbb{R}^2$, $r = \sqrt{x^2 + y^2}$ and $(x, y) = r\omega$ with

$$\omega = (\cos\theta, \sin\theta); \quad \theta \in (-\pi, \pi).$$

Consider the functions

$$\psi_0(x, y) = C_0(1 + r^2)^{3/2}e^{-r^2/2},$$
$$\psi_1(x, y) = C_1(r^2 - 2)(1 + r^2)^{3/2}e^{-r^2/2}(x, y),$$
$$\psi_2(x, y) = C_2(-4 + 6r^4 - 2r^6)(1 + r^2)^{3/2}e^{-r^2/2},$$

and

$$\psi_3(\underline{x}) = C_3(1 - 16r^2 + 21r^4 + 15r^6 + r^8)(1 + r^2)^{3/2}e^{-r^2/2}(x, y),$$

where the C_j's ($j = 0, 1, 2, 3$) are constants.
1) Compute the constants C_j's ($j = 0, 1, 2, 3$) in such a way we get normalized functions on the circle S^1.
2) Compute the Fourier transform of the ψ_j's, $j = 1, 2, 3, 4$.
3) Deduce possible wavelets on the circle S^1.

Exercise 6.

Let, for $(x, y, z) \in \mathbb{R}^3$, $\rho = \sqrt{x^2 + y^2 + z^2}$ and $(x, y, z) = \rho\omega$ with

$$\omega = e_1\cos\theta\sin\phi + e_2\sin\theta\sin\phi + e_3\cos\phi; \quad \theta \in (-\pi, \pi), \ \phi \in (0, \pi).$$

Consider the functions

$$\psi_1(\rho, \theta, \phi) = C_1(1 + \rho^2)^{3/2}e^{-\rho^2/2},$$
$$\psi_2(\rho, \theta, \phi) = C_2(1 + \rho^2)^{3/2}e^{-\rho^2/2}(\rho^2 - 2)\rho\omega,$$
$$\psi_3(\rho, \theta, \phi) = C_3(1 + \rho^2)^{3/2}e^{-\rho^2/2}(-6 - \rho^2 + 7\rho^4 - 2\rho^6)\rho\omega,$$

and

$$\psi_4(\rho, \theta, \phi) = C_4(1 + \rho^2)^{3/2}e^{-\rho^2/2}(1 - 16\rho^2 + 18\rho^4 + 15\rho^6 + \rho^8)\rho\omega,$$

where the C_j's ($j = 0, 1, 2, 3$) are constants.
1) Compute the constants C_j's ($j = 1, 2, 3, 4$) in such a way we get normalized functions on the sphere S^2.
2) Compute the Fourier transform of the ψ_j's, $j = 1, 2, 3, 4$.
3) Deduce possible wavelets on the sphere S^2.

Exercise 7.

Denote $\mathbb{T} = \mathbb{R}/\mathbb{Z}$ and consider $f \in L^1(\mathbb{T})$ such that $f(x) = \sin(2\pi(\gamma x + \theta))$, where $\gamma, \theta \in \mathbb{R}$ are fixed and $\gamma \in \mathbb{Z} \setminus \{0\}$. Let the generalized Haar wavelet be

$$\psi = \mathbf{1}_{[-\frac{1}{2},0)} + \mathbf{1}_{[0,\frac{1}{2})}.$$

Prove that

$$W_\psi f(a, b) = \frac{2}{\gamma\pi}\sin^2(\frac{\pi\gamma a}{2})\cos(2\pi(\gamma b + \theta)), \tag{3.3}$$

for all $(a, b) \in \mathbb{R}^+ \times \mathbb{R}$.

CHAPTER 4

Clifford wavelets

4.1 INTRODUCTION

The English mathematician William Kingdon Clifford introduced in 1878 a general set of algebras, which he called geometric algebras and which are now attached to his name as Clifford algebras. The interest in multivariate analysis using Clifford algebras started to grow in the last century. Since then, great works on Clifford analysis referring different classes of special functions have appeared.

Clifford analysis also offers some general context of Fourier one in signal/image processing by applying real, complex and quaternion numbers. In matrix spaces, Clifford algebras may be constructed using Pauli and Dirac matrices, which makes them suitable physical space such as Minkowski space-time and thus a unifying language for both mathematics and physics (see [83], [84]).

Clifford analysis offers a function theory, which is a higher-dimensional analogue of the theory of holomorphic functions in the complex plane, centered around the notion of monogenic functions. Indeed, monogenic function theory is considered generalization of the holomorphic function theory in the complex plane to higher dimensions and a refinement of the harmonic analysis based on the Laplace operator's factorizations. The construction of spherical monogenic functions has been studied for decades with different methods. Recently, orthogonal monogenic bases are developed for spheroidal reference domains.

Many works have related spheroidal monogenic functions to wavelets in the context of Clifford analysis. Therefore, classical wavelet theory has been constructed in the framework of Clifford analysis (see [75], [79], [80], [82], [83], [86]).

Clifford analysis, in its most basic form, is a refinement of harmonic analysis in higher-dimensional Euclidean spaces. By introducing the so-called Dirac operator, researchers introduced the notion of monogenic functions extending holomorphic ones. In this context, different concepts of real and complex analysis have been extended to the Clifford case, such as Fourier transform extended to Clifford-Fourier transform and derivation of functions. For example, Clifford-Fourier transform is related to or expressed in terms of an exponential operator. For the even-dimensional case, it yields a kernel based on Bessel functions. Similar to the classical Fourier transform, here also the new kernel satisfies a system of differential equations (see [79], [80], [82], [83], [86], [190]).

In [88], the authors survey the historical development of quaternion and Clifford-Fourier transforms and wavelets. Basic concepts have been revisited, and mathematical formulations have been enlightened. Hypercomplex Fourier transforms and wavelets have been revisited with overviews on quaternion Fourier transforms, Clifford-Fourier transforms and quaternion and Clifford wavelets.

Clifford algebra is characterized by additional facts as it provides a simpler model of mathematical objects compared to vector algebra. It also permits a simplification in the notations of mathematical expressions such as plane and volume segments in two and three as well as higher dimensions by using a coordinate-free representation. Such representation is characterized by one important feature resumed in the fact that the motion of an object may be described with respect to a coordinate frame defined on the object itself. This means that it permits use of a self-coordinate system related to the object in hand (see [75], [187], [286]).

Compared to spheroidal functions, wavelets are characterized by scale invariance of approximation spaces. Clifford algebra is one mathematical object that owns this characteristic. Recall that multiplication of real numbers scales their magnitudes according to their position in or out from the origin. However, multiplication of the imaginary part of a complex number performs a rotation: it is a multiplication that goes round and round instead of in and out. So, a multiplication of spherical elements by each other results in an element of the sphere. Again, repeated multiplication of the imaginary part results in orthogonal components. Thus, we need a coordinate system that results always in the object, a concept that we will see again and again in the algebra. In other words, Clifford algebra generalizes to higher dimensions by the same exact principles applied at lower dimensions, by providing an algebraic entity for scalars, vectors, bivectors and trivectors, and there is no limit to the number of dimensions it can be extended to. More details on Clifford algebra—its origins, history and developments—may be found in [2], [181], [182], [183], [184], [187], [318], [286].

From the applied point of view, Clifford algebras/analysis has been the object of many applications, especially in image processing. For example, in [92, 96, 250] algorithms and methods based on Clifford algebras have been developed for the aim of segmentation and color alterations. Clifford algebras have also been shown to be good framework for the geometry of images. See [44, 337]. In [223], a Marr wavelet model has been applied on samples of intensity values for each pixel of image to estimate the probability density function of the pixel intensity. See also [360, 223, 342]. In [90] and [91], the authors introduced new definition for general geometric Fourier transform covering some Clifford cases and showed possible applications in image processing.

4.2 DIFFERENT CONSTRUCTIONS OF CLIFFORD ALGEBRAS

Clifford analysis appeared as a generalization of the complex analysis and Hamiltonians. It extended complex calculus to some type of finite-dimensional associative algebra known as Clifford algebra endowed with suitable operations as well as inner products and norms. It is now applied widely in a variety of fields including geometry and theoretical physics. Clifford analysis offers a functional theory extending the theory of holomorphic functions of one complex variable.

4.2.1 Clifford original construction

The original idea of construction due to Clifford is based on the exterior Grassmann algebra $\bigwedge \mathbb{R}^m$ associated with the linear space $\mathbb{R}^m$. Consider an orthonormal basis $\{e_1, e_2, \cdots, e_n\}$ of $\mathbb{R}^m$ such as the canonical basis. The basis of the Clifford algebra $\bigwedge \mathbb{R}^m$ is composed of all possible k-exterior products of the elements $\{e_1, e_2, \cdots, e_n\}$, $0 \leq k \leq n$. We obtain $e_\emptyset = 1$, $e_{i_1 i_2 \ldots i_k} = e_{i_1} \wedge e_{i_2} \vdots \wedge e_{i_k}$, $1 \leq k \leq n$ by assuming that $e_i \wedge e_j = -e_j \wedge e_i$ for $i \neq j$ and $e_i \wedge e_i = 1$, $1 \leq i \leq n$. By this way, we obtain an associative non-commutative algebra on $\mathbb{R}$ with dimension 2^n denoted by $\mathbb{R}_m$.

Besides, Clifford himself has used a quite different way in [168] by putting $e_i \wedge e_i = -1$ and thus obtaind what he called the anti-Euclidean space $\mathbb{R}^{0,n}$ and the associated algebra.

4.2.2 Quadratic form-based construction

Clifford algebras have been introduced by using quadratic forms such as the construction developed in [331]. Let (V, Q) be an n-dimensional quadratic space and B be the bilinear form associated with the quadratic form Q. Consider next an associative algebra $\mathcal{A}$ with the following rules on addition and multiplication:

$$x^2 = Q(x) \quad \text{and} \quad xy + yx = 2B(x, y).$$

Let next $\{e_1, e_2, \cdots, e_n\}$ be an orthonormal basis of V (orthonormality relatively to the bilinear form B). We may obtain a structure of Clifford algebra for $\mathcal{A}$. More precisely, the Clifford algebra may be introduced as follows (see, for example, [186, 288]).

Let V be an n-dimensional real vector space and $Q : V \longrightarrow \mathbb{R}$ a quadratic form on V. Let also $\mathcal{A}$ be a real associative algebra with identity $\mathbf{1}_\mathcal{A}$ such that

   i. $\mathbb{R}$ and V are linear subspaces of $\mathcal{A}$.

   ii. For all $\underline{v} \in V$, we have $v^2 = vv = B(v, v) = Q(v)$.

   iii. The algebra $\mathcal{A}$ is generated as a real algebra by 1 and V.

Then, the algebra $\mathcal{A}$ is the Clifford algebra associated with the quadratic space (V, Q).

Already with the quadratic forms, we may observe Clifford algebras as a quotient result of suitable vector space and suitable quadratic form. This was the subject of [158] and resembles somehow the original construction due to Clifford and is based on exterior product. The construction here will instead apply the tensor product.

Let V be a vector space over the real field $\mathbb{R}$ with $dim(V) = n$ and $Q : V \longrightarrow \mathbb{R}$, a quadratic form. Consider next the tensor algebra of V denoted by $T(V) = \bigoplus_{k=0}^{\infty} \otimes^k V$.

Consider next the two-sided ideal $\mathcal{I}(Q)$ generated by the elements of the form

$\underline{x} \otimes \underline{x} - Q(\underline{x})\mathbf{1}_V$, where $\underline{x} \in V$ and $\mathbf{1}_V$ is the identity element for the multiplication in V. Then, the quotient $Cl(V, Q) = T(V)/\mathcal{I}(Q)$ is called the Clifford algebra associated with (V, Q).

4.2.3 A standard construction

We now present a standard construction of Clifford algebra issued from the Euclidean space $\mathbb{R}^m$ (or $\mathbb{C}^m$), which is the simplest one used in the majority of works dealing with this subject. See [288]. Consider an orthogonal basis of $\mathbb{R}^m$ (or $\mathbb{C}^m$) such as the canonical one and let A be the 2^n-dimensional linear space generated by the set

$$\{1, e_{i_1 \cdots i_k} = e_{i_1} e_{i_2} \cdots e_{i_k} | 1 \leq i_1 < i_2 < \cdots < i_k \leq n \text{ and } 1 \leq k \leq n\},$$

where a multiplication is defined by

$$e_i e_j + e_j e_i = 2\delta_{ij},$$

where δ_{ij} stands for the Kronecker symbol. We get here a Clifford algebra, which will be denoted by $\mathbb{R}_n$ (or $\mathbb{C}_n$ for the complex case). More precisely, we have in the literature the following definition.

Definition 4.1 *The Clifford algebra associated with $\mathbb{R}^m$, endowed with the usual Euclidean metric, is the extension of $\mathbb{R}^m$ to a unitary, associative algebra denoted by $\mathbb{R}_m$ over the reals, for which*

1. $x^2 = -|x|^2$ for any $x \in \mathbb{R}^m$,

2. $\mathbb{R}_n$ is generated (as an algebra) by $\mathbb{R}^m$,

3. $\mathbb{R}_n$ is not generated (as an algebra) by any proper subspace of $\mathbb{R}^m$.

Theorem 6 *The Clifford algebra $\mathbb{R}_m$ is uniquely determined up to an algebra isomorphism.*

The proof is based on the following preliminary result.

Lemma 4.2 *Denote for $I = (i_1, i_2, \ldots, i_l)$, $0 \leq l \leq n$, $e^I = (-1)^l e_{i_l} e_{i_{l-1}} \ldots e_{i_1}$ and for $a = (a_1, a_2, \ldots, a_n) \in \mathbb{R}^m$, $a_I = a_{i_1} a_{i_2} \ldots a_{i_l}$. We have*

$$2^{-n} \sum_I{}' e_I a e^I = \begin{cases} a_\emptyset & \text{if } n \text{ is even,} \\ a_\emptyset + a_{(1,2,\ldots,n)} e_{(1,2,\ldots,n)} & \text{if } n \text{ is odd.} \end{cases}$$

Proof. Denote $|.|$ the cardinality function. It is straightforward that for all $I, J \subseteq \{1, 2, \ldots, n\}$,

$$e_I e_J = (-1)^{|I||J|-|I \cap J|} e_J e_I.$$

As a result, whenever $x = \sum_J{}' x_J e_J \in \mathbb{R}_m$, we get

$$\sum_I{}' e_I x e^I = \sum_{I,J}{}' x_J e_I e_J e^I = \sum_{I,J}{}' (-1)^{|I||J|-|I \cap J|} x_J e_J.$$

Furthermore, for a fixed $J \subseteq \{1, 2, ..., n\}$,

$$\sideset{}{'}\sum_I (-1)^{|I||J|-|I \cap J|} = \sum_{i=0}^{|J|} \sum_{j=0}^{n-|J|} \sideset{}{'}\sum_{|I|=i+j, |I \cap J|=i} (-1)^{(i+j)|J|-i}$$

$$= \sum_{i=0}^{|J|} \sum_{j=0}^{n-|J|} \sideset{}{'}\sum_{|I|=i+j, |I \cap J|=i} (-1)^{j|J|}(-1)^{i|J|-1} C_{|J|}^i C_{n-|J|}^j$$

$$= (1 + (-1)^{|J|})^{n-|J|}(1 + (-1)^{|J|-1})^{|J|}$$

$$= 0$$

unless either n is odd and $J = \{1, 2, ..., n\}$, or $J = \emptyset$. In these cases, we end up with 2^n.

Proof of Theorem 6. It suffices to show that the set $(e_I)_{0 \leq |I| = l \leq n}$ is linearly independent. In other words, whenever the sum $\sideset{}{'}\sum_I a_I e_I$ is zero, necessarily $a_I = 0$ for all I. Using the minimality of $\mathbb{R}_n$, we observe that 1 and $e_{1,2,...,n}$ are linearly independent. Hence, $a_\emptyset = 0$. Next, by replacing a with $e^l a$, for arbitrary I, the result follows.

In the sequel, it will also be useful to embed $\mathbb{R}^{m+1}$ into $\mathbb{R}_m$ by identifying $(x_0, x) \in \mathbb{R}^{m+1} = \mathbb{R} \oplus \mathbb{R}^m$ with $x_0.e_0 + x \in \mathbb{R}^m$ (note that, by Assertion 1 in Definition 4.1, $1 \notin \mathbb{R}^m$), and call these elements Clifford vectors.

Proposition and Definition 4.3 • $\forall x \in \mathbb{R}^m \setminus \{0\}$, *it has a multiplicative inverse given by*

$$x^{-1} = \frac{\overline{x}}{|x|^2}.$$

• *The multiplicative group generated by all Clifford invertible vectors in $\mathbb{R}_m$ is called the Clifford group.*

Definition 4.4 • *For $a \in \mathbb{R}_n$, the real part of a is defined by*

$$Re(a) = a_\emptyset.$$

• *$\mathbb{R}_m$ is endowed with the natural Euclidean metric*

$$|a| = \sqrt{Re(a\overline{a})} = \sqrt{Re(\overline{a}a)}.$$

Lemma 4.5 • *For all $x, y \in \mathbb{R}_m$, we have*

$$|xy| \leq 2^{n-1}|x||y|$$

• *For all $x, y \in \mathbb{R}_m$, the quality*

$$|xy| = |x||y|$$

yields that at least one of x, y belongs to the Clifford group of $\mathbb{R}_m$.

4.3   GRADUATION IN CLIFFORD ALGEBRAS

We will focus, from now on, on the real Clifford algebra $\mathbb{R}_m$ (respectively, its complexification $\mathbb{C}_m$ if necessary) associated with the real (respectively, complex) Euclidean space $\mathbb{R}^m$ (respectively, $\mathbb{C}^m$).

The concept of graduation in Clifford algebra is an important task as it permits observation of the whole algebra in the form of a direct sum of subalgebras where the operations, especially multiplication, are easier and simpler.

Recall that an element $a \in \mathbb{R}_m$ is written in the form $a = \sum_A a_A e_A$, where A is an arbitrary ordered multi-index set $A = \{(i_1 i_2 \cdots i_k)\}$ with $1 \leq i_1 < i_2 < \cdots < i_k \leq n$ and where $k = |A|$ is the cardinality of A. For $|A| = 0$, we set $e_\emptyset = 1$. This way the element a may be written as

$$a = \sum_{k=0}^{n} \sum_{k=|A|} a_A e_A.$$

The subspace $\mathbb{R}_m^k = span_{\mathbb{R}} \{e_A, |A| = k\}$ is known as subspace of grade k. The subspace of grade 0 is the field $\mathbb{R}$ whose elements are called scalars, the subspace of grade 1 is the vector space $\mathbb{R}^m$ composed of vectors, the subspace of grade 2 is composed of the so-called bivectors. This may be continued for any grade $k \leq n$. This last grade gives the so-called pseudo-scalars. We obtain the decomposition

$$\mathbb{R}_m = \mathbb{R}_m^0 \oplus \mathbb{R}_m^1 \oplus .... \oplus \mathbb{R}_m^n$$

An element $a \in \mathbb{R}_n$ will be written as

$$
\begin{aligned}
a = \;& \underset{scalars}{a_\emptyset 1} + \underset{vectors}{(a_1 e_1 + ... + a_n e_n)} \\
& + \underset{bivectors}{(a_{12} e_{12} + a_{13} e_{13} + ...a_{ij} e_{ij} + ... + a_{n-1n} e_{n-1} e_n)} \\
& + ... + \underset{pseudo-scalar}{(a_{123...n} e_{123....n})}.
\end{aligned}
$$

Also, the element a may be written as $a = \sum_{k=0}^{n} [a]_k$, where $[a]_k$ is the projection of a on $\mathbb{R}_m^k$. It is straightforward that the projector $[.]_k : \mathbb{R}_m \longrightarrow \mathbb{R}_m^k$ for $k = 0, 1, \cdots, n$ satisfies

i. $[[a]_k]_k = [a]_k, \forall a \in \mathbb{R}_m$.

ii. $[\lambda a]_k = \lambda [a]_k = [a]_k \lambda, \forall a \in \mathbb{R}_m$ and $\lambda \in \mathbb{R}$.

iii. $[a + b]_k = [a]_k + [b]_k, \forall a, b \in \mathbb{R}_m$ and $\lambda \in \mathbb{R}$.

To finish with this review, we recall finally that the Clifford algebra $\mathbb{R}_m$ is $\mathbb{Z}_2$-grader in the sense that it may be split as

$$\mathbb{R}_m = \underset{k \text{ even}}{\bigoplus \mathbb{R}_m^k} \oplus \underset{k \text{ odd}}{\bigoplus \mathbb{R}_m^k}.$$

We usually denote

$$\mathbb{R}_m^+ = \bigoplus_{k \text{ even}} \mathbb{R}_m^k$$

and

$$\mathbb{R}_m^- = \bigoplus_{k \text{ odd}} \mathbb{R}_m^k.$$

Any element a may be written consequently as $a = [a]^+ + [a]^-$, where $[a]^\pm \in \mathbb{R}_m^\pm$. Moreover, we have the inclusions

$$\left.\begin{matrix} \mathbb{R}_m^+ \mathbb{R}_m^+ \\ \mathbb{R}_m^- \mathbb{R}_m^- \end{matrix}\right\} \subset \mathbb{R}_m^+ \quad \text{and} \quad \left.\begin{matrix} \mathbb{R}_m^+ \mathbb{R}_m^- \\ \mathbb{R}_m^- \mathbb{R}_m^+ \end{matrix}\right\} \subset \mathbb{R}_m^-.$$

Definition 4.6 *The center of the Clifford algebra $\mathbb{R}_m$ is*

$$\mathcal{Z}(\mathbb{R}_m) := \{a \in \mathbb{R}_m, \ ab = ba, \ \forall b \in \mathbb{R}_m\}.$$

We may observe easily that

$$\mathcal{Z}(\mathbb{R}_m) = \begin{cases} \mathbb{R} & \text{for } n \text{ even,} \\ \mathbb{R} \oplus \mathbb{R}e_{123\cdots n} & \text{for } n \text{ odd.} \end{cases}$$

4.4   SOME USEFUL OPERATIONS ON CLIFFORD ALGEBRAS

4.4.1   Products in Clifford algebras

Although there are different constructions of Clifford algebras, they intersect in many characteristics such as the different products. Indeed, we will see in the present section that tensor, exterior and Clifford products yield each other.

Let $\underline{x} = \sum_{i=1}^{n} e_i x_i$ and $\underline{y} = \sum_{i=1}^{n} e_i y_i$ be two elements in $\mathbb{R}^m$. We recall that the exterior is defined by

$$\underline{x} \wedge \underline{y} = \sum_{i<j} e_i e_j (x_i y_j - x_j y_i).$$

Figure 4.1 illustrates the exterior (wedge) product. Moreover, an interior product in the Clifford algebra $\mathbb{R}_n$ compatible with Definition 4.1, may be defined as

$$\underline{x} \cdot \underline{y} = -\sum_{j=1}^{n} x_j y_j = - <\underline{x}, \underline{y}>_{\mathbb{R}^m}$$

These two products combined yield the Clifford product

$$\underline{x}\underline{y} = \underline{x} \cdot \underline{y} + \underline{x} \wedge \underline{y}.$$

Otherwise, we observe conversely that

$$\underline{x} \cdot \underline{y} = \frac{1}{2}(\underline{x}\underline{y} + \underline{y}\underline{x}) \quad \text{and} \quad \underline{x} \wedge \underline{y} = \frac{1}{2}(\underline{x}\underline{y} - \underline{y}\underline{x}).$$

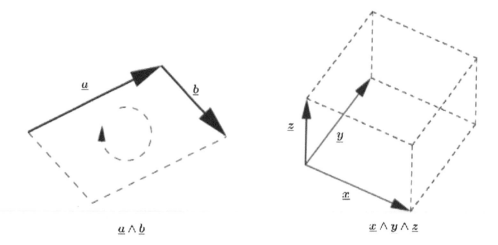

Figure 4.1 Wedge product of vectors.

4.4.2 Involutions on a Clifford algebra

There are many types of involutions that may be defined on Clifford algebras. The first one, known as the main involution or grade involution, is usually denoted by the tilde symbol~and constitutes an extension of vector reflection relatively to the origin in $\mathbb{R}^m$ to the whole algebra $\mathbb{R}_n$. It is defined on the basis elements e_A by

$$\widetilde{e_A} = (-1)^{|A|} e_A,$$

where as usual A is a multi-index. As a consequence, we get

$$\widetilde{\lambda} = \lambda \quad \text{and} \quad \widetilde{\underline{x}} = -\underline{x}, \ \forall \lambda \in \mathbb{R}, \ \forall \underline{x} \in \mathbb{R}^m.$$

More generally, we have

   i. $\widetilde{a} = [a]^+ - [a]^-, \ \forall a \in \mathbb{R}_m.$

   ii. $\widetilde{\widetilde{a}} = a, \ \forall a \in \mathbb{R}_m.$

   iii. $\widetilde{(a+b)} = \widetilde{a} + \widetilde{b}, \ \forall a, b \in \mathbb{R}_m.$

   iv. $\widetilde{(ab)} = \widetilde{a}\widetilde{b}, \ \forall a, b \in \mathbb{R}_m.$

The second involution type operation is known as the **reversion**, denoted by the symbol *, and is defined on the basis elements by

$$e_A^* = (-1)^{\frac{|A|(|A|-1)}{2}} e_A.$$

As a consequence, we get

   i. $\lambda^* = \lambda, \ \forall \lambda \in \mathbb{R}.$

   ii. $\underline{x}^* = \underline{x}, \ \forall \underline{x} \in \mathbb{R}^m.$

iii. $a^* = [a]^+ + [a]^-$, $\forall a \in \mathbb{R}_m$.

iv. $a^{**} = a$, $\forall a \in \mathbb{R}_m$.

v. $(a + b)^* = a^* + b^*$, $\forall a, b \in \mathbb{R}_m$.

vi. $(ab)^* = b^* a^*$, $\forall a, b \in \mathbb{R}_m$.

Sometimes we also need to introduce a Clifford conjugation, which consists of a composition of the main involution and the reversion. It is defined on the basis elements as

$$\overline{e_A} = (-1)^{\frac{|A|(|A|+1)}{2}} e_A.$$

As a result, we obtain

i. $\overline{\lambda} = \lambda$, $\forall \lambda \in \mathbb{R}$.

ii. $\overline{\underline{x}} = -\underline{x}$, $\forall \underline{x} \in \mathbb{R}^m$.

iii. $\overline{\overline{a}} = a$, $\forall a \in \mathbb{R}_m$.

iv. $\overline{(a + b)} = \overline{a} + \overline{b}$, $\forall a, b \in \mathbb{R}_m$.

v. $\overline{(ab)} = \overline{b}\,\overline{a}$, $\forall a, b \in \mathbb{R}_m$.

We achieve now the involutions with the complex case $\mathbb{C}_n$, where we deal with the complex Clifford conjugation. Recall here that an element $z \in \mathbb{C}_m$ can be written in the form $z = a + ib$, where $a, b \in \mathbb{R}_m$. The complex conjugation will be

$$z^\dagger = \overline{a} - i\overline{b}.$$

We may show easily that

$$(uv)^\dagger = u^\dagger v^\dagger \quad \text{and} \quad (\alpha u + \beta v)^\dagger = \alpha^c u^\dagger + \beta^c v^\dagger,$$

where $u, v \in \mathbb{C}_m$ and $\alpha, \beta \in \mathbb{C}$ and where the symbol $.^c$ stands for the complex conjugate in $\mathbb{C}$.

To finish with the operations on the Clifford algebra, we recall now the norm on Clifford algebra $\mathbb{R}_m$ or its complexification $\mathbb{C}_m$. We define an inner product on $\mathbb{R}_m$ as

$$\langle a, b \rangle = \left[a\overline{b} \right]_0 = [b\overline{a}]_0.$$

The associated norm will be defined for elements $a = \sum_A a_A e_A \in \mathbb{R}_m$ by

$$|a|^2 = \overline{a}a = \overline{a}a = \sum_A a_A^2.$$

It is easy to check that

i. $|a + b| \leq |a| + |b|$, $\forall a, b \in \mathbb{R}_m$.

ii. $|ab| \neq |a||b|$ in general.

iii. $|ab| \leq 2^n |a||b|$, $\forall a, b \in \mathbb{R}_m$.

iv. $|a\underline{x}| = |a||\underline{x}|$, $\forall \underline{x} \in \mathbb{R}^m$.

4.5 CLIFFORD FUNCTIONAL ANALYSIS

In this section, we propose to review the basic concepts of functional analysis on Clifford algebras. Recall that Clifford analysis constitutes in its basic form an extension of harmonic analysis in higher-dimensional Euclidean spaces. Different concepts of real and complex analysis will be reviewed. In this context, we will use for many times the Euclidean space $\mathbb{R}^{m+1}$ as a subalgebra in the Clifford algebra $\mathbb{R}_m$.

The most important concept in Clifford algebra-valued functions is the concept of monogenicity. One of the definitions starts by introducing a concept of differentiability for a pair of functions.

Definition 4.7 *Let f, g be two locally bounded, Clifford-valued functions defined in an open domain $\Omega \subseteq \mathbb{R}^{m+1}$. We introduce the Clifford derivative of the ordered pair (f, g) at a point $X \in \Omega$ by*

$$D(f|g)(\underline{x}) = \lim_{Q \downarrow \underline{x}} \frac{\int_{\partial Q} f \, \eta \, g \, d\sigma}{\int_Q dV}$$

for any domain Q such that $\underline{x} \in A \subset \Omega$ and where $d\sigma$ and dV are, respectively, the Lebesgue measures on ∂Q and Q. η is the outward normal vector to ∂Q.

More specifically, the pair (f, g) is Clifford differentiable at $\underline{x}$ if there exists an element $c \in \mathbb{R}^{m+1}$ such that, for any $\epsilon > 0$, there exists an open neighborhood $U \subseteq \Omega$ of X such that

$$\left| \int_{\partial Q} f \, n \, g \, d\sigma - c Vol(Q) \right| < \epsilon Vol(Q)$$

for all cube Q of $\mathbb{R}^{m+1}$ with $\underline{x} \in Q \subset U$ and such that, for some a priori fixed positive number C, $Vol(Q) \geq C diam(Q)^{m+1}$. Here, n is the outward unit normal to the boundary of the cube Q, $d\sigma$ is the surface measure of ∂Q and Vol stands for the usual Lebesgue measure in $\mathbb{R}^{m+1}$.

Definition 4.8 *The ordered pair (f, g) is said to be absolutely continuous on Ω if for any cube $Q \subset \Omega$ and any $\epsilon > 0$ there exists $\delta > 0$ such that*

$$\sum_{i \in J} \left| \int_{\partial Q_i} f \, \eta \, g \, d\sigma \right| \leq \epsilon$$

for any finite subdivision $(Q_i)_{i \in I}$ of Q, and any subset $J \subset I$ for which $\sum_{i \in J} Vol(Q_i) \leq \delta$.

Furthermore, we may go further and discuss a Stokes analogue formula. Let Ω be a bounded Lipschitz domain in $\mathbb{R}^{m+1}$ and let f, g be two Clifford algebra-valued continuous functions on Ω such that $D(f|g)$ is also continuous on Ω. We have

$$\int_{\partial \Omega} f \, \eta \, g \, d\sigma = \int \int_\Omega D(f|g) \, dV.$$

Besides, if the pair (f, g) is differentiable, it is easy to check the analogue of Leibnitz rule

$$D(f|g) = D(f|1)g + fD(l|g).$$

Denote $Dg = D(l|g)$ and $fD = D(f|1)$. It is straightforward that

$$\int_{\partial\Omega} f\, \eta\, g d\sigma = \int\int_{\Omega} (fD)g + f(Dg) dV.$$

We also set

$$\overline{D}f = \overline{D(\overline{f}|1)} \quad \text{and} \quad f\overline{D} = \overline{D(1|\overline{f})}.$$

It should be pointed out that at any point of differentiability $\underline{a} \in \Omega$ of the Clifford algebra-valued function $f = \sum_I' f_I e_I$, we may check that

$$(Df)(\underline{a}) = \sum_I' \sum_{j=0}^{n} \frac{\partial f_I}{\partial x_j}(\underline{a}) e_j e_I$$

and

$$(fD)(\underline{a}) = \sum_I' \sum_{j=0}^{n} \frac{\partial f_I}{\partial x_j}(\underline{a}) e_I e_j.$$

Notice that, by linearity considerations, it suffices to treat the case of a scalar-valued function f. We can also assume that the point of differentiability is the origin of the system. In this latter case, expanding f into its first-order Taylor series around the origin

$$f(\underline{x}) = f(0) + \sum_j x_j(\partial_j f)(0) + o(|\underline{a}|), \underline{a} = \sum_j e_j x_j \in \mathbb{R}^{m+1}$$

and using the easily checked fact that

$$\int_{\partial Q} x_j \, \eta \, d\sigma = e_j Vol(Q), \quad \forall j,$$

the derivative expressions above may be proved.

Finally, we may also derive a factorization for the Laplace operator Δ on $\mathbb{R}^{m+1}$ as

$$\Delta = D\overline{D} = \overline{D}D.$$

Definition 4.9 *A function f is said to be left monogenic (respectively, right monogenic, or two-sided monogenic) if $Df = 0$ (respectively, $fD = 0$, or $Df = fD = 0$).*

It is, in fact, possible to relate the concept of monogenicity to the differential operators known in real analysis such as the divergence and the curl. Indeed, let $F = u_0 - \sum_{j=1}^{m} u_j e_j$ be a $\mathbb{R}^{m+1}$-valued function defined on an open set Ω of $\mathbb{R}^{m+1}$. To show that F is left (right) monogenic, it is equivalent to prove that the $(m+1)$-tuple $U = (u_j)_{j=0}^{n}$ satisfies

$$div\, U = 0 \quad \text{and} \quad curl\, U = 0 \text{ on } \Omega.$$

Equivalently, we may show that the 1-form $\omega = u_0 dx_0 - u_l dx_1 - \cdots - u_m dx_m$ satisfies

$$d^*\omega = 0 \quad \text{and} \quad d^*\omega = 0 \quad \text{on } \Omega,$$

where d and d^* are the exterior differentiation operator and its formal transpose, respectively.

In addition, if the domain Ω is simply connected, we may show easily that F is monogenic if and only if there exists a unique (modulo an additive constant) real-valued harmonic function U on Ω such that $(u_j)_{j=0}^n = grad U$ in Ω (i.e. $F = \overline{D}U$) .

Monogenic functions are also introduced as generalizations of holomorphic functions, which are solutions of the so-called *Cauchy-Riemann operator*. For a function $F : \Omega \longrightarrow \mathbb{C}$, Ω, being a bounded domain in $\mathbb{C} \simeq \mathbb{R}^2$ such that $F(z) = F(x+iy) = U(x,y) + iV(x,y)$ and being holomorphic, is equivalent to

$$\partial_z f = 0 \iff \partial_x U = \partial_y V \quad \text{and} \quad \partial_y U = -\partial_x V. \tag{4.1}$$

The Cauchy-Riemann operator ∂_z will be replaced by the Dirac operator $\partial_{\underline{x}}$ on the Clifford algebra $\mathbb{R}_m$ or its complexification $\mathbb{C}_m$.

We will discuss in the remaining part the extension of functional analysis to the Clifford framework in more details and in a simple way especially for non-mathematicians.

The first important function that should be discussed is the famous exponential function in the framework of Clifford analysis. Formally, the exponential function is defined by means of the series

$$\exp(\underline{x}) = \sum_{k=0}^{\infty} \frac{\underline{x}^k}{k!}, \quad \underline{x} \in \mathbb{R}_m.$$

Recall that for $\underline{x} \in \mathbb{R}_m$, we have $|\underline{x}^k| \le 2^m |\underline{x}|^k$. We thus get a convergent series on the whole Clifford algebra $\mathbb{R}_m$. Moreover, we may show easily that

$$|\exp \underline{x}| \le \exp(2^{m-1}|\underline{x}|).$$

Our main result concerning this function is the following.

Theorem 7 $\exp : \mathbb{R}^{m+1} \longrightarrow \mathbb{R}^{m+1} \setminus \{0\}$ *is well-defined and onto. Furthermore, for all integer $N \ge 1$ and all $\underline{u} \in \mathbb{R}_m$, the equation $\underline{x}^N = \underline{u}$ has a solution in $\mathbb{R}_m$.*

In the latter, we will consider only functions with values on a Clifford algebra but where the variable is a vector in $\mathbb{R}^m$ taken as a part of the Clifford algebra. We put $\underline{x} = (x_0, \underline{x}) \in \mathbb{R}^{m+1}$ with $\underline{x} = \sum_{i=1}^{n} e_i x_i \in \mathbb{R}^m$. We will also write

$$\underline{x} = e_0 x_0 + \sum_{i=1}^{n} e_i x_i = \sum_{i=0}^{n} e_i x_i.$$

The vector space $\mathbb{R}^m$ can be seen as the hyperplane

$$\mathbb{R}^m \simeq \{\underline{x} = (x_0, \underline{x}) \in \mathbb{R}^{m+1} : x_0 = 0\}.$$

A function f defined on the vector space $\mathbb{R}^m$ and taking values in the Clifford algebra $\mathbb{R}_m$ or $\mathbb{C}_m$ will be expressed as

$$f(\underline{x}) = \sum_A e_A f_A(\underline{x}), \tag{4.2}$$

where f_A are complex-valued functions and $A \subset \{1, 2, \cdots, n\}$. We also recall that its conjugate $\overline{f}$ is

$$\overline{f}(\underline{x}) = \sum_A \overline{e_A} f_A(\underline{x})$$

whenever the functions f_A take values in $\mathbb{R}$. However, the conjugate will be

$$f^\dagger(\underline{x}) = \sum_A e_A^\dagger f_A(\underline{x})^\dagger$$

for a function with values in the complex Clifford algebra $\mathbb{C}_m$.

The continuity and derivability of f are to be taken component-wise. We thus denote for an open domain $\Omega \subset \mathbb{R}^m$

$$\mathcal{C}^{(r)}(\Omega, \mathbb{R}_m) := \{f : \Omega \longrightarrow \mathbb{R}_m; f = \sum_A f_A e_A \text{ with } f_A \in \mathcal{C}^{(r)}(\Omega, \mathbb{R})\}$$

$$\mathcal{C}^{(r)}(\Omega, \mathbb{C}_m) := \{f : \Omega \longrightarrow \mathbb{C}_m; f = \sum_A f_A e_A \text{ with } f_A \in \mathcal{C}^{(r)}(\Omega, \mathbb{C})\}.$$

The Clifford-valued function f belongs to the Lebesgue space $L^p(\Omega, \mathbb{R}_m, dV(\underline{x}))$ if all the components $f_A \in L^p(\Omega, \mathbb{R}, dx), 1 \leq p < \infty$, with the norm

$$|f|_p = 2^n \sum_A \left(\int_\Omega |f_A|^p \, dV(\underline{x}) \right)^{1/p}$$

being equivalent to the norm

$$\|f\|_p = \left(\int_\Omega |f|^p dV(\underline{x}) \right)^{1/p},$$

where $dV(\underline{x})$ stands for the Lebesgue measure on $\mathbb{R}^m$. We say that $f, g \in L^p(\mathbb{R}^m, \mathbb{R}_m, dV(\underline{x})$ are equal if the set $\{\underline{x} \in \mathbb{R}^m; f(\underline{x}) \neq g(\underline{x})\}$ is Lebesgue measure zero. The inner product on $L^2(\mathbb{R}^m, \mathbb{R}_m, dV(\underline{x}))$ is given by

$$\langle f, g \rangle_{L^2(\mathbb{R}^m, \mathbb{R}_m, dV(\underline{x}))} = \int_{\mathbb{R}^m} \overline{f(\underline{x})} g(\underline{x}) dV(\underline{x}). \tag{4.3}$$

Using the decomposition (4.2), we get

$$\langle f, g \rangle_{L^2(\mathbb{R}^m, dV(\underline{x}))} = \sum_{A,B} \int_{\mathbb{R}^m} \overline{f_A(\underline{x})} g_B(\underline{x}) \overline{e_A} e_B.$$

The inner product (4.3) satisfies the Cauchy-Schwartz inequality

$$\left| < f, g >_{L^2(\mathbb{R}^m, \mathbb{R}_m, dV(\underline{x}))} \right| \leq \|f\|_{L^2(\mathbb{R}^m, \mathbb{R}_m, dV(\underline{x}))} \|g\|_{L^2(\mathbb{R}^m, \mathbb{R}_m, dV(\underline{x}))}. \tag{4.4}$$

We now introduce the Dirac operator which generalizes the Cauchy-Riemann one for the complex case. It is defined by

$$\partial_{\underline{x}} = \sum_{k=1}^{n} e_k \partial_k, \tag{4.5}$$

where for $k = 1, 2, ..., m$, $\partial_k = \dfrac{\partial}{\partial_{x_k}}$. It may be seen also as a general form of the Weyl operator as

$$D_{\boldsymbol{x}} = \partial_0 + \partial_{\underline{x}}, \tag{4.6}$$

where analogously $\partial_0 = \dfrac{\partial}{\partial_{x_0}}$, known also as the Fueter-Delanghe operator (see [185, 210]). Such an operator acts from the left on Clifford-valued functions as

$$\partial_{\underline{x}} f(\underline{x}) = \sum_{i=1}^{n} e_i \partial_i f(\underline{x}) = \sum_{i,A} e_i e_A \partial_i f_A(\underline{x})$$

and on the right as

$$f \partial_{\underline{x}}(\underline{x}) = \sum_{i=1}^{n} \partial_i f(\underline{x}) e_i = \sum_{i,A} e_A e_i \partial_i f_A(\underline{x}).$$

In the same way, we get

$$D_{\boldsymbol{x}} f = \partial_0 f + \partial_{\underline{x}} f \quad \text{and} \quad f D_{\boldsymbol{x}} = f \partial_0 + f \partial_{\underline{x}}.$$

In the sequel, we will need for many times the conjugate of the Dirac operator

$$\overline{\partial}_{\underline{x}} = \sum_{k=1}^{n} \overline{e_k} \partial_k. \tag{4.7}$$

As for the real and complex classical cases of functional analysis, we will here also discuss the expressions of Dirac operator by means of the spherical coordinates. This will be used widely in the construction of Clifford wavelets. For this aim, we need to recall the spin groups. It consists of the subgroup of $\mathbb{R}_n^+$ composed of the products of even numbers of elements, where the spinor norm equals ± 1. In other words,

$$Spin(p, q) = \{s \in \mathbb{R}_{p,q}; s = \prod_{j=1}^{2l} \underline{\omega}_j \text{ with } \underline{\omega}_j^2 = \pm 1, 1 \le j \le 2l\}. \tag{4.8}$$

Such group acts on the Clifford-valued functions as follows:

$$s \in Spin(n) \to L_s : f(\underline{x}) \to s f(\overline{s}\underline{x}s)\overline{s}.$$

Definition 4.10 *A partial differential operator with constant coefficients is called spin-invariant if it commutes with L_s.*

Immediately, we notice that the Dirac operator is one of the differential operators that are *spin*-invariant.

Proposition 4.11 *The Dirac operator satisfies $\partial_{\underline{x}} L_s = L_s \partial_{\underline{x}}$ and thus it is spin-invariant.*

Proof 4.1 *As the spinor s is independent of $\underline{x}$, we may write*

$$\partial_{\underline{x}} L_s f(\underline{x}) = \partial_{\underline{x}} \left(s f(\overline{s}\underline{x}s)\overline{s} \right) = s \partial_{\underline{x}} \left(f(\overline{s}\underline{x}s) \right) \overline{s}.$$

Now, again, because of the independence of the spinor s of $\underline{x}$, we get

$$\partial_{\underline{x}} \left(f(\overline{s}\underline{x}s) \right) = (\partial_{\underline{x}} f)(\overline{s}\underline{x}s).$$

Consequently,

$$\partial_{\underline{x}} L_s f(\underline{x}) = L_x \partial_{\underline{x}} f(\underline{x}).$$

Definition 4.12 *A function $f \in C^1(\Omega, \mathbb{R}_m)$ is called left-monogenic (respectively, right-monogenic) on Ω iff*

$$D_{\boldsymbol{x}} f = (\partial_{x_0} + \partial_{\underline{x}}) f(x_0 + \underline{x}) = 0.$$

Respectively, $f D_{\boldsymbol{x}} = 0$.

Example 8 *The function $E : \mathbb{R}^m \setminus \{0\} \longrightarrow \mathbb{R}_m$ defined by*

$$E(\underline{x}) = \frac{\underline{x}}{|\underline{x}|^n},$$

where $|\underline{x}|^2 = \sum_{k=1}^{n} |x_k|^2$ is the Euclidean norm, is both left- and right-monogenic on $\mathbb{R}^m \setminus \{0\}$.

Since the Clifford product is not commutative, the two notions are not equivalent. The definition 4.12 is equivalent to a linear system of 2^n first order PDE.

Example 9 *For the simple case $n = 1$ on the Clifford algebra $\mathbb{R}_1 \simeq \mathbb{C}$, the left monogenicity may be written as*

$$D_{\boldsymbol{x}} f(x_0, x_1) = [\partial_0 + i\partial_1] f(x_0, x_1) = 0.$$

Writing f in the form $f(x_0, x_1) = u(x_0, x_1) + iv(x_0, x_1)$, we get

$$\partial_0 U = -\partial_1 V = 0 \quad and \quad \partial_0 V = \partial_1 U.$$

which is the Cauchy-Riemann conditions for a complex-valued function to be holomorphic.

Furthermore, the two-dimensional Laplacian may be obtained as the product of D and its complex conjugate $\overline{D}$: $\Delta_2 = D\overline{D}$.

We now return to the object of the section and continue to exploit the *spin* group action on the Dirac operator. We write

$$\underline{x} = r\underline{\eta}, \ \underline{\eta} = \sum_{i=0}^{n} e_i \eta_i, \ \eta_i = \frac{x_i}{r}, \ i = 0, 1, \cdots, n, \ r^2 = \sum_{i=0}^{n} |x_i|^2.$$

Using the spherical coordinates, the Dirac operator and its conjugate may be written as

$$\partial_{\underline{x}} = \underline{\eta}(\partial_r + \frac{1}{r}\partial_{\underline{\eta}}) \quad \text{and} \quad \overline{\partial_{\underline{x}}} = \overline{\underline{\eta}}(\partial_r + \frac{1}{r}\partial_{\underline{\eta}}),$$

where $\partial_{\underline{\eta}}$ is the so-called spherical Dirac operator acting on the sphere $\mathcal{S}^m$ and expressed by

$$\partial_{\underline{\eta}} = \underline{x} \wedge \partial_{\underline{x}}.$$

Consider next

$$\Gamma_{\underline{\eta}} = \overline{\underline{\eta}}\partial_{\underline{\eta}} \quad \text{and} \quad \Gamma_{\underline{\eta}}^\star = \underline{\eta}\overline{\partial_{\underline{\eta}}} \qquad (4.9)$$

called also the spherical Dirac operators, and their adjoints

$$\widetilde{\Gamma_{\underline{\eta}}} = \overline{\partial_{\underline{\eta}}}\underline{\eta} \quad \text{and} \quad \widetilde{\Gamma_{\underline{\eta}}^\star} = \partial_{\underline{\eta}}\overline{\underline{\eta}}. \qquad (4.10)$$

We obtain

$$\partial_{\underline{x}} = \underline{\eta}\left(\partial_r + \frac{1}{r}\Gamma_{\underline{\eta}}\right) = (\partial_r + \frac{1}{r}\widetilde{\Gamma_{\underline{\eta}}^\star})\underline{\eta}$$

and its conjugate

$$\overline{\partial_{\underline{x}}} = \left(\partial_r + \frac{1}{r}\widetilde{\Gamma_{\underline{\eta}}}\right)\overline{\underline{\eta}}.$$

As a result, we obtain the decomposition of the Laplacian Δ_m as

$$\Delta_m = \partial_r^2 + \frac{n}{r}\partial_r + \Delta_{\underline{\eta}},$$

where $\Delta_{\underline{\eta}}$ is the well-known Laplace-Beltrami operator.

Considering next the identity operator $\boldsymbol{I}$, we obtain a simplified form of the Laplace-Beltrami operator

$$\Delta_{\underline{\eta}} = \left((n-1)\,\boldsymbol{I} - \Gamma_{\underline{\eta}}^\star\right)\Gamma_{\underline{\eta}}^\star.$$

Denote now $\boldsymbol{E} = \sum_{i=0}^{n} x_i \partial_i = r\partial_r$, the Euler operator. We my check easily that

$$\underline{x}D_{\underline{x}} = \boldsymbol{E} - \underline{x} \wedge \partial_{\underline{x}} + (x_0 \partial_{\underline{x}} - \underline{x}\partial_0).$$

On the other hand, observe that

$$\underline{x}D_{\underline{x}} = \boldsymbol{E} + \Gamma_{\underline{\eta}}$$

and that

$$\underline{x}\partial_{\underline{x}} + \partial_{\underline{x}}\underline{x} = -2\boldsymbol{E} - n.$$

We get the simple expression

$$\Gamma_{\underline{\eta}} = \underline{x} \wedge \partial_{\underline{x}} = -\sum_{i<j} e_{ij}(x_i \partial_j - x_j \partial_i).$$

$\Gamma_{\underline{\eta}}$ is called the **angular Dirac operator** and $L_{ij} = x_i \partial_j - x_j \partial_i$ for $i, j = 1, 2, \cdots, n$ are the **angular momentum operators**.

4.6 EXISTENCE OF MONOGENIC EXTENSIONS

In this section, we propose to review the problem of existence of monogenic extensions for analytic (harmonic, holomorphic) functions to monogenic ones and thus introduce some methods to generate monogenic functions.

Let $f : \Omega \subset \mathbb{R}^m \longrightarrow \mathbb{R}$ be an analytic (harmonic) function. Does there exists a Clifford algebra-valued function g, defined on Ω, such that its real value $[g(.)]_0$ is equal to f,

$$f(\underline{x}) = [g(\underline{x})]_0 \, , \forall \underline{x} \in \Omega \, ?$$

Different methods have been developed in the literature to tackle this problem. The following result yields a monogenic extension on star-shaped domains.

Theorem 10 *Let $\Omega \subset \mathbb{R}^m$ be an open and star-shaped domain relative to the origin and $f : \Omega \to \mathbb{R}$ be harmonic. Then, the function*

$$F(\underline{x}) = f(\underline{x})e_0 + \int_0^1 t^{n-1}\overline{\partial_{\underline{x}}f(t\underline{x})}\underline{x}dt - \left[\int_0^1 t^{n-1}\overline{\partial_{\underline{x}}f(t\underline{x})} \times dt\right]_0$$

is left monogenic in Ω and its scalar part is precisely the function f.

Proof 4.2 *Left as an exercise to the reader.*

A second recent method to generate monogenic functions from analytic ones is called Cauchy-Kowalevski extension (CK-extension). Such a method is based essentially on resolving some systems of differential equations based on the Dirac operator. See [75, 76, 78, 186, 341].

Let f be an analytic function on an open set $\Omega \subset \mathbb{R}^m$. A monogenic extension may be generated by resolving the system

$$\begin{cases} D_{\underline{x}}F(x_0, \underline{x}) = 0 \text{ in } \mathbb{R}^{m+1}, \\ F(0, \underline{x}) = f(\underline{x}). \end{cases}$$

One method to realize such an extension is by applying the exponential operator $e^{-x_0\partial_{\underline{x}}}$ by setting

$$F(x_0, \underline{x}) = e^{-x_0\partial_{\underline{x}}}f(\underline{x}) = \sum_{k=0}^{\infty} \frac{(-x_0)^k}{k!}\partial_{\underline{x}}^k f(\underline{x}). \tag{4.11}$$

Formally, we have

$$\begin{aligned} D_{\boldsymbol{x}}\left(e^{-x_0\partial_{\underline{x}}}f(\underline{x})\right) &= (\partial_0 + \partial_{\underline{x}})\left(e^{-x_0\partial_{\underline{x}}}f(\underline{x})\right) \\ &= \partial_0\left(e^{-x_0\partial_{\underline{x}}}f(\underline{x})\right) + \partial_{\underline{x}}\left(e^{-x_0\partial_{\underline{x}}}f(\underline{x})\right) \\ &= -\partial_{\underline{x}}\left(e^{-x_0\partial_{\underline{x}}}f(\underline{x})\right) + \partial_{\underline{x}}\left(e^{-x_0\partial_{\underline{x}}}f(\underline{x})\right) \\ &= 0 \end{aligned}$$

Example 11 *We propose to compute the CK-extension of the function* $f : (\underline{x}) = e^{\underline{x}^2/2}$, $\underline{x} \in \mathbb{R}^m$. *Let*

$$F(x_0, \underline{x}) = e^{-x_0 \partial_{\underline{x}}} e^{\underline{x}^2/2} = \sum_{k=0}^{\infty} \frac{(-x_0)^k}{k!} \partial_{\underline{x}}^k e^{\underline{x}^2/2} = e^{\underline{x}^2/2} \sum_{k=0}^{\infty} \frac{x_0^k}{k!} H_k(\underline{x}),$$

where the H_k *are the Clifford-Hermite polynomials*

$$\begin{aligned} H_0(\underline{x}) &= 1, & H_1(\underline{x}) &= \underline{x} \\ H_2(\underline{x}) &= \underline{x}^2 + n, & H_3(\underline{x}) &= \underline{x}^3 + (n+2)\underline{x} \end{aligned}$$

and

$$H_4(\underline{x}) = \underline{x}^4 + 2(n+2)\underline{x}^2 + n(n+2).$$

We may check that the H_k *are mutually orthogonal in the sense that* $k < l$

$$\int_{\mathbb{R}^m} [H_l(\underline{x})]^{\dagger} H_k(\underline{x}) \exp(\frac{\underline{x}^2}{2}) dV(\underline{x}) = 0 \tag{4.12}$$

for $k \neq l$. *More details may be found in [24, 25, 77, 78].*

Definition 4.13 *A monogenic function* f *is called* axial monogenic *if*

$$f(x_0, \underline{x}) = A(x_0, r) + \underline{\eta} B(x_0, r),$$

where $r > 0$ *and* $\underline{\eta}$ *are the spherical coordinates of* $\underline{x}$ *and* A, B *are two scalar-valued functions satisfying the so-called Vekua system*

$$\partial_{x_0} A - \partial_r B = \frac{n-1}{r} B \quad and \quad \partial_{x_0} B + \partial_r A = 0.$$

See, for instance, [171].

In the next part, we propose to review some useful integral transforms such as Cauchy integral, Stokes theorem and Cauchy-Clifford integral formula.

Let $\Omega \subset \mathbb{R}^m$ be an open set. Let also $U \subset \Omega$ be a compact and orientable piecewise differentiable bounded domain in $\mathbb{R}^m$ with Lipschitzian boundary ∂U. The surface element on ∂U will be denoted by

$$d\sigma_{\underline{x}} = \sum_{j=1}^{n} (-1)^j e_j dx_{[j]},$$

where $dx_{[j]} = dx_1 \wedge dx_2 \wedge \cdots dx_{j-1} \wedge dx_{j+1} \wedge \cdots \wedge dx_n$. Let $\eta(\underline{x})$ denote the unit outward pointing normal-vector to the boundary ∂U at $\underline{x} \in \partial U$. We know that

$$d\sigma_{\underline{x}} = \eta(\underline{x}) dS(\underline{x}),$$

with $dS(\underline{x})$ being the surface element on ∂U. The following result is a variant of the Stokes theorem in the Clifford framework.

Theorem 12 (Clifford-Stokes Theorem) *Let $\Omega \subset \mathbb{R}^m$ be an open set. Let also $U \subset \Omega$ be a compact and orientable piecewise differentiable bounded domain in $\mathbb{R}^m$ with Lipschitzian boundary ∂U. Let $f, g \in \mathcal{C}^1(\Omega)$. Then*

$$\int_{\partial U} f(\underline{x})d\sigma_{\underline{x}}g(\underline{x}) = \int_U [(\partial_{\underline{x}}f)g + f(\partial_{\underline{x}}g)]dV(\underline{x}).$$

In particular, for $f \equiv 1$ we have

$$\int_{\partial U} d\sigma_{\underline{x}}g(\underline{x}) = \int_U \partial_{\underline{x}}g(\underline{x})dV(\underline{x}).$$

Moreover, whenever f is left-monogenic and g is right-monogenic on Ω then

$$\int_{\partial U} g(\underline{x})d\sigma_{\underline{x}}f(\underline{x}) = 0.$$

Proof 4.3 *Left as an exercise to the reader.*

Theorem 13 *Let f be left-monogenic on an open $U \subset \Omega$ as in Theorem 12 and let $\underline{y} \in U$. Then*

$$f(\underline{y}) = \frac{1}{a_n} \int_{\partial U} E(\underline{x} - \underline{y})\eta(\underline{x})f(\underline{x})d\sigma_{\underline{x}}.$$

Similarly, whenever f is right-monogenic, we have

$$f(\underline{y}) = \frac{1}{a_n} \int_{\partial U} f(\underline{x})\eta(\underline{x})E(\underline{x} - \underline{y})d\sigma_{\underline{x}},$$

where $E(\underline{x}) = \frac{1}{\omega_m}\frac{\underline{x}}{|\underline{x}|^m}$ is the fundamental solution (Green function) of the Dirac operator

$$\partial_{\underline{x}}E(\underline{x}) = \delta(\underline{x}).$$

$\omega_m = \frac{2\pi^{\frac{m}{2}}}{\Gamma(\frac{m}{2})}$ *is the area of the unit sphere in $\mathbb{R}^m$.*

Proof 4.4 *We only prove the result for the left-monogenic function. We consider the sphere $S^{n-1}(\underline{y}, r)$ for $r > 0$ chosen small enough such that the disc whose boundary is $S^{n-1}(\underline{y}, r)$ is included in U. Applying Theorem 12 on E and f, we get*

$$\int_{\partial U} E(\underline{x} - \underline{y})\eta(\underline{x})f(\underline{x})d\sigma_{\underline{x}} = \int_{S^{n-1}(\underline{y},r)} E(\underline{x} - \underline{y})\eta(\underline{x})f(\underline{x})d\sigma_{\underline{x}}.$$

Observing that $S^{n-1}(\underline{y}, r)$, we immediately have

$$\eta(\underline{x}) = \frac{\underline{x}}{|\underline{x}|} = \frac{\underline{y} - \underline{x}}{|\underline{x} - \underline{y}|} = \frac{\underline{y} - \underline{x}}{r}.$$

Consequently, on $S^{n-1}(\underline{y}, r)$,

$$E(\underline{x} - \underline{y})\eta(\underline{x}) = \frac{\underline{x} - \underline{y}}{r^n}\frac{\underline{y} - \underline{x}}{r} = \frac{r^2}{r^{n+1}} = \frac{1}{r^{n-1}}.$$

As a result,

$$\int_{S^{n-1}(\underline{y},r)} E(\underline{x}-\underline{y})\eta(\underline{x})f(\underline{x})d\sigma_{\underline{x}} = \int_{S^{n-1}(\underline{y},r)} \frac{1}{r^{n-1}}f(\underline{x})d\sigma_{\underline{x}}$$

$$= \int_{S^{n-1}(\underline{y},r)} \frac{f(\underline{x})-f(\underline{y})+f(\underline{y})}{r^{n-1}}d\sigma_{\underline{x}}$$

$$= \int_{S^{n-1}(\underline{y},r)} \frac{f(\underline{x})-f(\underline{y})}{r^{n-1}}d\sigma_{\underline{x}} + \int_{S^{n-1}(\underline{y},r)} f(\underline{y})d\sigma_{\underline{x}}$$

$$= \int_{S^{n-1}(\underline{y},r)} \frac{f(\underline{x})-f(\underline{y})}{\left|\underline{x}-\underline{y}\right|^{n-1}}d\sigma_{\underline{x}} + f(\underline{y})\int_{S^{n-1}(\underline{y},r)} d\sigma_{\underline{x}}$$

$$= \frac{1}{a_n}\int_{S^{n-1}(\underline{y},r)} \frac{f(\underline{x})-f(\underline{y})}{\left|\underline{x}-\underline{y}\right|^{n-1}}d\sigma_{\underline{x}} + \frac{f(\underline{y})}{a_n}\int_{S^{n-1}} d\sigma_{\underline{x}}$$

$$= \int_{S^{n-1}(\underline{y},r)} \frac{f(\underline{x})-f(\underline{y})}{\left|\underline{x}-\underline{y}\right|^{n-1}}d\sigma_{\underline{x}} + a_n f(\underline{y}).$$

By continuity, we have

$$\lim_{r\to 0}\int_{S^{n-1}(\underline{y},r)} \frac{f(\underline{x})-f(\underline{y})}{\left|\underline{x}-\underline{y}\right|^{n-1}}d\sigma_{\underline{x}} = 0,$$

Hence, we finally obtain

$$f(\underline{y}) = \frac{1}{\omega_m}\int_{\partial U} E(\underline{x}-\underline{y})\eta(\underline{x})f(\underline{x})d\sigma_{\underline{x}}.$$

4.7   CLIFFORD-FOURIER TRANSFORM

In this section, we propose to review some basic concepts of the Clifford-Fourier transform. Recall that the classical Fourier transform can be seen as the operator exponential

$$\mathcal{F} = \exp\left(-i\frac{\pi}{2}\mathcal{H}\right) = \sum_{k=0}^{\infty} \frac{1}{k!}\left(-i\frac{\pi}{2}\right)^k \mathcal{H}^k, \tag{4.13}$$

where $\mathcal{H}$ is the scalar-valued operator

$$\mathcal{H} = \frac{-1}{2}\left(\Delta_n + \underline{x}^2 + n\right) \tag{4.14}$$

called Hermite operator. A first characterization of these operators is the following dealing with the invariance under the Fourier transform.

Proposition 4.14 *The operators* $\mathcal{H}$ *and* $\exp(-i\frac{\pi}{2}\mathcal{H})$ *are Fourier invariant in the sense that*

$$\widehat{\mathcal{H}(f)} = \mathcal{H}(\widehat{f})$$

and

$$\widehat{\exp(-i\frac{\pi}{2}\mathcal{H})(f)} = \exp(-i\frac{\pi}{2}\mathcal{H})(\widehat{f}).$$

Proof 4.5 *We have*

$$\widehat{\mathcal{H}(f)}(\underline{\xi}) = \frac{1}{(2\pi)^{\frac{n}{2}}} \int_{\mathbb{R}^m} \frac{-1}{2} (\Delta_n f(\underline{x}) + \underline{x}^2 f(\underline{x}) + n f(\underline{x})) e^{-i\underline{x}\cdot\underline{\xi}} dV(\underline{x})$$

$$= \frac{-1}{2} \Big[\frac{1}{(2\pi)^{\frac{n}{2}}} \int_{\mathbb{R}^m} \partial_{\underline{x}}^2 f(\underline{x}) e^{-i\underline{x}\cdot\underline{\xi}} dV(\underline{x}) + \frac{1}{(2\pi)^{\frac{n}{2}}} \int_{\mathbb{R}^m} \underline{x}^2 f(\underline{x}) e^{-i\underline{x}\cdot\underline{\xi}} dV(\underline{x})$$

$$+ n \frac{1}{(2\pi)^{\frac{n}{2}}} \int_{\mathbb{R}^m} f(\underline{x}) e^{-i\underline{x}\cdot\underline{\xi}} dV(\underline{x}) \Big]$$

$$= \frac{-1}{2} \Big[\underline{\xi}^2 \widehat{f}(\underline{\xi}) + \partial_{\underline{\xi}}^2 \widehat{f}(\underline{\xi}) + n \widehat{f}(\underline{\xi}) \Big]$$

$$= \mathcal{H}(\widehat{f})(\underline{\xi}).$$

We now investigate the second point. We have

$$\widehat{\exp(-i\frac{\pi}{2}\mathcal{H})(f)}(\underline{\xi}) = \frac{1}{(2\pi)^{\frac{n}{2}}} \int_{\mathbb{R}^m} \exp(-i\frac{\pi}{2}\mathcal{H})(f)(\underline{x}) e^{-i\underline{x}\cdot\underline{\xi}} dV(\underline{x})$$

$$= \frac{1}{(2\pi)^{\frac{n}{2}}} \int_{\mathbb{R}^m} \left\{ \sum_{k=0}^{\infty} \frac{(-i\frac{\pi}{2})^k}{k!} \mathcal{H}^k(f)(\underline{x}) \right\} e^{-i\underline{x}\cdot\underline{\xi}} dV(\underline{x})$$

$$= \sum_{k=0}^{\infty} \frac{(-i\frac{\pi}{2})^k}{k!} \left\{ \frac{1}{(2\pi)^{\frac{n}{2}}} \int_{\mathbb{R}^m} \mathcal{H}^k(f)(\underline{x}) e^{-i\underline{x}\cdot\underline{\xi}} dV(\underline{x}) \right\}$$

$$= \sum_{k=0}^{\infty} \frac{(-i\frac{\pi}{2})^k}{k!} \widehat{\mathcal{H}^k(f)}(\underline{\xi}).$$

Now, by iterating on the parameter k and using the first point above, we get

$$\widehat{\mathcal{H}^k(f)} = \mathcal{H}\{\widehat{\mathcal{H}^{k-1}(f)}\}\cdots = \mathcal{H}^k \widehat{\mathcal{H}^0(f)} = \mathcal{H}^k \widehat{f}.$$

As a result,

$$\widehat{\exp(-i\frac{\pi}{2}\mathcal{H})(f)}(\underline{\xi}) = \sum_{k=0}^{\infty} \frac{(-i\frac{\pi}{2})^k}{k!} \mathcal{H}^k \widehat{f}(\underline{\xi}) = \exp(-i\frac{\pi}{2}\mathcal{H}\widehat{f}).$$

To extend the Fourier transform to Clifford algebras, we start by exploiting the Hermite operators. The first step consists in splitting the Laplace operator into product of Dirac operators to obtain Clifford-valued operators. See, for example, [81, 85, 86, 190]. Denote

$$\Phi_1 = \frac{1}{2} (\partial_{\underline{x}} - \underline{x}) (\partial_{\underline{x}} + \underline{x}) \quad \text{and} \quad \Phi_2 = \frac{1}{2} (\partial_{\underline{x}} + \underline{x}) (\partial_{\underline{x}} - \underline{x}).$$

The operators Φ_1 and Φ_2 may be expressed otherwise as in the following proposition.

Proposition 4.15 Φ_1 *and* Φ_2 *satisfy the following assertions:*

i. $\Phi_1 = \mathcal{H} + \Gamma_{\underline{x}}$.

ii. $\Phi_2 = \mathcal{H} - \Gamma_{\underline{x}} + n$.

iii. $\Phi_1 + \Phi_2 = 2\left(\mathcal{H} + \frac{n}{2}\right).$

iv. $\Phi_1 - \Phi_2 = 2\left(\Gamma_{\underline{x}} - \frac{n}{2}\right).$

Proof 4.6 *We recall firstly that* $\Gamma_{\underline{x}} = \frac{-1}{2}(\underline{x}\partial_{\underline{x}} - \partial_{\underline{x}}\underline{x} - n).$ *Consequently,*

$$
\begin{aligned}
\Phi_1 &= \frac{1}{2}\left(\partial_{\underline{x}} - \underline{x}\right)\left(\partial_{\underline{x}} + \underline{x}\right) \\
&= \frac{1}{2}\left(\partial_{\underline{x}}^2 + \partial_{\underline{x}}\underline{x} - \underline{x}\partial_{\underline{x}} - \underline{x}^2\right) \\
&= \frac{1}{2}\left(\partial_{\underline{x}}^2 - \underline{x}^2 - n\right) - \frac{1}{2}\left(\underline{x}\partial_{\underline{x}} - \partial_{\underline{x}}\underline{x} - n\right) \\
&= \mathcal{H} + \Gamma_{\underline{x}}.
\end{aligned}
$$

Similarly, we have

$$
\begin{aligned}
\Phi_2 &= \frac{1}{2}\left(\partial_{\underline{x}} + \underline{x}\right)\left(\partial_{\underline{x}} - \underline{x}\right) \\
&= \frac{1}{2}\left(\partial_{\underline{x}}^2 - \underline{x}^2 - n + \underline{x}\partial_{\underline{x}} - \partial_{\underline{x}}\underline{x} + n\right) \\
&= \mathcal{H} + \frac{1}{2}\left(\underline{x}\partial_{\underline{x}} - \partial_{\underline{x}}\underline{x} + 2n - n\right) \\
&= \mathcal{H} - \Gamma_{\underline{x}} + n.
\end{aligned}
$$

The assertions iii and iv follow from i and ii.

Proposition 4.16 Φ_1 *and* Φ_2 *are Fourier-invariant, in the sense that*

$$\widehat{\Phi_1 f} = \Phi_1\widehat{f} \quad and \quad \widehat{\Phi_2 f} = \Phi_2\widehat{f}.$$

Proof 4.7 *We prove firstly that*

$$\widehat{\Gamma_{\underline{x}}(f)} = \Gamma_{\underline{\xi}}(\widehat{f}).$$

Indeed, we have

$$
\begin{aligned}
\widehat{\Gamma_{\underline{x}}(f)}(\underline{\xi}) &= \frac{-1}{2}\left(\widehat{\underline{x}\partial_{\underline{x}}f} - \widehat{\partial_{\underline{x}}\underline{x}f} - n\widehat{f}\right)(\underline{\xi}) \\
&= \frac{-1}{2}\left(\partial_{\underline{\xi}}\widehat{\partial_{\underline{x}}f} - \underline{\xi}\widehat{\underline{x}f} - n\widehat{f}\right)(\underline{\xi}) \\
&= \frac{-1}{2}\left(\partial_{\underline{\xi}}\underline{\xi} - \underline{\xi}\partial_{\underline{\xi}} - n\right)\widehat{f}(\underline{\xi}) \\
&= \Gamma_{\underline{\xi}}(\widehat{f})(\underline{\xi}).
\end{aligned}
$$

Consequently,

$$\widehat{\Phi_1 f} = \widehat{\mathcal{H}f} + \widehat{\Gamma_{\underline{x}}f} = \Phi_1\widehat{f}.$$

Similarly for Φ_2.

Now, we are able to define the Clifford-Fourier transform.

Definition 4.17 *The Clifford-Fourier transform is the couple $\mathcal{F} = (\mathcal{F}_+, \mathcal{F}_-)$ of exponential operators*

$$\mathcal{F}_+ = \exp(-i\frac{\pi}{2}\mathcal{H}_+) \ \text{and} \ \mathcal{F}_- = \exp(-i\frac{\pi}{2}\mathcal{H}_-).$$

To join the classical form of the Fourier transform and to obtain a one operator $\mathcal{F}$ from the pair $(\mathcal{F}_+, \mathcal{F}_-)$, researchers considered the rule

$$\mathcal{F}^2 = \mathcal{F}_+\mathcal{F}_-.$$

By letting

$$\mathcal{H} = \frac{1}{2}(\mathcal{H}_+ + \mathcal{H}_-),$$

we come back to the classical form $\mathcal{F} = \exp(-i\frac{\pi}{2}\mathcal{H})$. To join the operators Φ_1 and Φ_2, a simple choice may be

$$\mathcal{H}_+ = \Phi_1 - \frac{n}{2} \quad \text{and} \quad \mathcal{H}_- = \Phi_2 - \frac{n}{2}.$$

Another alternative proposed in [75, 190] consists of taking

$$\mathcal{H}_+ = \Phi_1 \quad \text{and} \quad \mathcal{H}_- = \Phi_2 - n.$$

As a consequence, we may define an integral representation for the new Clifford-Fourier transform as

$$\mathcal{F}_+[f](\underline{\xi}) = \frac{1}{(2\pi)^{\frac{n}{2}}} \int_{\mathbb{R}^m} \exp\left(-i\frac{\pi}{2}\left(\Gamma_{\underline{\xi}} - \frac{n}{2}\right)\right) e^{-i\underline{x}\cdot\underline{\xi}} f(\underline{x})dV(\underline{x})$$

$$\mathcal{F}_-[f](\underline{\xi}) = \frac{1}{(2\pi)^{\frac{n}{2}}} \int_{\mathbb{R}^m} \exp\left(i\frac{\pi}{2}\left(\Gamma_{\underline{\xi}} - \frac{n}{2}\right)\right) e^{-i\underline{x}\cdot\underline{\xi}} f(\underline{x})dV(\underline{x}). \quad (4.15)$$

In terms of the pair of operators, one may set

$$\mathcal{F}_+^{\frac{1}{2}} = \exp\left(-i\frac{\pi}{4}\mathcal{H}_+\right) \quad \text{and} \quad \mathcal{F}_-^{\frac{1}{2}} = \exp\left(-i\frac{\pi}{4}\mathcal{H}_-\right)$$

and thus put

$$\mathcal{F} = \mathcal{F}_+^{\frac{1}{2}}\mathcal{F}_-^{\frac{1}{2}}.$$

Hereafter, we discuss some properties of the Clifford-Fourier transform.

Proposition 4.18 *For two Clifford algebra-valued functions f and g and $a, b \in \mathbb{C}_m$, we have*

$$\mathcal{F}_\pm[fa + gb] = \mathcal{F}_\pm[f]\,a + \mathcal{F}_+[g]\,b.$$

For $\lambda > 0$, we have

$$\mathcal{F}_\pm[f(\lambda\bullet)](\underline{\xi}) = \frac{1}{\lambda^n}\mathcal{F}_\pm[f]\left(\frac{\underline{\xi}}{\lambda}\right).$$

Proof 4.8 *The first assertion is obvious. For the second assertion, we will show firstly that*

$$\mathcal{F}\left[f(\lambda\bullet)\right](\underline{\xi}) = \frac{1}{\lambda^n}\mathcal{F}\left[f\right](\frac{\underline{\xi}}{\lambda}).$$

We have

$$\mathcal{F}\left[f(\lambda\bullet)\right](\underline{\xi}) = \frac{1}{(2\pi)^{\frac{n}{2}}}\int_{\mathbb{R}^m} f(\lambda\underline{x})e^{-i\underline{x}\cdot\underline{\xi}}dV(\underline{x}).$$

By setting $\underline{y} = \lambda\underline{x}$, we get

$$\mathcal{F}\left[f(\lambda\bullet)\right](\underline{\xi}) = \frac{1}{\lambda^n}\frac{1}{(2\pi)^{\frac{n}{2}}}\int_{\mathbb{R}^m} f(\underline{y})e^{-i\underline{x}\cdot\frac{\underline{\xi}}{\lambda}}dV(\underline{y}) = \frac{1}{\lambda^n}\widehat{f}(\frac{\underline{\xi}}{\lambda}).$$

It follows that

$$\mathcal{F}_+\left[f(\lambda\bullet)\right](\underline{\xi}) = \exp\left(-i\frac{\pi}{2}\left(\Gamma_{\underline{\xi}} - \frac{n}{2}\right)\right)\frac{1}{\lambda^n}\widehat{f}(\frac{\underline{\xi}}{\lambda})$$

and

$$\mathcal{F}_-\left[f(\lambda\bullet)\right](\underline{\xi}) = \exp\left(i\frac{\pi}{2}\left(\Gamma_{\underline{\xi}} - \frac{n}{2}\right)\right)\frac{1}{\lambda^n}\widehat{f}(\frac{\underline{\xi}}{\lambda}).$$

Observe next that $\Gamma_{\underline{x}} = \Gamma_{\frac{\underline{x}}{\lambda}}$. As a result,

$$\mathcal{F}_+\left[f(\lambda\bullet)\right](\underline{\xi}) = \exp\left(-i\frac{\pi}{2}\left(\Gamma_{\frac{\underline{\xi}}{\lambda}} - \frac{n}{2}\right)\right)\frac{1}{\lambda^n}\widehat{f}(\frac{\underline{\xi}}{\lambda}) = \frac{1}{\lambda^n}\mathcal{F}_+\left[f\right](\frac{\underline{\xi}}{\lambda}).$$

Similarly for $\mathcal{F}_-\left[f(\lambda\bullet)\right](\underline{\xi})$.

Proposition 4.19

$$\mathcal{F}_{\pm}\left[\bullet f(\bullet)\right](\underline{\xi}) = \mp(\mp i)^n\partial_{\underline{\xi}}\mathcal{F}_+\left[f\right](\underline{\xi}).$$

Proof 4.9 *We have*

$$\mathcal{F}_+\left[\bullet f(\bullet)\right](\underline{\xi}) = \exp\left(-i\frac{\pi}{2}\left(\Gamma_{\underline{\xi}} - \frac{n}{2}\right)\right)\mathcal{F}\left[\bullet f(\bullet)\right](\underline{\xi})$$

$$= i\exp\left(-i\frac{\pi}{2}\left(\Gamma_{\underline{\xi}} - \frac{n}{2}\right)\right)\partial_{\underline{\xi}}\mathcal{F}\left[f\right](\underline{\xi})$$

$$= i\left\{\sum_{k=0}^{\infty}\frac{(-i\frac{\pi}{2})^k}{k!}\left(\Gamma_{\underline{\xi}} - \frac{n}{2}\right)^k\right\}\partial_{\underline{\xi}}\mathcal{F}\left[f\right](\underline{\xi}). \tag{4.16}$$

Observe next that

$$\partial_{\underline{\xi}}\Gamma_{\underline{\xi}} + \Gamma_{\underline{\xi}}\partial_{\underline{\xi}} = (n-1)\partial_{\underline{\xi}}$$

and that $\Gamma_{\underline{\xi}}\partial_{\underline{\xi}} = \partial_{\underline{\xi}}\left[n - 1 - \Gamma_{\underline{\xi}}\right]$. We deduce that

$$\left(\Gamma_{\underline{\xi}} - \frac{n}{2}\right)\partial_{\underline{\xi}} = \partial_{\underline{\xi}}\left(\frac{n-2}{2} - \Gamma_{\underline{\xi}}\right).$$

Consequently,

$$\left(\Gamma_{\underline{\xi}} - \frac{n}{2}\right)^k \partial_{\underline{\xi}} = \partial_{\underline{\xi}} \left(\frac{n-2}{2} - \Gamma_{\underline{\xi}}\right)^k.$$

As a result, we obtain

$$\mathcal{F}_+ \left[\bullet f(\bullet)\right](\underline{\xi}) = i \left\{ \sum_{k=0}^{\infty} \frac{(-i\frac{\pi}{2})^k}{k!} \left(\Gamma_{\underline{\xi}} - \frac{n}{2}\right)^k \right\} \partial_{\underline{\xi}} \mathcal{F}[f](\underline{\xi})$$

$$= i \left\{ \sum_{k=0}^{\infty} \frac{(-i\frac{\pi}{2})^k}{k!} \left(\Gamma_{\underline{\xi}} - \frac{n}{2}\right)^k \partial_{\underline{\xi}} \right\} \mathcal{F}[f](\underline{\xi})$$

$$= i \left\{ \sum_{k=0}^{\infty} \frac{(-i\frac{\pi}{2})^k}{k!} \partial_{\underline{\xi}} \left(\frac{n-2}{2} - \Gamma_{\underline{\xi}}\right)^k \right\} \mathcal{F}[f](\underline{\xi})$$

$$= i \partial_{\underline{\xi}} \exp\left(i\frac{\pi}{2}\left(\Gamma_{\underline{\xi}} - \frac{n}{2} + 1\right)\right) \mathcal{F}[f](\underline{\xi})$$

$$= i \exp\left(i\frac{\pi}{2}\right) \partial_{\underline{\xi}} \exp\left(i\frac{\pi}{2}\left(\Gamma_{\underline{\xi}} - \frac{n}{2}\right)\right) \mathcal{F}[f](\underline{\xi})$$

$$= -\partial_{\underline{\xi}} \mathcal{F}_- [f](\underline{\xi}).$$

The same techniques hold for $\mathcal{F}_-$.

In fact, we may have more general rules.

Proposition 4.20 *For $k \in \mathbb{N}$, we have*

$$\mathcal{F}_+ \left[\underline{x}^{2k} f\right](\underline{\xi}) = (-1)^k \partial_{\underline{\xi}}^{2k} \mathcal{F}_+ [f](\underline{\xi})$$

$$\mathcal{F}_+ \left[\underline{x}^{2k+1} f\right](\underline{\xi}) = -(-1)^k \partial_{\underline{\xi}}^{2k+1} \mathcal{F}_- [f](\underline{\xi}).$$

In fact, the explicit form of the kernel of (4.15) is already a difficult problem. We know just few cases where the expression is simplified such as the case where $n = 2$ in which

$$\mathcal{F}_+[f](\underline{\xi}) = \frac{1}{2\pi} \int_{\mathbb{R}^2} \exp(\underline{\xi} \wedge \underline{x}) f(\underline{x}) dV(\underline{x})$$

and

$$\mathcal{F}_-[f](\underline{\xi}) = \frac{1}{2\pi} \int_{\mathbb{R}^2} \exp(\underline{x} \wedge \underline{\xi}) f(\underline{x}) dV(\underline{x}).$$

For even dimensions, in general, a first step has been developed in [86] in terms of the Bessel function. For example, for $n = 4$ we have the kernels

$$K_+(\underline{x}, \underline{\xi}) = \sqrt{\frac{\pi}{2}} |\underline{x} \wedge \underline{\xi}|^{-1/2} \left((1 + \underline{x} \cdot \underline{\xi}) J_{1/2}(|\underline{x} \wedge \underline{\xi}|) + \frac{(\underline{\xi} \wedge \underline{x})}{|\underline{x} \wedge \underline{\xi}|} J_{3/2}(|\underline{x} \wedge \underline{\xi}|)(\underline{x} \cdot \underline{\xi}) \right)$$

$$K_-(\underline{x}, \underline{\xi}) = \sqrt{\frac{\pi}{2}} |\underline{x} \wedge \underline{\xi}|^{-1/2} \left((1 - \underline{x} \cdot \underline{\xi}) J_{1/2}(|\underline{x} \wedge \underline{\xi}|) + \frac{(\underline{\xi} \wedge \underline{x})}{|\underline{x} \wedge \underline{\xi}|} J_{3/2}(|\underline{x} \wedge \underline{\xi}|)(\underline{x} \cdot \underline{\xi}) \right).$$

Recently, a general form has been developed in [183] for all even dimensions.

4.8 CLIFFORD WAVELET ANALYSIS

Clifford wavelets or wavelets on Clifford algebras are the last variants of wavelet functions developed by researchers in order to overcome many problems that are not well investigated by classical transforms. The challenge in such concepts is not the wavelet functions themselves but the structure of Clifford algebras and their flexibility to include different forms of vector analysis in the same time.

In the present section, we propose to develop two main methods to construct Clifford wavelets. The first one is based on *spin* groups and thus includes the factor of rotations in the wavelet analysis provided with the translation and dilatation factors. This method generalizes in some sense the first attempt in developing multidimensional wavelets such as Cauchy ones. See J. P. Antoine and his team works [11, 12, 13].

The second part is concerned with the development of wavelet analysis from monogenic functions, mainly polynomials. These constitute an extension of orthogonal polynomials to the case of Clifford algebras. Recall that orthogonal polynomials are widely applied in wavelet theory on Euclidean spaces. Extensions to the case of Clifford framework are few, especially the works of Ghent University group on Clifford analysis, Brackx et al. and recently more general extensions due to Arfaoui, Ben Mabrouk, Cattani in [24, 25].

4.8.1 Spin-group based Clifford wavelets

As in the classical cases of wavelets on Euclidean spaces, we seek some properties that must be satisfied for a Clifford algebra-valued function to be a mother wavelet.

Definition 4.21 (Clifford Wavelet) *Let* $\psi \in L^1 \cap L^2(\mathbb{R}^m, \mathbb{R}_m, dV(\underline{x}))$ *be such that* $\psi\psi \in L^1 \cap L^2(\mathbb{R}^m, \mathbb{R}_m, dV(\underline{x}))$. *The function* ψ *is said to be a Clifford mother wavelet iff the following assertions hold simultaneously.*

  i. $\widehat{\psi}(\underline{\xi}) \left[\widehat{\psi}(\underline{\xi})\right]^\dagger$ *is scalar.*

  ii. The admissibility condition

$$\mathcal{A}_\psi = (2\pi)^n \int_{\mathbb{R}^m} \frac{\widehat{\psi}(\underline{\xi}) \left[\widehat{\psi}(\underline{\xi})\right]^\dagger}{|\underline{\xi}|^n} dV(\underline{\xi}) < \infty.$$

The function ψ is said to be admissible and $\mathcal{A}_\psi$ is its admissibility constant. We notice here also that being admissible as a mother wavelet the function ψ should satisfy some oscillation property such as

$$\widehat{\psi}(\underline{0}) = 0 \iff \int_{\mathbb{R}^m} \psi(\underline{x}) dV(\underline{x}).$$

Example 14 *A generalized Clifford Mexican hat wavelet has been provided in [78] by considering*

$$\psi(\underline{x}) = \exp(\frac{1}{2}\underline{x}^2) H_n(\underline{x}) = (-1)^n \partial_{\underline{x}} \exp(\frac{\underline{x}^2}{2}),$$

where H_n is the Clifford version of the Hermite polynomial of degree n. We get

$$\widehat{\psi}(\underline{\xi}) = (2\pi)^{\frac{n}{2}}(-i)^n \underline{\xi}^n \exp(\frac{\underline{\xi}^2}{2}).$$

Consequently, its admissibility constant will be

$$\mathcal{A}_\psi = (2\pi)^n \int_{\mathbb{R}^m} \frac{\left|\widehat{\psi}(\underline{\xi})\right|^2}{\left|\underline{\xi}\right|^n} dV(\underline{\xi})$$

$$= (2\pi)^{2n} \int_{\mathbb{R}^m} \left|\exp(\frac{\underline{\xi}^2}{2})\right|^2 dV(\underline{\xi})$$

$$= (2\pi)^n \omega_m \int_0^\infty r^{m-1} e^{-r^2} dr < \infty.$$

Now, starting with an admissible Clifford mother wavelet, we generate a whole set of wavelets by the action of general group of translations, dilations and spin-rotations, which generalizes the affine group applied in the case of real wavelets.

For $(a, \underline{b}, s) \in \mathbb{R}^+ \times \mathbb{R}^m \times Spin(n)$, denote

$$\psi^{a,\underline{b},s}(\underline{x}) = \frac{1}{a^{\frac{n}{2}}} s\psi(\frac{\overline{s}(\underline{x}-\underline{b})s}{a})\overline{s}.$$

It is straightforward that whenever ψ is admissible, the copies $\psi^{a,\underline{b},s}$ are also admissible and satisfy precisely

$$\mathcal{A}_{\psi^{a,\underline{b},s}} = \frac{a^{n/2}}{(2\pi)^n} \mathcal{A}_\psi < \infty.$$

We will now see that these copies permit further to approximate any function in $L^2(\mathbb{R}^m, \mathbb{R}_m, dV(\underline{x}))$.

Proposition 4.22 *The set $\Lambda_\psi = \{\psi^{a,\underline{b},s} : a > 0, \underline{b} \in \mathbb{R}^m, s \in Spin(n)\}$ is dense in $L^2(\mathbb{R}^m, \mathbb{R}_m, dV(\underline{x}))$.*

Proof 4.10 *Let $f \in L^2(\mathbb{R}^m, \mathbb{R}_m, dV(\underline{x}))$ be an analyzed function such that*

$$< \psi^{a,\underline{b},s}, f >_{L^2(\mathbb{R}^m,\mathbb{R}_m,dV(\underline{x}))} = 0, \ \forall a > 0, \ \underline{b} \in \mathbb{R}^m \ and \ s \in Spin(n).$$

We shall prove that $f = 0$. Indeed, we already know from Parseval identity of the Clifford-Fourier transform that

$$< \psi^{a,\underline{b},s}, f >_{L^2(\mathbb{R}^m,\mathbb{R}_m,dV(\underline{x}))} = < \widehat{\psi^{a,\underline{b},s}}, \widehat{f} >_{L^2(\mathbb{R}^m,\mathbb{R}_m,dV(\underline{x}))} = 0.$$

On the other hand, observe that

$$< \widehat{\psi^{a,\underline{b},s}}, \widehat{f} >_{L^2(\mathbb{R}^m,\mathbb{R}_m,dV(\underline{x}))} = a^{\frac{n}{2}} \int_{\mathbb{R}^m} e^{i\underline{b}\cdot\underline{\xi}} s \left[\widehat{\psi}(a\overline{s}\underline{\xi}s)\right]^\dagger \overline{s}\widehat{f}(\underline{\xi}) dV(\underline{\xi}) = 0.$$

This yields necessarily that

$$s \left[\widehat{\psi}(a\overline{s}\underline{\xi}s)\right]^\dagger \overline{s}\widehat{f}(\underline{\xi}) = 0.$$

Now, as for fixed $\underline{\xi} \neq 0$ in $\mathbb{R}^m$, we already have

$$\left\{ a\bar{s}\underline{\xi}s, a > 0 \text{ and } s \in Spin(n) \right\} = \mathbb{R}^m.$$

It results that

$$\hat{f} = 0 \text{ and so } f = 0.$$

We now introduce the Clifford wavelet continuous transform of functions.

Definition 4.23 *The Clifford wavelet continuous transform of a function f in $L^2(\mathbb{R}^m, \mathbb{R}_m, dV(\underline{x}))$ with respect to an admissible mother wavelet ψ is*

$$T_\psi\left[f\right](a, \underline{b}, s) = < \psi^{a, \underline{b}, s}, f >_{L^2(\mathbb{R}^m, \mathbb{R}_m, dV(\underline{x}))} = \int_{\mathbb{R}^m} \left[\psi^{a, \underline{b}, s}(\underline{x})\right]^\dagger f(\underline{x}) dV(\underline{x}).$$

Such transform possesses, as for other integral transforms of functions, many useful properties as stated in the following proposition.

Proposition 4.24 *The Clifford continuous wavelet transform is*

  i. Translation-invariant, in the sense that

$$T_\psi\left[f(\bullet - \underline{c}\right](a, \underline{b}, s) = T_\psi\left[f\right](a, \underline{b} - \underline{c}, s).$$

  ii. Dilation-invariant, in the sense that

$$T_\psi\left[\frac{1}{\lambda^{\frac{n}{2}}} f(\frac{\bullet}{\lambda})\right](a, \underline{b}, s) = T_\psi\left[f\right](\frac{a}{\lambda}, \frac{\underline{b}}{\lambda}, s).$$

  iii. Spin-rotation-invariant, in the sense that

$$T_\psi\left[L_t f\right](a, \underline{b}, s) = t T_\psi\left[f\right](a, \bar{t}\underline{b}t, \bar{t}s)\bar{t}.$$

Proof 4.11 *i. Observe that*

$$T_\psi\left[f(\bullet - \underline{c}\right](a, \underline{b}, s) = \frac{1}{a^{\frac{n}{2}}} \int_{\mathbb{R}^m} s \left[\psi\left(\frac{\bar{s}(\underline{x} - \underline{b})s}{a}\right)\right]^\dagger \bar{s} f(\underline{x} - \underline{c}) dV(\underline{x}).$$

$$= \frac{1}{a^{\frac{n}{2}}} \int_{\mathbb{R}^m} s \left[\psi\left(\frac{\bar{s}(\underline{y} - (\underline{b} - \underline{c}))s}{a}\right)\right]^\dagger \bar{s} f(\underline{y}) dV(\underline{y})$$

$$= T_\psi\left[f\right](a, \underline{b} - \underline{c}, s).$$

ii. We have

$$T_\psi\left[\frac{1}{\lambda^{\frac{n}{2}}} f(\frac{\bullet}{\lambda})\right](a, \underline{b}, s) = \frac{1}{a^{\frac{n}{2}}} \int_{\mathbb{R}^m} s \left[\psi\left(\frac{\bar{s}(\underline{x} - \underline{b})s}{a}\right)\right]^\dagger \bar{s} \frac{1}{\lambda^{\frac{n}{2}}} f(\frac{\underline{x}}{\lambda}) dV(\underline{x})$$

$$= (\frac{\lambda}{a})^{\frac{n}{2}} \int_{\mathbb{R}^m} s \left[\psi\left(\frac{\bar{s}(\lambda\underline{y} - \underline{b})s}{a}\right)\right]^\dagger \bar{s} f(\underline{y}) dV(\underline{y})$$

$$= \frac{1}{(\frac{a}{\lambda})^{\frac{n}{2}}} \int_{\mathbb{R}^m} s \left[\psi\left(\frac{\bar{s}(\underline{y} - \frac{\underline{b}}{\lambda})s}{\frac{a}{\lambda}}\right)\right]^\dagger \bar{s} f(\underline{y}) dV(\underline{y})$$

$$= T_\psi\left[f\right](\frac{a}{\lambda}, \frac{\underline{b}}{\lambda}, s).$$

iii. *Observe that*

$$T_\psi \left[L_t f \right] (a, \underline{b}, s) = \frac{1}{a^{\frac{n}{2}}} \int_{\mathbb{R}^m} s \left[\psi \left(\frac{\overline{s}(x - \underline{b})s}{a} \right) \right]^\dagger \overline{s} t f(\overline{t} \underline{x} t) \overline{t} dV(x)$$

$$= \frac{1}{a^{\frac{n}{2}}} \int_{\mathbb{R}^m} s \left[\psi \left(\frac{\overline{s}(t y \overline{t} - \underline{b})s}{a} \right) \right]^\dagger \overline{s} t f(\overline{t} y t) \overline{t} dV(y)$$

$$= \frac{1}{a^{\frac{n}{2}}} \int_{\mathbb{R}^m} s \left[\psi \left(\frac{\overline{s}(t y \overline{t} - \underline{b})s}{a} \right) \right]^\dagger \overline{s} t f(\overline{t} y t) \overline{t} dV(y)$$

$$= t \frac{1}{a^{\frac{n}{2}}} \int_{\mathbb{R}^m} \{\overline{t}s\} \left[\psi \left(\frac{\overline{s} t (y - \overline{t} \underline{b} t) \overline{t} s}{a} \right) \right]^\dagger \{\overline{s}t\} f(\overline{t} y t) \overline{t} dV(y) \overline{t}$$

$$= t T_\psi \left[f \right] (a, \overline{t} \underline{b} t, \overline{t}s) \overline{t}$$

We now propose to develop Parseval-Plancherel-type rules, which are the most important formula in wavelet theory as they permit reconstruction of functions from their wavelet transforms.

We firstly introduce an inner product relative to the continuous Clifford wavelet transform. Let

$$\mathcal{H}_\psi = \left\{ T_\psi \left[f \right], \ f \in L^2(\mathbb{R}^m, \mathbb{R}_m, dV(\underline{x})) \right\}$$

be the image of $L^2(\mathbb{R}^m, \mathbb{R}_m, dV(\underline{x}))$ relative to the operator T_ψ. We define the inner product by

$$\left[T_\psi \left[f \right], T_\psi \left[g \right] \right] = \frac{1}{A_\psi} \int_{Spin(n)} \int_{\mathbb{R}^m} \int_{\mathbb{R}^+} (T_\psi \left[f \right] (a, \underline{b}, s))^\dagger T_\psi \left[g \right] (a, \underline{b}, s) \frac{da}{a^{n+1}} dV(\underline{b}) ds,$$

where ds stands for the Haar measure on $Spin(n)$.

Proposition 4.25 $T_\psi : L^2(\mathbb{R}^m, \mathbb{R}_m, dV(\underline{x})) \longrightarrow \mathcal{H}_\psi$ *is an isometry.*

Proof 4.12 *We shall prove that*

$$\left[T_\psi \left[f \right], T_\psi \left[g \right] \right] = < f, g >_{L^2(\mathbb{R}^m, \mathbb{R}_m, dV(\underline{x}))} . \tag{4.17}$$

Denote

$$\Phi_\psi(a, s, \underline{\xi}) \left[f \right] (-\underline{b}) = \left[\left[\widehat{\psi}(a \overline{s} \underline{\xi} s) \right]^\dagger \overline{s} \widehat{f}(\underline{\xi}) \right] (-\underline{b})$$

and

$$\Phi_\psi(a, s, \underline{\xi}) \left[g \right] (-\underline{b}) = \left[\left[\widehat{\psi}(a \overline{s} \underline{\xi} s) \right]^\dagger \overline{s} \widehat{g}(\underline{\xi}) \right] (-\underline{b}).$$

We immediately obtain

$$T_\psi \left[f \right] (a, \underline{b}, s) = a^{\frac{n}{2}} s (2\pi)^{\frac{n}{2}} \widehat{\Phi_\psi(a, \underline{\xi}, s)} \left[f \right] (-\underline{b})$$

and

$$T_\psi \left[g \right] (a, \underline{b}, s) = a^{\frac{n}{2}} s (2\pi)^{\frac{n}{2}} \widehat{\Phi_\psi(a, \underline{\xi}, s)} \left[g \right] (-\underline{b}).$$

Applying Parseval formula, we get

$$\left\langle \Phi_\psi \widehat{(a, \bullet, s)}\, [f], \Phi_\psi \widehat{(a, \bullet, s)}\, [g] \right\rangle = \left\langle \Phi_\psi (a, \bullet, s)\, [f], \Phi_\psi (a, \bullet, s)\, [g] \right\rangle.$$

Consequently,

$$[T_\psi\, [f], T_\psi\, [g]]$$

$$= \frac{1}{(2\pi)^n \mathcal{A}_\psi} \int_{Spin(n)} \int_{\mathbb{R}^+} \left\{ \int_{\mathbb{R}^m} (\Phi_\psi(a, \underline{\xi}, s)\, [f]\, (\underline{\xi}))^\dagger \Phi_\psi(a, \underline{\xi}, s)\, [g]\, (\underline{\xi}) dV(\underline{b}) \right\} \frac{da}{a} ds$$

$$= \frac{1}{(2\pi)^n \mathcal{A}_\psi} \int_{Spin(n)} \int_{\mathbb{R}^+} \left\{ \int_{\mathbb{R}^m} \left[\left(\left[\widehat{\psi}(a\overline{s}\underline{\xi}s) \right]^\dagger \overline{s} \widehat{f}(\underline{\xi}) \right)^\dagger \left[\widehat{\psi}(a\overline{s}\underline{\xi}s) \right]^\dagger \overline{s} \widehat{g}(\underline{\xi}) dV(\underline{\xi}) \right\} \frac{da}{a} ds \right.$$

$$= \frac{1}{(2\pi)^n \mathcal{A}_\psi} \int_{Spin(n)} \int_{\mathbb{R}^+} \left\{ \int_{\mathbb{R}^m} \left[\widehat{f}(\underline{\xi}) \right]^\dagger s \widehat{\psi}(a\overline{s}\underline{\xi}s) \left[\widehat{\psi}(a\overline{s}\underline{\xi}s) \right]^\dagger \overline{s} \widehat{g}(\underline{\xi}) dV(\underline{\xi}) \right\} \frac{da}{a} ds$$

$$= \frac{1}{(2\pi)^n \mathcal{A}_\psi} \int_{\mathbb{R}^m} \left[\widehat{f}(\underline{\xi}) \right]^\dagger \left\{ \int_{Spin(n)} \int_{\mathbb{R}^+} s \widehat{\psi}(a\overline{s}\underline{\xi}s) \left[\widehat{\psi}(a\overline{s}\underline{\xi}s) \right]^\dagger \overline{s} \frac{da}{a} ds \right\} \widehat{g}(\underline{\xi}) dV(\underline{\xi}).$$

Observing now that

$$\int_{Spin(n)} \int_{\mathbb{R}^+} s\widehat{\psi}(a\overline{s}\underline{\xi}s) \left[\widehat{\psi}(a\overline{s}\underline{\xi}s) \right]^\dagger \overline{s} \frac{da}{a} ds = \frac{\mathcal{A}_\psi}{(2\pi)^n}, \tag{4.18}$$

we get immediately

$$[T_\psi\, [f], T_\psi\, [g]] = \frac{1}{(2\pi)^n \mathcal{A}_\psi} \int_{\mathbb{R}^m} \left[\widehat{f}(\underline{\xi}) \right]^\dagger \left\{ \int_{\mathcal{S}^{n-1}} \int_{\mathbb{R}^+} \Gamma(t, \nu) dS(\underline{\nu}) \frac{da}{t} \right\} \widehat{g}(\underline{\xi}) dV(\underline{\xi}),$$

where we denoted $\Gamma(t, \nu) = \widehat{\psi}(t\underline{\nu}) \left[\widehat{\psi}(t\underline{\nu}) \right]^\dagger$. *Otherwise, by taking* $\underline{u} = t\underline{\nu}$, *we obtain*

$$[T_\psi\, [f], T_\psi\, [g]] = \frac{1}{(2\pi)^n \mathcal{A}_\psi} \int_{\mathbb{R}^m} \left[\widehat{f}(\underline{\xi}) \right]^\dagger \left\{ \int_{\mathbb{R}^m} \frac{\widehat{\psi}(\underline{u}) \left[\widehat{\psi}(\underline{u}) \right]^\dagger}{|\underline{u}|^n} dV(\underline{u}) \right\} \widehat{g}(\underline{\xi}) dV(\underline{\xi})$$

$$= \int_{\mathbb{R}^m} \left[\widehat{f}(\underline{\xi}) \right]^\dagger \widehat{g}(\underline{\xi}) dV(\underline{\xi})$$

$$= < \widehat{f}, \widehat{g} >$$

$$= < f, g >. \tag{4.19}$$

Corollary 4.26 *The operator*

$$T_\psi : L^2(\mathbb{R}^m, \mathbb{R}_m, dV(\underline{x})) \longrightarrow L_2(\mathbb{R}_+ \times \mathbb{R}^m \times Spin(n), \frac{1}{\mathcal{A}_\psi} \frac{da\, dV(\underline{b})ds}{a^{n+1}})$$

is an isometry. More precisely, we have the Parseval-Plancherel equality

$$\int_{Spin(n)} \int_{\mathbb{R}^m} \int_{\mathbb{R}^+} (T_\psi\, [f]\, (a, \underline{b}, s))^2 \frac{da}{a^{n+1}} dV(\underline{b})ds = \mathcal{A}_\psi \|f\|_2^2.$$

As a result of the last corollary and as in the real case, we have here a Clifford wavelet reconstruction formula.

Proposition 4.27 *For all $f \in L^2(\mathbb{R}^m, \mathbb{R}_m, dV(\underline{x}))$, we have*

$$f(\underline{x}) = \frac{1}{A_\psi} \int\limits_{Spin(n)} \int\limits_{\mathbb{R}^m} \int\limits_{\mathbb{R}^+} \psi^{a,\underline{b},s}(\underline{x}) T_\psi[f](a,\underline{b},s) \frac{da}{a^{n+1}} dV(\underline{b}) ds$$

in $L^2(\mathbb{R}^m, \mathbb{R}_m, dV(\underline{x}))$.

Proof 4.13 *Let f and g be two square integrable Clifford-valued functions with Clifford wavelet transforms $T_\psi[f]$ and $T_\psi[g]$, respectively. From 4.25 and using 4.23, we have*

$$< f, g >_{L^2(\mathbb{R}^m, \mathbb{R}_m, dV(\underline{x}))}$$

$$= \frac{1}{\mathcal{A}_\psi} \int\limits_{Spin(n)} \int\limits_{\mathbb{R}^m} \int\limits_{\mathbb{R}^+} (T_\psi[f](a,\underline{b},s))^\dagger T_\psi[g](a,\underline{b},s) \frac{da}{a^{n+1}} dV(\underline{b}) ds$$

$$= \frac{1}{\mathcal{A}_\psi} \int\limits_{Spin(n)} \int\limits_{\mathbb{R}^m} \int\limits_{\mathbb{R}^+} [T_\psi[f](a,\underline{b},s)]^\dagger T_\psi[g](a,\underline{b},s) \frac{da}{a^{n+1}} dV(\underline{b}) ds$$

$$= \frac{1}{\mathcal{A}_\psi} \int\limits_{Spin(n)} \int\limits_{\mathbb{R}^m} \int\limits_{\mathbb{R}^+} < \psi^{a,\underline{b},s}, f >^\dagger_{L^2(\mathbb{R}^m, \mathbb{R}_m, dV(\underline{x}))} T_\psi[g](a,\underline{b},s) \frac{da}{a^{n+1}} dV(\underline{b}) ds$$

$$= \frac{1}{\mathcal{A}_\psi} \int\limits_{Spin(n)} \int\limits_{\mathbb{R}^m} \int\limits_{\mathbb{R}^+} < f, \psi^{a,\underline{b},s} >_{L^2(\mathbb{R}^m, \mathbb{R}_m, dV(\underline{x}))} T_\psi[g](a,\underline{b},s) \frac{da}{a^{n+1}} dV(\underline{b}) ds$$

$$= < f, \frac{1}{\mathcal{A}_\psi} \int\limits_{Spin(n)} \int\limits_{\mathbb{R}^m} \int\limits_{\mathbb{R}^+} \psi^{a,\underline{b},s}(\underline{x}) T_\psi[g](a,\underline{b},s) \frac{da}{a^{n+1}} dV(\underline{b}) ds >_{L^2(\mathbb{R}^m, \mathbb{R}_m, dV(\underline{x}))}.$$

Then

$$g(\underline{x}) = \frac{1}{\mathcal{A}_\psi} \int\limits_{Spin(n)} \int\limits_{\mathbb{R}^m} \int\limits_{\mathbb{R}^+} \psi^{a,\underline{b},s}(\underline{x}) T_\psi[g](a,\underline{b},s) \frac{da}{a^{n+1}} dV(\underline{b}) ds.$$

The last result in this section deals with reproducing kernels relative to the continuous Clifford wavelet transform.

Theorem 15 *A function $F(a,\underline{b},s) \in L_2\left(\mathbb{R}_+ \times \mathbb{R}^m \times Spin(n), \mathcal{A}_\psi^{-1} a^{-(n+1)} da dV(\underline{b}) ds\right)$ is the Clifford wavelet transform of a square integrable function f iff*

$$F(a,\underline{b},s) = \frac{1}{C_\psi} \int_{Spin(n)} \int_{\mathbb{R}^m} \int_0^{+\infty} \left(K_\psi(a,\underline{b},s;\tilde{a},\underline{\tilde{b}},\tilde{s})\right)^\dagger F(\tilde{a},\underline{\tilde{b}},\tilde{s}) \frac{d\tilde{a}}{\tilde{a}^{n+1}} dV(\underline{\tilde{b}}) d\tilde{s},$$

where $K_\psi(a,\underline{b},s;\tilde{a},\underline{\tilde{b}},\tilde{s}) = T_\psi[\psi^{a,\underline{b},s}](\tilde{a},\underline{\tilde{b}},\tilde{s}) = < \psi^{\tilde{a},\underline{\tilde{b}},\tilde{s}}, \psi^{a,\underline{b},s} >$ is the reproducing kernel.

Proof 4.14 *Left to the readers as an exercise.*

4.8.2 Monogenic polynomial-based Clifford wavelets

In this section, we propose to review a second method to construct wavelets on Clifford algebras. The idea is based on the so-called monogenic polynomials, which constitute an extension of orthogonal polynomials on Clifford algebras. There are, in fact, many types of such polynomials and different associated wavelets have been obtained. We will focus here on just one way due to the well-known Gegenbauer polynomials, known also as ultraspheroidal polynomials. Other classes may be developed by the readers by similar techniques. Furthermore, we may refer to [24, 25, 21, 20, 75, 79, 80, 82, 83, 86, 87, 76, 77, 78, 81, 85] for other existing classes of polynomials and associated wavelets in both the classical context and the Clifford one.

Gegenbauer polynomials are defined on the orthogonality interval $I =]-1, 1[$ relatively to the weight function

$$\omega_p(x) = (1 - x^2)^{p - \frac{1}{2}}$$

via Rodrigues rule as

$$G_m^p(x) = \frac{(-1)^m \Gamma(p + \frac{1}{2})\Gamma(m + 2p)}{2^m m! \Gamma(2p)\Gamma(p + m + \frac{1}{2})} \frac{1}{\omega_p(x)} \frac{d^m \omega_{p+m}}{dx^m}(x).$$

Denote next

$$R_m^p = \frac{(-1)^m \Gamma(p + m + \frac{1}{2})\Gamma(p + \frac{1}{2})\Gamma(m + 2p)}{2^m m! \Gamma(2p)}$$

and

$$a_{m,p}^k = \frac{(-1)^k C_m^k}{\Gamma(p + m + \frac{1}{2} - k)\Gamma(p + \frac{1}{2} + k)}.$$

By splitting

$$\omega_p(x) = (1 - x)^{p - \frac{1}{2}}(1 + x)^{p - \frac{1}{2}}$$

and applying Leibnitz derivation rule, we get the explicit form

$$G_m^p(x) = R_m^p \sum_{k=0}^{m} a_{m,p}^k (1 - x)^{m-k}(1 + x)^k.$$

For $m = 0$ and $m = 1$, this gives, respectively,

$$G_0^p(x) = 1 \quad \text{and} \quad G_1^p(x) = 2px.$$

For $m = 2$, we get

$$G_2^p(x) = 2p(p + 1)\left[x^2 - \frac{1}{2p + 2}\right]$$

and for $m = 3$,

$$G_3^p(x) = \frac{4}{3}p(p + 1)(p + 2)\left[x^3 - \frac{3}{2p + 4}x\right].$$

Gegenbauer polynomials G_m^p may also be introduced via the induction rule stated for $p \geq \frac{-1}{2}$ by

$$mG_m^p(x) = 2x(m + p - 1)G_{m-1}^p(x) - (m + 2p - 2)G_{m-2}^p(x), \qquad (4.20)$$

with initial polynomials G_0^p and G_1^p as above.

Gegenbauer wavelets are examples of wavelets on the interval and depend on four parameters j, n, m, p. The parameter $j \in \mathbb{N}$ is related to the level of resolution, $n \in \{1, 2, 3, ..., 2^{j-1}\}$ is related to the translation, $m = 0, 1, 2, ..., M - 1,\quad M > 0$, is the degree of the Gegenbauer polynomial, and finally a reel parameter $p > -\dfrac{1}{2}$ is related to the order of Gegenbauer polynomials. The Gegenbauer mother wavelet is defined on $[0, 1)$ by $\psi^{m,p}(x) = G_m^p(x)$. Next, the translation-dilation copies of $\psi^{m,p}$ are defined by

$$\psi_{j,n}^{m,p}(x) = \begin{cases} \dfrac{1}{\sqrt{L_m^p}} 2^{\frac{j}{2}} G_m^p(2^j x - 2n + 1) & , \quad \dfrac{2n-2}{2^j} \leq t < \dfrac{2n}{2^j}, \\ 0 & , \quad \text{elsewhere,} \end{cases}$$

Remark 4.28 • *For $p = \dfrac{1}{2}$, we get Legendre wavelets.*

- *For $p = 0$ and $p = 1$, we obtain the Chebyshev wavelet of first and second kind, respectively.*

To obtain the mutual orthogonality of Gegenbauer wavelets $\psi_{j,n}^{m,p}$, the weight function associated with the Gegenbauer polynomials has to be dilated and translated as for the Gegenbauer wavelets. Thus, we obtain a translation-dilation copy of the weight ω as

$$\omega_{j,n}(x) = \omega(2^j x - 2n + 1) = (1 - (2^j x - 2n + 1)^2)^{p - \frac{1}{2}}.$$

At fixed level of resolution, we get

$$\omega_{j,n}(x) = \begin{cases} \omega_{j,1}(x) & , \quad 0 \leq x < \dfrac{1}{2^{j-1}}, \\ \omega_{j,2}(x) & , \quad \dfrac{1}{2^{j-1}} \leq x < \dfrac{2}{2^{j-1}}, \\ \omega_{j,3}(x) & , \quad \dfrac{2}{2^{j-1}} \leq x < \dfrac{3}{2^{j-1}}, \\ \vdots & \\ \omega_{j,2^{j-1}}(x) & , \quad \dfrac{2^{j-1}-1}{2^{j-1}} \leq x < 1. \end{cases}$$

According to such wavelets, a function $f \in L^2[0, 1)$ may be expressed in terms of a series

$$f = \sum_{j=1}^{\infty} \sum_{n \in \mathbb{Z}} C_{j,n}^{m,p} \psi_{j,n}^{m,p}, \tag{4.21}$$

where the coefficient $C_{j,}^{m,p}$ are the so-called wavelet coefficients given by

$$C_{j,n}^{m,p} = <f, \psi_{j,n}^{m,p}> = \int_0^1 \omega_{j,n}(x) \psi_{j,n}^{m,p}(x) f(x) \, dx.$$

For more details, we may refer to [315], [316], [325], [333].

In this section, we propose to revisit the Gegenbauer polynomials associated with the real weight function $\omega(x) = (1 + x^2)^\alpha$, $\alpha \in \mathbb{R}$, extended to the context of Clifford algebra-valued polynomials. Consider the Clifford algebra-valued weight function

$$\omega_\alpha(\underline{x}) = (1 + |\underline{x}|^2)^\alpha, \ \alpha \in \mathbb{R}.$$

The general Clifford-Gegenbauer polynomials, denoted by $G_{\ell,m,\alpha}(\underline{x})$, are generated by the CK-extension $F^*(t, \underline{x})$ defined by

$$F^*(t, \underline{x}) = \sum_{\ell=0}^{\infty} \frac{t^\ell}{\ell!} G_{\ell,m,\alpha}(\underline{x}) \, \omega_{\alpha-\ell}(\underline{x}); \ \ t \in \mathbb{R}, \ \ \underline{x} \in \mathbb{R}_m.$$

As for the real case of orthogonal polynomials, we impose a left monogenic property on F^* in $\mathbb{R}^{m+1}$ to obtain a recursive relation on the general Clifford-Gegenbauer polynomials $G_{\ell,m,\alpha}$. Hence, since F^* is monogenic, we have

$$(\partial_t + \partial_{\underline{x}})F^*(t, \underline{x}) = 0. \tag{4.22}$$

The first part related to the time derivative is evaluated as

$$\partial_t F^*(t, \underline{x}) = \sum_{\ell=0}^{\infty} \frac{t^\ell}{\ell!} G_{\ell+1,m,\alpha}(\underline{x}) \, \omega_{\alpha-\ell-1}(\underline{x}).$$

For the second part $\partial_{\underline{x}} F^*(t, \underline{x})$, we shall use the following technical lemma.

Lemma 4.29 *For all $n \in \mathbb{N}$, we have*

$$\partial_{\underline{x}}(\underline{x}^n) = \gamma_{n,m} \underline{x}^{n-1},$$

where

$$\gamma_{n,m} = \begin{cases} -n & \textit{if } n \textit{ is even.} \\ -(m+n-1) & \textit{if } n \textit{ is odd.} \end{cases}$$

So, now, observing that

$$\partial_{\underline{x}}(\underline{x}) = -m, \quad \partial_{\underline{x}}(\underline{x}^2) = -2\underline{x} \quad \text{and} \quad \partial_{\underline{x}}(|\underline{x}|^2) = 2\underline{x},$$

we get

$$\partial_{\underline{x}} F^*(t, \underline{x}) = \sum_{\ell=0}^{\infty} \frac{t^\ell}{\ell!} \left(\partial_{\underline{x}} G_{\ell,m,\alpha}(\underline{x}) \omega_{\alpha-\ell}(\underline{x}) + G_{\ell,m,\alpha}(\underline{x}) \partial_{\underline{x}} \omega_{\alpha-\ell}(\underline{x}) \right).$$

Observing again that

$$\partial_{\underline{x}} \omega_{\alpha-\ell}(\underline{x}) = 2(\alpha - \ell)\underline{x} \, \omega_{\alpha-\ell-1}(\underline{x}),$$

the monogenicity property (4.22) leads to the recurrence relation

$$G_{\ell+1,m,\alpha}(\underline{x})\omega_{\alpha-\ell-1}(\underline{x}) + \omega_{\alpha-\ell}(\underline{x})\partial_{\underline{x}} G_{\ell,m,\alpha}(\underline{x})$$
$$+2(\alpha - \ell)\omega_{\alpha-\ell-1}(\underline{x})\underline{x} G_{\ell,m,\alpha}(\underline{x}) = 0,$$

or equivalently

$$G_{\ell+1,m,\alpha}(\underline{x}) = -2(\alpha - \ell)\underline{x}G_{\ell,m,\alpha}(\underline{x}) - (1 + |\underline{x}|^2)\partial_{\underline{x}}G_{\ell,m,\alpha}(\underline{x}). \qquad (4.23)$$

Starting from $G_{0,m,\alpha}(\underline{x}) = 1$, we obtain as examples

$$G_{1,m,\alpha}(\underline{x}) = -2\alpha\underline{x},$$

$$G_{2,m,\alpha}(\underline{x}) = 2\alpha[(2(\alpha - 1) + m)\underline{x}^2 - m],$$

$$G_{3,m,\alpha}(\underline{x}) = [-4\alpha((2\alpha - 1) + m)(\alpha - 1)]\underline{x}^3 + 4\alpha(\alpha - 1)(m + 2)\underline{x}.$$

The Clifford-Gegenbauer polynomials may also be introduced via the Rodrigues formula subject to the next proposition.

Proposition 4.30

$$G_{\ell,m,\alpha}(\underline{x}) = (-1)^\ell \, \omega_{\ell-\alpha}(\underline{x})\partial_{\underline{x}}^\ell(\, \omega_\alpha(\underline{x})).$$

Proof. We proceed by recurrence on ℓ. For $\ell = 1$, we have

$$\partial_{\underline{x}}\, \omega_\alpha(\underline{x}) = 2\alpha\, \underline{x}\, \omega_{\alpha-1}(\underline{x}) = (-1)(-2\alpha\underline{x})\, \omega_{\alpha-1}(\underline{x}) = (-1)\, \omega_{\alpha-1}(\underline{x})G_{1,m}^\alpha(\underline{x}),$$

which means that

$$G_{1,m}^\alpha = (-1)\, \omega_{1-\alpha}(\underline{x})\partial_{\underline{x}}\, \omega_\alpha(\underline{x}).$$

For $\ell = 2$, we get

$$\begin{aligned}
\partial_{\underline{x}}^{(2)}\, \omega_\alpha(\underline{x}) &= 2\alpha[2(\alpha - 1)\underline{x}^2(1 - \underline{x}^2)^{\alpha-2} - m(1 - \underline{x}^2)^{\alpha-1}] \\
&= (-1)^2\omega_{\alpha-2}[2\alpha[2(\alpha - 1) + m]\underline{x}^2 - m] \\
&= (-1)^2\omega_{\alpha-2}(\underline{x})\, G_{2,m,\alpha}(\underline{x}).
\end{aligned}$$

Hence,

$$G_{2,m}^{\alpha,\beta} = (-1)^2\omega_{2-\alpha}(\underline{x})\partial_{\underline{x}}^{(2)}\omega_\alpha(\underline{x}).$$

So, assume the recurrence hypothesis

$$G_{\ell,m}^{\alpha,\beta}(\underline{x}) = (-1)^\ell\omega_{\ell-\alpha}(\underline{x})\partial_{\underline{x}}^{(\ell)}\omega_\alpha(\underline{x}).$$

Denote

$$\Im(\underline{x}) = -2(\alpha - \ell)\underline{x}(-1)^\ell\omega_{\ell-\alpha}(\underline{x})\partial_{\underline{x}}^{(\ell)}\omega_\alpha(\underline{x}),$$

and

$$\Re(\underline{x}) = (1 + |\underline{x}|^2)(-1)^\ell 2(\ell - \alpha)\underline{x}\omega_{\ell-\alpha-1}(\underline{x})\, \partial_{\underline{x}}^\ell\omega_\alpha(\underline{x}).$$

From (4.23) and the recurrence hypothesis above, we obtain

$$\begin{aligned}
G_{\ell+1,m}^{\alpha,\beta}(\underline{x}) &= -2(\alpha - \ell)\underline{x}(-1)^\ell\omega_{\ell-\alpha}(\underline{x})\partial_{\underline{x}}^{(\ell)}\omega_\alpha(\underline{x}) \\
&\quad -(1 + |\underline{x}|^2)\partial_{\underline{x}}[(-1)^\ell\omega_{\ell-\alpha}(\underline{x})\partial_{\underline{x}}^{(\ell)}\omega_\alpha(\underline{x})] \\
&= \Im(\underline{x}) - \Re(\underline{x}) - (1 + |\underline{x}|^2)(-1)^\ell\omega_{\ell-\alpha}(\underline{x})\partial_{\underline{x}}^{(\ell+1)}\omega_\alpha(\underline{x}).
\end{aligned}$$

Simple calculus yields that

$$
\begin{aligned}
\Re(\underline{x}) &= (1+|\underline{x}|^2)(-1)^\ell 2(\ell-\alpha)\underline{x}\omega_{\ell-\alpha-1}(\underline{x})\,\partial_{\underline{x}}^\ell\omega_\alpha(\underline{x}) \\
&= (-1)^\ell 2(\ell-\alpha)\underline{x}\omega_{\ell-\alpha}(\underline{x})\,\partial_{\underline{x}}^\ell\omega_\alpha(\underline{x}) \\
&= \Im(\underline{x}).
\end{aligned}
$$

Hence,

$$
\begin{aligned}
G_{\ell+1,m}^{\alpha,\beta}(\underline{x}) &= -(1+|\underline{x}|^2)(-1)^\ell\omega_{\ell-\alpha}(\underline{x})\partial_{\underline{x}}^{(\ell+1)}\omega_\alpha(\underline{x}) \\
&= (-1)^{\ell+1}\omega_{\ell-\alpha+1}(\underline{x})\partial_{\underline{x}}^{(\ell+1)}\omega_\alpha(\underline{x}).
\end{aligned}
$$

Recently, a generalized class of Clifford-Gegenbauer polynomials and wavelets has been developed in [21] based on the 2-parameter Clifford weight function

$$
\omega_{\alpha,\beta}(\underline{x}) = (1-|\underline{x}|^2)^\alpha(1+|\underline{x}|^2)^\beta.
$$

Denote such polynomials by $Z_{\ell,m}^{\alpha,\beta}(\underline{x})$ and the CK-extension F^* can be expressed as

$$
F^*(t,\underline{x}) = \sum_{\ell=0}^\infty \frac{t^\ell}{\ell!} Z_{\ell,m}^{\alpha,\beta}(\underline{x})\,\omega_{\alpha-\ell,\beta-\ell}(\underline{x}).
$$

From the monogenicity relation $(\partial_t+\partial_{\underline{x}})F^*(t,\underline{x}) = 0$, the authors proved the following result.

Proposition 4.31 *The 2-parameter Clifford-Gegenbauer polynomials $Z_{\ell,m}^{\alpha,\beta}$ satisfy the recurrence relation*

$$
\begin{aligned}
Z_{\ell+1,m}^{\alpha,\beta}(\underline{x}) &= [2(\alpha-\ell)\underline{x}(1-\underline{x}^2) - 2(\beta-\ell)\underline{x}\,(1+\underline{x}^2)]Z_{\ell,m}^{\alpha,\beta}(\underline{x}) \\
&\quad - \omega_{1,1}(\underline{x})\partial_{\underline{x}}(Z_{\ell,m}^{\alpha,\beta}(\underline{x})).
\end{aligned}
$$

For example, starting with $Z_{0,m}^{\alpha,\beta}(\underline{x}) = 1$, a simple calculation yields that

$$
Z_{1,m}^{\alpha,\beta}(\underline{x}) = 2(\alpha-\beta)\underline{x} - 2(\alpha+\beta)\underline{x}^3.
$$

For $\ell = 1$, we get

$$
\begin{aligned}
Z_{2,m}^{\alpha,\beta}(\underline{x}) &= 2(\alpha-\beta)m + [4\alpha(\alpha-1)+4\beta(\beta-1)-8\alpha\beta-2(\alpha+\beta)m]\underline{x}^2 \\
&\quad + [8\beta(\beta-1)-8\alpha(\alpha-1)+2(\beta-\alpha)m]\underline{x}^4 \\
&\quad + [4\alpha(\alpha-1)+4\beta(\beta-1)+8\alpha\beta+2(\beta+\alpha)m]\underline{x}^6.
\end{aligned}
$$

For $\ell = 2$, we obtain

$$
\begin{aligned}
Z_{3,m}^{\alpha,\beta}(\underline{x}) &= [[4(\alpha-\beta)^2-4(\alpha+\beta)]m+8\alpha(\alpha-1)+8\beta(\beta-1)-8\alpha\beta]\underline{x} \\
&\quad + [-26\alpha(\alpha-1)+40\beta(\beta-1)-16\alpha\beta \\
&\qquad\qquad -4(\alpha+\beta)+4(\beta-\alpha)[\alpha+\beta-2]]\underline{x}^3 \\
&\quad + [16\beta(\beta-1)(\alpha-\beta)-16\alpha(\alpha-1)(\alpha-\beta)-4(\alpha-\beta)^2 m] \\
&\quad - [4\alpha(\alpha-1)+4\beta(\beta-1)-8\alpha\beta-2(\alpha+\beta)m][2(\alpha+\beta-2)]\underline{x}^5 \\
&\quad + [8\alpha(\alpha-1)+8\beta(\beta-1)+16\alpha\beta+4(\beta+\alpha)m](\alpha-\beta) \\
&\quad - 2[8\beta(\beta-1)-8\alpha(\alpha-1)+2(\beta-\alpha)m](\alpha+\beta)\underline{x}^7 \\
&\quad - [4\alpha(\alpha-1)+4\beta(\beta-1)+8\alpha\beta+2(\beta+\alpha)m[2\alpha+2\beta-2]\underline{x}^9.
\end{aligned}
$$

Remark 4.32 $Z_{\ell,m}^{\alpha,\beta}(\underline{x})$ *is a polynomial of degree 3ℓ in $\underline{x}$.*

Proposition 4.33 *The 2-parameter Clifford-Gegenbauer polynomials $Z_{\ell,m}^{\alpha,\beta}$ may be obtained via the Rodrigues formula*

$$Z_{\ell,m}^{\alpha,\beta}(\underline{x}) = (-1)^\ell \, \omega_{\ell-\alpha,\ell-\beta}(\underline{x}) \, \partial_{\underline{x}}^\ell [(1 + \underline{x}^2)^\alpha (1 - \underline{x}^2)^\beta].$$

Proof. For $\ell = 0$, the situation is obvious. For $\ell = 1$, we have

$$\begin{aligned}
\partial_{\underline{x}}(\omega_{\alpha,\beta}(\underline{x})) &= -2\alpha\underline{x}\,\omega_{\alpha-1,\beta}(\underline{x}) + 2\beta\underline{x}\,\omega_{\alpha,\beta-1}(\underline{x}) \\
&= (-1)\,\omega_{\alpha-1,\beta-1}(\underline{x})[2\alpha\underline{x}(1 - \underline{x}^2) - 2\beta\underline{x}(1 + \underline{x}^2)] \\
&= (-1)\,\omega_{\alpha-1,\beta-1}(\underline{x})Z_{1,m}^{\mu,\alpha}(\underline{x}).
\end{aligned}$$

Thus,

$$Z_{1,m}^{\alpha,\beta}(\underline{x}) = (-1)\omega_{1-\alpha,1-\beta}(\underline{x})\partial_{\underline{x}}(\omega_{\alpha,\beta}(\underline{x})).$$

For $\ell = 2$, we get

$$\begin{aligned}
&\partial_{\underline{x}}^2(\omega_{\alpha,\beta}(\underline{x})) \\
&= -2\alpha[\partial_{\underline{x}}(\underline{x}(1 + \underline{x}^2)^{\alpha-1})(1 - \underline{x}^2)^\beta + \underline{x}(1 + \underline{x}^2)^{\alpha-1}\partial_{\underline{x}}(1 - \underline{x}^2)^\beta] \\
&\quad +2\beta[\partial_{\underline{x}}(\underline{x}(1 + \underline{x}^2)^\alpha)(1 - \underline{x}^2)^{\beta-1} + \underline{x}(1 + \underline{x}^2)^\alpha\partial_{\underline{x}}(1 - \underline{x}^2)^{\beta-1}] \\
&= 2m\alpha\omega_{\alpha-1,\beta}(\underline{x}) + 4\alpha(\alpha - 1)\underline{x}^2\omega_{\alpha-2,\beta}(\underline{x}) \\
&\quad -4\alpha\beta\underline{x}^2\omega_{\alpha-1,\beta-1}(\underline{x}) - 2m\beta\,\omega_{\alpha,\beta-1}(\underline{x}) \\
&\quad -4\alpha\beta\underline{x}^2\omega_{\alpha-1,\beta-1}(\underline{x}) + 4\beta(\beta - 1)\underline{x}^2\omega_{\alpha,\beta-2}(\underline{x}) \\
&= (-1)^2\omega_{\alpha-2,\beta-2}(\underline{x})[2m\alpha\omega_{1,2}(\underline{x}) \\
&\quad +4\alpha(\alpha - 1)\underline{x}^2(1 - \underline{x}^2)^2 - 4\alpha\beta\underline{x}^2\omega_{1,1}(\underline{x}) \\
&\quad -2m\beta\omega_{2,1}(\underline{x}) - 4\alpha\beta\underline{x}^2\omega_{1,1}(\underline{x}) + 4\beta(\beta - 1)\underline{x}^2(1 + \underline{x}^2)^2] \\
&= (-1)^2\omega_{\alpha-2,\beta-2}(\underline{x})Z_{2,m}^{\alpha,\beta}(\underline{x}).
\end{aligned}$$

Then

$$Z_{2,m}^{\alpha,\beta}(\underline{x}) = (-1)^2\omega_{2-\alpha,2-\beta}(\underline{x})\partial_{\underline{x}}(\omega_{\alpha,\beta}(\underline{x})).$$

Now assume that

$$Z_{\ell,m}^{\alpha,\beta}(\underline{x}) = (-1)^\ell \, \omega_{\ell-\alpha,\ell-\beta}(\underline{x}) \, \partial_{\underline{x}}^\ell \, \omega_{\alpha,\beta}(\underline{x}).$$

Denote

$$\begin{aligned}
\wp &= [2(\alpha - \ell)\underline{x}(1 - \underline{x}^2) - 2(\beta - \ell)\underline{x}\,(1 + \underline{x}^2)] \\
&\quad [(-1)^\ell \, \omega_{\ell-\alpha,\ell-\beta}(\underline{x}) \, \partial_{\underline{x}}^\ell \, \omega_{\alpha,\beta}(\underline{x})]
\end{aligned}$$

and

$$\begin{aligned}
\aleph &= (-1)^\ell \, \omega_{1,1}(\underline{x})\,[2(\alpha - \ell)\underline{x}\,\omega_{\ell-\alpha-1,\ell-\beta}(\underline{x}) \\
&\quad -2(\beta - \ell)\underline{x}\,\omega_{\ell-\alpha,\ell-\beta-1}(\underline{x})]\partial_{\underline{x}}^\ell \, \omega_{\alpha,\beta}(\underline{x}).
\end{aligned}$$

Then, we derive that

$$\begin{aligned}
Z_{\ell+1,m}^{\alpha,\beta}(\underline{x}) &= [2(\alpha - \ell)\underline{x}(1 - \underline{x}^2) - 2(\beta - \ell)\underline{x}\,(1 + \underline{x}^2)]Z_{\ell,m}^{\alpha,\beta}(\underline{x}) \\
&\quad - \omega_{1,1}(\underline{x})\partial_{\underline{x}}Z_{\ell,m}^{\alpha,\beta}(\underline{x}) \\
&= \wp - \aleph - (-1)^\ell \, \omega_{\ell-\alpha+1,\ell-\beta+1}(\underline{x}) \, \partial_{\underline{x}}^{\ell+1}\omega_{\alpha,\beta}(\underline{x}).
\end{aligned}$$

Otherwise, we have

$$
\begin{aligned}
\aleph &= (-1)^{\ell}[2(\alpha - \ell)\underline{x}\,\omega_{\ell-\alpha,\ell-\beta+1}(\underline{x}) \\
&\qquad -2(\beta - \ell)\underline{x}\,\omega_{\ell-\alpha+1,\ell-\beta-1}(\underline{x})]\partial_{\underline{x}}^{\ell}\omega_{\alpha,\beta}(\underline{x}) \\
&= (-1)^{\ell}[2(\alpha - \ell)\underline{x}(1 - \underline{x}^2) \\
&\qquad -2(\beta - \ell)\underline{x}(1 + \underline{x}^2)]\,\omega_{\ell-\alpha,\ell-\beta}(\underline{x})\partial_{\underline{x}}^{\ell}\omega_{\alpha,\beta}(\underline{x}) \\
&= \wp.
\end{aligned}
$$

Hence,

$$
Z_{\ell+1,m}^{\alpha,\beta}(\underline{x}) = (-1)^{\ell+1}\,\omega_{\ell-\alpha+1,\ell-\beta+1}(\underline{x})\,\partial_{\underline{x}}^{\ell+1}(\omega_{\alpha,\beta}(\underline{x})).
$$

The following orthogonality relation is proved.

Proposition 4.34 *Let*

$$
I_{\ell,t,p}^{\alpha,\beta} = \int_{\mathbb{R}^m} \underline{x}^{\ell}\, Z_{t,m}^{\alpha+p,\beta+p}(\underline{x})\,\omega_{\alpha,\beta}(\underline{x})\,dV(\underline{x}).
$$

For $4t < 1 - m - 2(\alpha + \beta)$, we have

$$
I_{\ell,t,t}^{\alpha,\beta} = 0.
$$

Proof. Denote

$$
I_{\ell,t} = \int_{\mathbb{R}^m} \underline{x}^{\ell}\,\partial_{\underline{x}}^{t}(\omega_{\alpha+t,\beta+t}(\underline{x})\,dV(\underline{x})).
$$

Using Stokes's theorem, we obtain

$$
\begin{aligned}
&\int_{\mathbb{R}^m} \underline{x}^{\ell} Z_{t,m}^{\alpha+t,\beta+t}(\underline{x})\omega_{\alpha,\beta}(\underline{x})dV(\underline{x}) \\
&= \int_{\mathbb{R}^m} \underline{x}^{\ell}(-1)^t \omega_{t-\alpha-t,t-\beta-t}(\underline{x})\partial_{\underline{x}}^{t}(\omega_{\alpha+t,\beta+t}(\underline{x}))\omega_{\alpha,\beta}(\underline{x})\,dV(\underline{x}) \\
&= (-1)^t \int_{\mathbb{R}^m} \underline{x}^{\ell}\,\partial_{\underline{x}}^{t}(\omega_{\alpha+t,\beta+t}(\underline{x})\,dV(\underline{x})) \\
&= (-1)^t \int_{\mathbb{R}^m} \underline{x}^{\ell}\,\partial_{\underline{x}}\partial_{\underline{x}}^{t-1}(\omega_{\alpha+t,\beta+t}(\underline{x}))\,dV(\underline{x})) \\
&= (-1)^t \left[\int_{\partial\mathbb{R}^m} \underline{x}^{\ell}\partial_{\underline{x}}^{t-1}\omega_{\alpha+t,\beta+t}(\underline{x})\,dV(\underline{x}) \right. \\
&\qquad\qquad \left. - \int_{\mathbb{R}^m} \partial_{\underline{x}}(\underline{x}^{\ell})\partial_{\underline{x}}^{t-1}\omega_{\alpha+t,\beta+t}(\underline{x})dV(\underline{x})\right].
\end{aligned}
$$

Denote already

$$
I = \int_{\partial\mathbb{R}^m} \underline{x}^{\ell}\partial_{\underline{x}}^{t-1}\omega_{\alpha+t,\beta+t}(\underline{x})\,dV(\underline{x})
$$

and

$$
II = \int_{\mathbb{R}^m} \partial_{\underline{x}}(\underline{x}^{\ell})\partial_{\underline{x}}^{t-1}\omega_{\alpha+t,\beta+t}(\underline{x})dV(\underline{x}).
$$

The integral I vanishes due to the assumption

$$
0 < t < \frac{1 - m - 2(\alpha + \beta)}{4}.
$$

Due to Lemma 4.29, the latter satisfies

$$II = \gamma_{l,m} \int_{\mathbb{R}^m} \underline{x}^{\ell-1} \partial_{\underline{x}}^{t-1} \omega_{\alpha+t,\beta+t}(\underline{x})\, dV(\underline{x}) = \gamma_{l,m} I_{\ell-1,t-1}.$$

Hence, we obtain

$$
\begin{aligned}
&\int_{\mathbb{R}^m} \underline{x}^{\ell} Z_{t,m}^{\alpha+t,\beta+t}(\underline{x}) \omega_{\alpha,\beta}(\underline{x}) dV(\underline{x}) \\
&= (-1)^{t+1} \gamma_{l,m} I_{\ell-1,t-1} \\
&= (-1)^{t+1} \gamma_{l,m} [(-1)^t \gamma_{l-1,m} I_{\ell-2,t-2}] \\
&= (-1)^{2t+1} \gamma_{l,m} \gamma_{l-1,m} I_{\ell-2,t-2} \\
&\quad\vdots \\
&= C(m,\ell,t) I_0 \\
&= 0,
\end{aligned}
$$

where

$$C(m,\ell,t) = (-1)^{ml+1} \prod_{k=0}^{m} \gamma_{k,m}.$$

Definition 4.35 *The generalized 2-parameter Clifford-Gegenbauer mother wavelet is defined by*

$$\psi_{\ell,m}^{\alpha,\beta}(\underline{x}) = Z_{\ell,m}^{\alpha+\ell,\beta+\ell}(\underline{x}) \omega_{\alpha,\beta}(\underline{x}) = (-1)^{\ell} \partial_{\underline{x}}^{(\ell)} \omega_{\alpha+\ell,\beta+\ell}(\underline{x}).$$

Furthermore, the wavelet $\psi_{\ell,m}^{\alpha,\beta}(\underline{x})$ has vanishing moments as is shown in the next proposition.

Proposition 4.36 *The following assertions hold.*

1. *For $0 < k < -m - \ell - 2(\alpha + \beta)$ and $k < \ell$, we have*

$$\int_{\mathbb{R}^m} \underline{x}^k \psi_{\ell,m}^{\alpha,\beta}(\underline{x}) dV(\underline{x}) = 0.$$

2. *Its Clifford-Fourier transform is*

$$\widehat{\psi_{\ell,m}^{\alpha,\beta}(\underline{u})} = (-i)^{\ell} \underline{\xi}^{\ell} (2\pi)^{\frac{m}{2}} \rho^{1-\frac{m}{2}+\ell} \int_0^{\infty} \widetilde{\omega}_{\alpha,\beta}^l(r) J_{\frac{m}{2}-1}(r\rho) dr,$$

where

$$\widetilde{\omega}_{\alpha,\beta}^l(r) = ((1-r^2)\varepsilon_r)^{\alpha+\ell}(1+r^2)^{\beta+\ell} r^{\frac{m}{2}}$$

with $\varepsilon_r = sign(1-r)$.

Proof. The first assertion is a natural consequence of Proposition 4.34. We prove the second. We have

$$
\begin{aligned}
\widehat{\psi}_{\ell,m}^{\alpha,\beta}(\underline{u}) &= \int_{\mathbb{R}^m} \psi_{\ell,m}^{\alpha,\beta}(\underline{x}) e^{-i\underline{x}.\underline{u}} \, dV(\underline{x}) \\
&= (-1)^{\ell} \int_{\mathbb{R}^m} \partial_{\underline{x}}^{\ell} \left(\omega_{\alpha+\ell,\beta+\ell}(\underline{x}) \right) e^{-i\underline{x}.\underline{u}} \, dV(\underline{x}) \\
&= (-1)^{\ell} \int_{\mathbb{R}^m} \omega_{\alpha+\ell,\beta+\ell}(\underline{x}) e^{-i\underline{x}.\underline{u}} (i\underline{u})^{\ell} \, dV(\underline{x}) \\
&= (-1)^{\ell} (i\underline{u})^{\ell} \int_{\mathbb{R}^m} \omega_{\alpha+\ell,\beta+\ell}(\underline{x}) \, e^{-i\underline{x}.\underline{u}} \, dV(\underline{x}) \\
&= (-1)^{\ell} (i\underline{u})^{\ell} \int_{\mathbb{R}^m} (1-|\underline{x}|^2)^{\alpha+\ell} (1+|\underline{x}|^2)^{\beta+\ell} e^{-i\underline{x}.\underline{u}} \, dV(\underline{x}) \\
&= (-1)^{\ell} (i\underline{u})^{\ell} \widehat{\omega_{\alpha+\ell,\beta+\ell}}(\underline{u}).
\end{aligned}
$$

This Fourier transform can be simplified by using the spherical co-ordinates. By definition, we have

$$
\widehat{\omega_{\alpha+\ell,\beta+\ell}}(\underline{u}) = \int_{\mathbb{R}^m} (1-|\underline{x}|^2)^{\alpha+\ell} (1+|\underline{x}|^2)^{\beta+\ell} e^{-i<\underline{x},\underline{u}>} \, dV(\underline{x}). \tag{4.24}
$$

Introducing spherical co-ordinates

$$
\underline{x} = r\underline{\omega}, \quad \underline{u} = \rho\underline{\xi}, \quad r = |\underline{x}|, \quad \rho = |\underline{u}|, \quad \underline{\omega} \in S^{m-1}, \underline{\xi} \in S^{m-1}
$$

(S^{m-1} is the unit sphere in $\mathbb{R}^m$), expression (4.24) becomes

$$
\widehat{\omega_{\alpha+\ell,\beta+\ell}}(\underline{u}) = \int_0^{\infty} \widetilde{\omega}_{\alpha,\beta}^{\ell}(r) \, r^{\frac{m}{2}-1} \, dr \int_{S^{m-1}} e^{-i<r\underline{\omega},\rho\underline{\xi}>} d\sigma(\underline{\omega}),
$$

where $d\sigma(\underline{\omega})$ stands for the Lebesgue measure on S^{m-1}.
Using the following lemma

Lemma 4.37

$$
\int_{S^{m-1}} e^{-i<r\underline{\omega},\rho\underline{\xi}>} d\sigma(\underline{\omega}) = \frac{(2\pi)^{\frac{m}{2}} J_{\frac{m}{2}-1}(r\rho)}{(r\rho)^{\frac{m}{2}-1}},
$$

where $J_{\frac{m}{2}-1}$ is the Bessel function of the first kind of order $\frac{m}{2} - 1$ and $d\sigma$ is the Lebesgue measure on the sphere S^{m-1}.

We obtain

$$
\widehat{\omega_{\alpha+\ell,\beta+\ell}}(\underline{u}) = (2\pi)^{\frac{m}{2}} \rho^{1-\frac{m}{2}+\ell} \int_0^{\infty} \widetilde{\omega}_{\alpha,\beta}^{\ell}(r) J_{\frac{m}{2}-1}(r\rho) dr.
$$

Consequently, we obtain the following expression for the Fourier transform of the (α, β)-Clifford-Jacobi wavelets

$$
\widehat{\psi_{\ell,m}^{\alpha,\beta}}(\underline{u}) = (-i)^{\ell} \underline{\xi}^{\ell} (2\pi)^{\frac{m}{2}} \rho^{1-\frac{m}{2}+\ell} \int_0^{\infty} \widetilde{\omega}_{\alpha,\beta}^{\ell}(r) J_{\frac{m}{2}-1}(r\rho) dr.
$$

Definition 4.38 *The copy of the generalized 2-parameter Clifford-Gegenbauer wavelet at the scale $a > 0$ and the position $\underline{b}$ is defined by*

$$\frac{\underline{b}}{a}\psi_{\ell,m}^{\alpha,\beta}(\underline{x}) = a^{-\frac{m}{2}}\psi_{\ell,m}^{\alpha,\beta}\big(\frac{\underline{x}-\underline{b}}{a}\big).$$

Definition 4.39 *The wavelet transform of a function f in L_2 according to the generalized 2-parameter Clifford-Gegenbauer wavelet at the scale a and the position $\underline{b}$ is*

$$C_{a,\underline{b}}(f) = \int_{\mathbb{R}^m} f(\underline{x})\frac{\underline{b}}{a}\psi_{\ell,m}^{\alpha,\beta}(\underline{x})dV(\underline{x}).$$

The following lemma guarantees that the candidate $\psi_{\ell,m}^{\alpha,\beta}$ is indeed a mother wavelet. Analogue result is already checked in [79]. The proof is based on the asymptotic behavior of Bessel functions.

Lemma 4.40 *The quantity*

$$\mathcal{A}_{\ell,m}^{\alpha,\beta} = \frac{1}{\omega_m}\int_{\mathbb{R}^m}|\widehat{\psi_{\ell,m}^{\alpha,\beta}}(\underline{x})|^2\frac{dV(\underline{x})}{|\underline{x}|^m}$$

is finite. (ω_m is the volume of the unit sphere S^{m-1} in $\mathbb{R}^m$.)

To state the final result dealing with the reconstruction formula relative to the constructed new wavelets, we introduce firstly the inner product

$$< C_{a,\underline{b}}(f), C_{a,\underline{b}}(g) >= \frac{1}{\mathcal{A}_{\ell,m}^{\alpha,\beta}}\int_{\mathbb{R}^m}\int_0^{+\infty}\overline{C_{a,\underline{b}}(f)}C_{a,\underline{b}}(g)\frac{da}{a^{m+1}}dV(\underline{b}).$$

We obtain the following result.

Theorem 16 *Any function $f \in L_2(\mathbb{R}_m)$ may be reconstructed by*

$$f(x) = \frac{1}{\mathcal{A}_{\ell,m}^{\alpha,\beta}}\int_{a>0}\int_{b\in\mathbb{R}^m}C_{a,\underline{b}}(f)\,\psi\left(\frac{\underline{x}-\underline{b}}{a}\right)\frac{da\,dV(\underline{b})}{a^{m+1}},$$

where the equality has to be understood in the L_2-sense.

The proof reposes on the following result.

Lemma 4.41 *It holds that*

$$\int_{a>0}\int_{b\in\mathbb{R}^m}\overline{C_{a,\underline{b}}(f)}C_{a,\underline{b}}(g)\frac{da\,dV(\underline{b})}{a^{m+1}} = \mathcal{A}_{\ell,m}^{\mu,\alpha}\int_{\mathbb{R}^m}f(\underline{x})\overline{g(\underline{x})}dV(\underline{x}).$$

Proof. Using the Clifford-Fourier transform, we observe that

$$C_{a,\underline{b}}(f)(\underline{b}) = a^{\frac{m}{2}}\widetilde{\widehat{f}(\underline{.})\widehat{\psi}(a\underline{.})}(\underline{b}),$$

where, $\widetilde{h}(\underline{u}) = h(-\underline{u})$, $\forall\,h$. Thus,

$$\overline{C_{a,\underline{b}}(f)}C_{a,\underline{b}}(g) = \overline{\left(\widehat{f}(\underline{.})a^{\frac{m}{2}}\widehat{\psi}(a\underline{.})\right)}(-\underline{b})\left(\widehat{g}(\underline{.})a^{\frac{m}{2}}\widehat{\psi}(a\underline{.})\right)(-\underline{b}).$$

Consequently,

$$
\begin{aligned}
< C_{a,\underline{b}}(f), C_{a,\underline{b}}(g) > &= \int_{a>0} \int_{\mathbb{R}^m} \overline{\widehat{f}(\underline{.})a^{\frac{m}{2}}\widehat{\psi}(a\underline{.})}\, \widehat{g}(\underline{.})a^{\frac{m}{2}}\widehat{\psi}(a\underline{.}) \frac{da\, dV(\underline{b})}{a^{m+1}} \\
&= \int_{a>0} \int_{\mathbb{R}^m} \overline{\widehat{f}(\underline{b})}\widehat{g}(\underline{b}) \frac{a^m |\widehat{\psi}(a\underline{b})|^2}{a^{m+1}}\, da\, dV(\underline{b}) \\
&= \mathcal{A}_{\ell,m}^{\mu,\alpha} \int_{\omega \in \mathbb{R}^m} \overline{\widehat{f}(\underline{b})}\widehat{g}(b)\, dV(\underline{b}) \\
&= \mathcal{A}_{\ell,m}^{\mu,\alpha} < \widehat{f}, \widehat{g} > \\
&= < f, g > .
\end{aligned}
$$

Proof of Theorem 16. It follows immediately from Lemma 4.41.

4.9   SOME EXPERIMENTATIONS

In this section, we reproduce in brief an analogue application as in [24] based on the following two-dimensional Clifford weight function

$$
\omega(\underline{x}) = (1 + |\underline{x}|^2)^{3/2} e^{-|\underline{x}|^2/2},
$$

where $\underline{x} = e_1 x_1 + e_2 x_2$, $x_1, x_2 \in \mathbb{R}$ and (e_1, e_2) is the canonical basis of $\mathbb{R}^2$ equipped with the product rule $e_i e_j + e_j e_i = 2\delta_{ij}$, where δ_{ij} stands for the Kronecker symbol. By denoting $r^2 = x_1^2 + x_2^2 = |\underline{x}|^2$, we get as examples the Clifford mother wavelets (non-normalized)

$$
\psi_{0,2}(\underline{x}) = (1 + r^2)^{3/2} e^{-r^2/2},
$$

$$
\psi_{1,2}(\underline{x}) = (r^2 - 2)(1 + r^2)^{3/2} e^{-r^2/2} \underline{x},
$$

$$
\psi_{2,2}(\underline{x}) = (-4 + 6r^4 - 2r^6)(1 + r^2)^{3/2} e^{-r^2/2},
$$

and

$$
\psi_{3,2}(\underline{x}) = (1 - 16r^2 + 21r^4 + 15r^6 + r^8)(1 + r^2)^{3/2} e^{-r^2/2} \underline{x}.
$$

These wavelets are represented in Figure 4.2. Some examples of these Clifford wavelets have been applied to some biosignals in [24, 25]. Interested readers may be referred to such references. We here conduct another different experimentation by conducting a wavelet decomposition of a 2D image by means of the Clifford wavelet $\psi_{1,2}(\underline{x})$ above a single level (1-level) decomposition. The illustrations in Figure 4.3 have been obtained from the child Mohamed Amine Ben Mabrouk's photo (son of the author Anouar Ben Mabrouk).

The image has been processed for a 1-level wavelet decomposition. The upper right-hand sub-figure illustrates the 1-level approximation of the analyzed original image. The second line sub-figures and the left-hand sub-figure are the components of the 1-level detail projection of the original image.

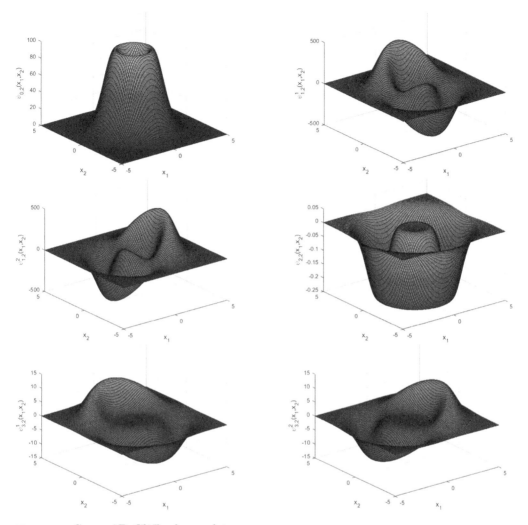

Figure 4.2 Some 2D Clifford wavelets.

The next experimentation concerns the application of 3D Clifford wavelets for 3D image processing tested on magnetic resonance images. We will apply here in the case where $m = 3$ the following 3D Clifford wavelets

$$\psi_{0,3}(\underline{x}) = (1 - \underline{x})\zeta(\rho),$$

$$\psi_{1,3}(\underline{x}) = \left[(\rho^4 - \rho^2 + 1) + (\rho^2 - 2)\underline{x}\right]\zeta(\rho),$$

$$\psi_{2,3}(\underline{x}) = \left[(\rho^6 + 4\rho^4 + 12\rho^2 + 5) + (-\rho^6 - \rho^4 - 2\rho^2 + 2)\underline{x}\right]\zeta(\rho)$$

$$\psi_{3,3}(\underline{x}) = \left[(\rho^{10} - 6\rho^6 + 35\rho^4 + 45\rho^2 + 1) + (\rho^8 - 2\rho^6 - 2\rho^4 + 2\rho^2 + 5)\underline{x}\right]\zeta(\rho),$$

$$\zeta(\rho) = (1 + \rho^2)^{3/2}\exp(-\frac{\rho^2}{2})$$

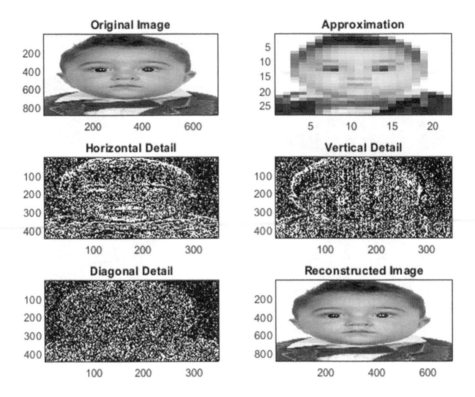

Figure 4.3 $\psi_{1,2}(\underline{x})$-Clifford wavelet 1-level decomposition of Mohamed Amine's photo.

where $\underline{x} = e_1 x_1 + e_2 x_2 + e_3 x_3$ by means of the canonical basis (e_1, e_2, e_3) of $\mathbb{R}^3$ equipped already with the extra operations $e_i e_j + e_j e_i = -2\delta_{ij}$ and where $\rho = |\underline{x}| = \sqrt{x_1^2 + x_2^2 + x_3^2}$. As in [24], we examined the method using in the present case the 3D Clifford wavelet $\psi_{3,3}(\underline{x})$ above. The experimentation has resulted in Figure 4.4.

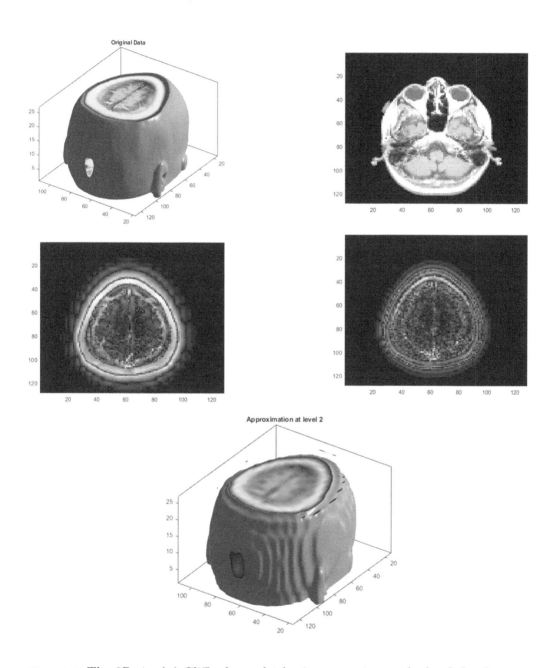

Figure 4.4 The 3D $\psi_{3,3}(\underline{x})$ Clifford wavelet brain processing at the level $J = 2$.

4.10 EXERCISES

Exercise 1.

Show $\forall \underline{x} \in \mathbb{R}^m$ and $\forall A_k \in \mathbb{R}_m$, a k-grade, the following assertions.

a. $\underline{x} \cdot A_k = [\underline{x}A_k]_{k-1} = \frac{1}{2}(\underline{x}A_k - (-1)^k A_k \underline{x})$.

b. $\underline{x} \wedge A_k = [\underline{x}A_k]_{k+1} = \frac{1}{2}(\underline{x}A_k + (-1)^k A_k \underline{x})$.

c. Generalize the previous assertions to

$$A_k \cdot B_l = [A_k B_l]_{|k-l|}, \text{ with } A_k \cdot B_l = 0 \text{ if } kl = 0$$
$$A_k \wedge B_l = [A_k B_l]_{|k+l|}$$

for all k-grade element A_k and l-grade element B_l in $\mathbb{R}_m$.

Exercise 2.

a. Show that the main involution and the reversion commute:

$$\widetilde{a^*} = \widetilde{a}^*, \quad \forall a \in \mathbb{R}_n.$$

b. Show that for all $a = \sum_{k=0}^{n} [a]_k \in \mathbb{R}_m = \bigoplus_{k=0}^{m} \mathbb{R}_m^k$, we have

$$\widetilde{a} = \sum_k (-1)^k [a]_k,$$
$$a^* = \sum_k (-1)^k \left([a]_{2k} + [a]_{2k+1}\right),$$
$$\overline{a} = \sum_k (-1)^k \left([a]_{2k} - [a]_{2k+1}\right).$$

Exercise 3.

Consider the Pauli matrices

$$\sigma_0 = \begin{pmatrix} 1 & 0 \\ 0 & 1 \end{pmatrix}, \quad \sigma_1 = \begin{pmatrix} 0 & 1 \\ 1 & 0 \end{pmatrix}, \quad \sigma_2 = \begin{pmatrix} 0 & -i \\ i & 0 \end{pmatrix}, \quad \sigma_3 = \begin{pmatrix} 1 & 0 \\ 0 & -1 \end{pmatrix}.$$

a. Compute the products $\sigma_i \sigma_j$ for all i, j.

b. Let $\mathbb{A}$ be the algebra spanned by σ_i, $i = 1, 2, 3, 4$. Prove that $\mathbb{A}$ is isomorphic to $\mathcal{M}_2(\mathbb{C})$.

Exercise 4.

Consider the Dirac matrices

$$\gamma_0 = \begin{pmatrix} \sigma_0 & 0 \\ 0 & -\sigma_0 \end{pmatrix}, \quad \gamma_1 = \begin{pmatrix} 0 & \sigma_1 \\ -\sigma_1 & 0 \end{pmatrix}, \quad \gamma_2 = \begin{pmatrix} 0 & \sigma_2 \\ -\sigma_2 & 0 \end{pmatrix}, \quad \gamma_3 = \begin{pmatrix} 0 & \sigma_3 \\ -\sigma_3 & 0 \end{pmatrix}.$$

a. Compute

$$\gamma_i\gamma_j + \gamma_j\gamma_i, \quad i,j = 0,1,2,3.$$

b. Characterize the Clifford algebra $\mathbb{A}$ spanned by σ_i, $i = 1,2,3,4$.

c. Prove that $\mathbb{A}$ is isomorphic to $\mathcal{M}_4(\mathbb{C})$.

Exercise 5.

Show that for a Clifford number $\underline{x} \in \mathbb{R}_m$

$$|\exp \underline{x}| = \exp(Re(\underline{x})).$$

Exercise 6.

Let f be a vector field on $\mathbb{R}_m$. Show that

a. $\operatorname{div} f = \partial_{\underline{x}} \cdot f = \dfrac{1}{2}(\partial_{\underline{x}} f + f\partial_{\underline{x}}).$

b. $\operatorname{curl} f = \partial_{\underline{x}} \wedge f = \dfrac{1}{2}(\partial_{\underline{x}} f - f\partial_{\underline{x}}).$

Exercise 7.

Show that

$$\partial_{\underline{x}}[f(\underline{x})]_+ = [\partial_{\underline{x}} f(\underline{x})]_- \quad \text{and} \quad \partial_{\underline{x}}[f(\underline{x})]_- = [\partial_{\underline{x}} f(\underline{x})]_+$$

Exercise 8.

Consider the Clifford-Hermite weight function

$$W(\underline{x}) = \exp(-\frac{|\underline{x}|^2}{2}).$$

Denote $F * (t, \underline{x})$ its CK-extension and $H_l(\underline{x})$ the Clifford-Hermite polynomials generated by $F * (t, \underline{x})$.
1) By applying the characteristics of $F * (t, \underline{x})$ show that

$$H_{l+1}(\underline{x}) - \underline{x}H_l(\underline{x}) + \partial_{\underline{x}}H_l(\underline{x}) = 0, \; \forall l \in \mathbb{N}.$$

2) Show that $H_l(\underline{x})$ satisfies the Rodrigues formula

$$H_l(\underline{x}) = (-1)^l \exp(\frac{|\underline{x}|^2}{2})\partial_{\underline{x}}^l \left[\exp(-\frac{|\underline{x}|^2}{2}) \right].$$

3) Prove finally the following mutual orthogonality for the polynomials $H_l(\underline{x})$:

$$\int_{\mathbb{R}^m} \exp(-\frac{|\underline{x}|^2}{2})\overline{H_l(\underline{x})}H_k(\underline{x})dV(\underline{x}) = \delta_{l,k}\gamma_l,$$

where $\gamma_{l,m}$ is a constant depending only on l and m (to be explicated).

Exercise 9.

We conserve the same assumptions and notations as in Exercise 8. Consider for $l \in \mathbb{N}$ the function

$$\psi_l(\underline{x}) = \exp(-\frac{|\underline{x}|^2}{2})H_l(\underline{x}).$$

1) Compute the Clifford-Fourier transform of ψ_l denote by $\mathcal{F}[\psi_l](\underline{\xi})$, $\forall \underline{\xi} \in \mathbb{R}^m$.
2) Compute the admissibility constant

$$A_l = \frac{(2\pi)^m}{\omega_m} \int_{\mathbb{R}^m} \frac{\mathcal{F}[\psi_l](\underline{\xi})\Big(\mathcal{F}[\psi_l](\underline{\xi})\Big)^\dagger}{|\underline{\xi}|^m} dV(\underline{\xi}),$$

where ω_m is the volume of the unit sphere $\mathbb{S}^m$ in $\mathbb{R}^m$.
3) Show finally that

$$\int_{\mathbb{R}^m} \underline{x}^j \psi_l(\underline{x}) dV(\underline{x}) = 0, \text{ for } j = 0, 1, \ldots, l-1.$$

Exercise 10.

Let $m \in \mathbb{N}$ decomposed into a sum $m = p + q$ in such a way the vector variable $\underline{x} \in \mathbb{R}^m = \mathbb{R}^{p+q}$ will be decomposed as a sum $\underline{x} = \underline{y} + \underline{z}$, with $\underline{y} \in \mathbb{R}^p$ and $\underline{z} \in \mathbb{R}^q$. Consider next the Clifford-Hermite weight

$$W(\underline{x}) = \exp(-\alpha\frac{|\underline{y}|^2}{2})\exp(-\beta\frac{|\underline{w}|^2}{2}), \ \alpha, \beta > 0.$$

Let $F_{\alpha,\beta} * (t, \underline{x})$ be its CK-extension and $H_l^{\alpha,\beta}(\underline{x})$ the Clifford-Hermite polynomials generated by $F_{\alpha,\beta} * (t, \underline{x})$.
By following the same steps as Exercise 8 and Exercise 9,
1) Provide a recurrence rule for the polynomials $H_l^{\alpha,\beta}(\underline{x})$.
2) Provide a Rodrigues formula for $H_l^{\alpha,\beta}(\underline{x})$.
3) Prove finally a mutual orthogonality relation for $H_l^{\alpha,\beta}(\underline{x})$.
4) Provide an analogue $\psi_l^{\alpha,\beta}(\underline{x})$ for the Clifford-Hermite wavelet.
5) Compute the Clifford-Fourier transform of the new wavelet $\psi_l^{\alpha,\beta}(\underline{x})$.
6) Compute its admissibility constant.
7) Show finally that $\psi_l^{\alpha,\beta}(\underline{x})$ has l vanishing moments.
8) Try to express $H_l^{1,1}(\underline{x})$ by means of $H_l(\underline{y})$ and $H_l(\underline{z})$.

CHAPTER 5

Quantum wavelets

5.1 INTRODUCTION

Quantum wavelets are special types of wavelet functions characterized by special properties that may not be satisfied by other functions. In the present context, our aim is to develop wavelet functions based on some special functions such as Bessel one. Bessel functions form an important class of special functions and are applied almost everywhere in mathematical physics. They are also known as cylindrical functions, or cylindrical harmonics, because of their strong link to the solutions of the Laplace equation in cylindrical coordinates.

The purpose of this chapter is to present some wavelet analysis in the framework of quantum theory (q-theory). We aim precisely to review details on q-Bessel functions introduced in the context of q-theory, which makes a variant of Bessel functions on the real line. Famous relations associated with wavelet transforms such as Plancherel/Parseval ones as well as reconstruction formula will be investigated.

A first step in wavelet/Bessel transform interconnection has been conducted in [216], where the theory of continuous wavelet transform has been extended to a class of generalized one associated with a class of singular differential operators. This class contains, in particular, the so-called Bessel function (see also [347]). Such an extension is natural as Bessel functions may yield good bases for functional spaces similar to known cases such as Fourier, Haar and orthogonal polynomials. Besides, Bessel functions are applied in various domains, especially partial differential equations, wave motion, diffusion, etc.

Quantum or q-theory, in fact, provides a discrete refinement of continuous harmonic analysis on suitable sub-spaces such as $\mathbb{R}_q$ composed of the discrete grid $\pm q^n$, $n \in \mathbb{Z}$, $q \in (0, 1)$, which is a dense subset in $\mathbb{R}$.

5.2 BESSEL FUNCTIONS

There are three main ways to introduce Bessel functions. The first one is due to a special type of second-order singular and linear differential equations where the Bessel functions form a generator set of solutions. The second one resembles orthogonal polynomials and provides Bessel functions as solutions of three-level recurrent singular and linear functional equation. The third one provides Bessel functions via the Rodriguez derivation formula.

Definition 5.1 *The Bessel equation is a linear differential equation of second order written in the form*

$$y" + \frac{1}{x}y' + \left(1 - \frac{v^2}{x^2}\right)y = 0,$$

where v is a positive constant.

Theorem 17 *Bessel's differential equation has a solution of the form*

$$J_v(x) = \left(\frac{x}{2}\right)^v \sum_{k\geq 0} \frac{(-1)^k}{k!\Gamma(v+k+1)} \left(\frac{x}{2}\right)^{2k}.$$

Γ *is the Gamma Euler's function. The function J_v is called Bessel function of the first kind of order v.*

Proof. Put

$$y = x^p \sum_{i\geq 0} a_i x^i = \sum_{i\geq 0} a_i x^{i+p},$$

where p is a real parameter. By replacing y and its derivatives in Definition 5.1, we get

$$a_0(p^2 - v^2) = 0, \qquad a_1((p+1)^2 - v^2) = 0$$

and

$$a_i((i+v)^2 - v^2) + a_{i-2} = 0, \ \forall\, i \geq 2.$$

For $p = v$, we get $a_1 = 0$ and

$$a_i = -\frac{a_{i-2}}{i(2v+i)}, \ \forall\, i \geq 2.$$

Hence, $a_{2k+1} = 0$ for all $k \geq 0$, and

$$a_{2k} = (-1)^k \frac{a_0}{2^{2k}k!(v+k)(v+k-1)\ldots(v+1)}, \ \forall k \geq 0.$$

Taking $a_0 = \dfrac{1}{2^v\Gamma(v+1)}$ and observing that

$$\Gamma(v+k+1) = (v+k)(v+k-1)\ldots(v+1)\Gamma(v+1),$$

we get

$$a_{2k} = (-1)^k \frac{1}{2^{2k+v}k!\Gamma(v+k+1)}, \ k \geq 0.$$

So as the theorem.

Remark 5.2 *1. For $p = -v$, the solution of Bessel's equation in Definition 5.1 is called Bessel's function of the first kind with order $-v$ and is denoted $J_{-v}(x)$ with*

$$J_{-v}(x) = \left(\frac{x}{2}\right)^{-v} \sum_{k \geq 0} \frac{(-1)^k}{k!\Gamma(k - v + 1)} \left(\frac{x}{2}\right)^{2k}.$$

2. The same solution can be obtained by choosing $p+1 = v$ in the proof of Theorem 17.

3. For $v = n \in \mathbb{N}$, we have $J_n = (-1)^n J_{-n}$.

4. The Bessel function of the second kind of order α denoted usually Y_α is given by

$$Y_\alpha(x) = \begin{cases} \dfrac{\cos(\pi\alpha)J_\alpha(x) - J_{-\alpha}(x)}{\sin(\pi\alpha)}, & for \; \alpha \notin \mathbb{Z} \\ \lim\limits_{v \to \alpha} \dfrac{\cos(\pi v)J_v(x) - J_{-v}(x)}{\sin(\pi v)}, & for \; \alpha \in \mathbb{Z}. \end{cases}$$

Proposition 5.3 *Let λ and μ be two different roots of the Bessel function $J_v(x)$. The following orthogonality property holds,*

$$\int_0^1 x \, J_v(\lambda x) J_v(\mu x) \, dx = 0.$$

Proof. Denote

$$y_{v,\lambda}(x) = J_v(\lambda x) \qquad and \qquad y_{v,\mu}(x) = J_v(\mu x).$$

Then, $y_{v,\lambda}$ and $y_{v,\mu}$ are solutions of the following Bessel-type differential equations

$$(xy'_{v,\lambda})'(x) + (\lambda^2 x - \frac{v^2}{x})y_{v,\lambda}(x) = 0 \tag{5.1}$$

and

$$(xy'_{v,\mu})'(x) + (\mu^2 x - \frac{v^2}{x})y_{v,\mu}(x) = 0. \tag{5.2}$$

Multiplying equation (5.1) by $y_{v,\mu}$ and equation (5.2) by $y_{v,\lambda}$ and integrating the difference on $(0, 1)$ we get

$$\int_0^1 \left[(xy'_{v,\lambda})'(x)y_{v,\mu}(x) - (xy'_{v,\mu})'(x)y_{v,\lambda}(x) \right] dx + (\lambda^2 - \mu^2) \int_0^1 xy_{v,\lambda}(x)y_{v,\mu}(x)dx = 0. \tag{5.3}$$

Next, an integration by parts in the first integral yields

$$\int_0^1 \left[(xy'_{v,\lambda})'(x)y_{v,\mu}(x) - (xy'_{v,\mu})'(x)y_{v,\lambda}(x) \right] dx = y'_{v,\lambda}(1)y_{v,\mu}(1) - y'_{v,\mu}(1)y_{v,\lambda}(1).$$

Recall now that

$$y_{v,\lambda}(1) = J_v(\lambda) = y_{v,\mu}(1) = J_v(\mu) = 0.$$

This implies that

$$\int_0^1 \Big[(xy'_{v,\lambda})'(x) y_{v,\mu}(x) - (xy'_{v,\mu})'(x) y_{v,\lambda}(x) \Big] dx = 0.$$

Hence, it remains in (5.3)

$$(\lambda^2 - \mu^2) \int_0^1 x \, J_v(\lambda x) J_v(\mu x) \, dx = 0.$$

Since $\lambda \neq \mu$, we get

$$\int_0^1 x \, J_v(\lambda x) J_v(\mu x) \, dx = 0.$$

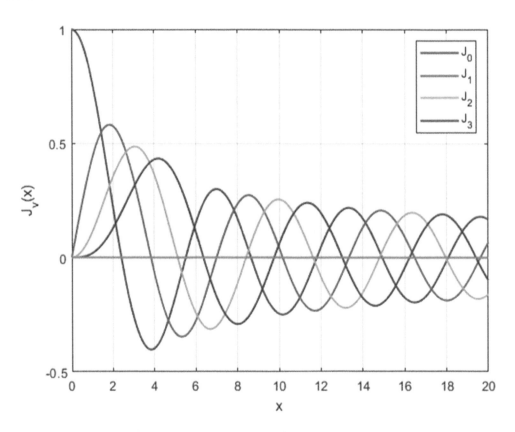

Figure 5.1 Graphs of Bessel functions J_v of the first kind for $v = 0, 1, 2, 3$.

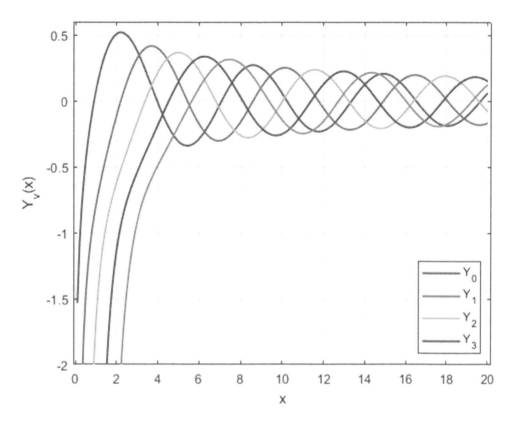

Figure 5.2 Graphs of Bessel functions Y_v of the second kind for $v = 0, 1, 2, 3$.

We now review the notions of Bessel and Fourier-Bessel transforms of functions. The readers may refer to [217] for more details. We denote the inner product in $L^2(\mathbb{R}^+, dx)$ by

$$\langle f, g \rangle = \int_0^\infty f(x) \, \overline{g(x)} \, dx$$

and the associated norm by $\|.\|_2$. Similarly, we denote the inner product in $L^2(\mathbb{R}^+, \xi d\xi)$ by

$$\langle f, g \rangle_\xi = \int_0^\infty f(\xi) \, \overline{g(\xi)} \, \xi d\xi$$

and the associated norm by $\|.\|_{\xi,2}$.

Definition 5.4 *Let $f \in L^2(\mathbb{R}^+, dx)$. The Bessel transform of f is defined by*

$$\mathcal{B}(f)(\xi) = \int_0^{+\infty} f(x) \sqrt{x} J_v(x\xi) \, dx, \ \forall \, \xi > 0,$$

where J_v is the Bessel function of first kind and index v.

We immediately have the following characteristics.

Proposition 5.5 *1. For all $f \in L^2(\mathbb{R}^+, dx)$, $\mathcal{B}(f) \in L^2(\mathbb{R}^+, \xi d\xi)$.*

2. *The Bessel transform $\mathcal{B}$ is invertible and*

$$\mathcal{B}^{-1}(g)(x) = \int_0^{+\infty} g(\xi)\sqrt{x}\,J_v(x\xi)\,\xi\,d\xi, \quad \forall g \in L^2(\mathbb{R}^+, \xi d\xi).$$

Proof. 1. Let f and g be in $L^2(\mathbb{R}^+, dx)$. We have

$$
\begin{aligned}
< \mathcal{B}(f), \mathcal{B}(g) >_\xi &= \int_0^{+\infty} \mathcal{B}(f)(\xi)\overline{\mathcal{B}(g)(\xi)}\xi d\xi \\
&= \int_{\mathbb{R}_+^3} \sqrt{x}\sqrt{y}f(x)\overline{g(y)}J_v(x\xi)J_v(y\xi)\xi dx\,dy\,d\xi \\
&= \int_{\mathbb{R}_+^2} \sqrt{x}\sqrt{y}f(x)\overline{g(y)}\frac{\delta(x-y)}{x}dx\,dy \\
&= \int_{\mathbb{R}_+} \sqrt{x}\sqrt{x}f(x)\overline{g(x)}\frac{1}{x}dx \\
&= < f, g > .
\end{aligned}
$$

So, taking $g = f$, we get

$$\|\mathcal{B}(f)\|_{\xi,2} = \|f\|_2,$$

which means that $\mathcal{B}$ is an isometry.

2. Denote the right-hand quantity by $\widetilde{\mathcal{B}}(f)$. We will prove that $\mathcal{B}(\widetilde{\mathcal{B}}(f)) = f$. Indeed,

$$
\begin{aligned}
\mathcal{B}(\widetilde{\mathcal{B}}(f))(\xi) &= \int_0^{+\infty} \widetilde{\mathcal{B}}(f)(x)\sqrt{x}J_v(x\xi)\,dx \\
&= \int_0^{+\infty}\int_0^{+\infty} f(\eta)\sqrt{x}J_v(x\eta)\,\eta\sqrt{x}J_v(x\xi)\,d\eta\,dx \\
&= \int_0^{+\infty} f(\eta)\eta\frac{\delta(\eta-\xi)}{\eta}d\eta = f(\xi).
\end{aligned}
$$

Definition 5.6 *The Fourier-Bessel transform of order v is defined by*

$$\mathcal{FB}(f)(\xi) = \int_0^\infty f(x)\,J_v(x\xi)\,xdx; \quad \forall\,f. \tag{5.4}$$

Lemma 5.7 *Fourier-Bessel transform and Bessel transform are related via the equality*

$$\mathcal{FB}(f)(\xi) = \mathcal{B}(\sqrt{.}f)(\xi).$$

Indeed, For $t > 0$, we have

$$
\begin{aligned}
\mathcal{B}(\sqrt{t}f)(\xi) &= \int_0^{+\infty} \sqrt{t}f(t)\sqrt{t}J_v(t\xi)\,dt \\
&= \int_0^{+\infty} f(t)\,t\,J_v(t\xi)\,dt \\
&= \int_0^\infty f(t)\,J_v(t\xi)\,tdt \\
&= \mathcal{FB}(f)(\xi).
\end{aligned}
$$

5.3   BESSEL WAVELETS

There are in literature several approaches to introduce Bessel wavelets. We refer, for instance, to [315, 317]. For $1 \le p < \infty$ and $\mu > 0$, denote

$$L_\sigma^p(\mathbb{R}_+) = \left\{ f : \mathbb{R}_+ \to \mathbb{R}; \ \|f\|_{p,\sigma}^p = \int_0^\infty |f(x)|^p \, d\sigma(x) < \infty \right\},$$

where $d\sigma(x)$ is the measure defined by

$$d\sigma(x) = \frac{x^{2\mu}}{2^{\mu-\frac{1}{2}} \, \Gamma(\mu + \frac{1}{2})} \chi_{[0,\infty[}(x) \, dx.$$

Denote also

$$j_\mu(x) = 2^{\mu-\frac{1}{2}} \, \Gamma(\mu + \frac{1}{2}) \, x^{\frac{1}{2}-\mu} \, J_{\mu-\frac{1}{2}}(x),$$

where $J_{\mu-\frac{1}{2}}(x)$ is the Bessel function of order $v = \mu - \frac{1}{2}$.
Denote next

$$D(x, y, z) = \int_0^\infty j_\mu(xt) \, j_\mu(yt) \, j_\mu(zt) \, d\sigma(t).$$

Definition 5.8 *For a 1-variable function f, we define a translation operator*

$$\tau_x f(y) = \widetilde{f}(x, y) = \int_0^\infty D(x, y, z) \, f(x) \, d\sigma(z), \quad \forall \, 0 < x, y < \infty,$$

and for a 2-variables function f, we define a dilation operator

$$D_a f(x, y) = \frac{1}{a^{2\mu+1}} \, f(\frac{x}{a}, \frac{y}{a}).$$

Proposition 5.9

$$\int_0^\infty j_\mu(zt) \, D(x, y, z) \, d\sigma(z) = j_\mu(xt) \, j_\mu(yt), \quad \forall \, 0 < x, y < \infty, \ 0 \le t < \infty$$

and

$$\int_0^\infty D(x, y, z) \, d\sigma(z) = 1.$$

Proof. We will develop the proof of the first point. The second is left to the reader as a simple exercise. So, simple observations permit us to write

$$\int_0^\infty j_\mu(zt) D(x, y, z) d\sigma(z)$$
$$= \frac{(xyt)^{1/2-\mu}}{C_\mu^2} \int_0^\infty \int_0^\infty s^{3/2-\mu} J_{1/2-\mu}(xs) J_{1/2-\mu}(ys) z J_{1/2-\mu}(xzt) J_{1/2-\mu}(zs) ds dz,$$

where $\dfrac{1}{C_\mu} = 2^{\mu-\frac{1}{2}} \, \Gamma(\mu + \frac{1}{2})$. Using Fubini's rule, we obtain

$$\int_0^\infty j_\mu(zt) D(x, y, z) d\sigma(z)$$
$$= \frac{(xyt)^{1/2-\mu}}{C_\mu^2} \int_0^\infty s^{3/2-\mu} J_{1/2-\mu}(xs) J_{1/2-\mu}(ys) \frac{\delta(t-s)}{s} ds$$
$$= \frac{(xyt)^{1/2-\mu}}{C_\mu^2} t^{3/2-\mu} J_{1/2-\mu}(xt) J_{1/2-\mu}(yt) \frac{1}{t}$$
$$= j_\mu(xt) j_\mu(yt).$$

Definition 5.10 *The Bessel wavelet copies $\Psi_{a,b}$ are defined from the Bessel mother wavelet $\Psi \in L^2_\sigma(\mathbb{R}_+)$ by*

$$\Psi_{a,b}(x) = D_a\tau_b\Psi(x) = \frac{1}{a^{2\mu+1}} \int_0^\infty D(\frac{b}{a}, \frac{x}{a}, z)\, \Psi(z)\, d\sigma(x), \quad \forall a, b \geq 0.$$

The continuous Bessel wavelet transform (CBWT) of a function $f \in L^2_\sigma(\mathbb{R}_+)$, at the scale a and the position b, is defined by

$$(B_\Psi f)(a,b) = \frac{1}{a^{2\mu+1}} \int_0^\infty \int_0^\infty f(t)\, \overline{\Psi}(z)\, D(\frac{b}{a}, \frac{t}{a}, z)\, d\sigma(z)\, d\sigma(t).$$

It is well known in Bessel wavelet theory that such a transform is a continuous function according to the variable (a, b). The following result is a variant of Parseval/Plancherel rules for the case of Bessel wavelet transforms.

Theorem 18 *[315] Let $\Psi \in L^2_\sigma(\mathbb{R}_+)$ and $f, g \in L^2_\sigma(\mathbb{R}_+)$. Then*

$$\int_0^\infty \int_0^\infty (B_\Psi f)(a,b)\, \overline{(B_\Psi g)}(a,b)\, \frac{d\sigma(a)\, d\sigma(b)}{a^{2\mu+1}} = \mathcal{A}_\Psi\, \langle f, g \rangle,$$

whenever

$$\mathcal{A}_\Psi = \int_0^\infty t^{-2\mu-1}\, |\widehat{\Psi}(t)|^2\, d\sigma(t) < \infty.$$

Proof. We have

$$(B_\Psi f)(a,b) = \int_0^{+\infty} f(t)\, \Psi_{a,b}(t)\, d\sigma(t)$$

$$= \frac{1}{a^{2\sigma+1}} \int_0^\infty \int_0^\infty f(t)\, \overline{\Psi}(z)\, D(\frac{b}{a}, \frac{t}{a}, z)\, d\sigma(z)\, d\sigma(t).$$

Now observe that

$$D(\frac{b}{a}, \frac{t}{a}, z) = \int_0^{+\infty} j_\mu(\frac{b}{a}u)\, j_\mu(\frac{t}{a}u)\, j_\mu(zu)\, d\sigma(u).$$

Hence,

$$(B_\Psi f)(a,b) = \frac{1}{a^{2\mu+1}} \int_{\mathbb{R}^3_+} f(t)\Psi(z)j_\mu(\frac{b}{a}u)j_\mu(\frac{t}{a}u)j_\mu(zu)d\sigma(u)d\sigma(z)d\sigma(t)$$

$$= \frac{1}{a^{2\mu+1}} \int_{\mathbb{R}^2_+} \widehat{f}(\frac{u}{a})\Psi(z)j_\mu(\frac{b}{a}u)j_\mu(zu)d\sigma(u)d\sigma(z)$$

$$= \frac{1}{a^{2\mu+1}} \int_{\mathbb{R}_+} \widehat{f}(\frac{u}{a})\widehat{\Psi}(u)j_\mu(\frac{b}{a}u)d\sigma(u)$$

$$= \int_{\mathbb{R}_+} \widehat{f}(\eta)\widehat{\Psi}(a\eta)j_\mu(b\eta)d\sigma(\eta)$$

$$= \left(\widehat{f}(\eta)\widehat{\Psi}(a\eta)\right)(b).$$

As a result,

$$\int_{\mathbb{R}_+^2} (B_\Psi f)(a,b)\overline{(B_\Psi g)}(a,b)\frac{d\sigma(a)}{a^{2\eta+1}}d\sigma(b)$$

$$= \int_{\mathbb{R}_+^2} \widehat{f}(\eta)\widehat{\Psi}(a\eta)\overline{\widehat{g}(\eta)}\overline{\widehat{\Psi}(a\eta)}d\sigma(\eta)\frac{d\sigma(a)}{a^{2\sigma+1}}$$

$$= \int_{\mathbb{R}_+^2} \widehat{f}(\eta)\overline{\widehat{g}(\eta)}\left(\int_{\mathbb{R}_+} |\widehat{\Psi}(a\eta)|^2\frac{d\sigma(a)}{a^{2\sigma+1}}\right)d\sigma(\eta)$$

$$= C_\Psi \int_{\mathbb{R}_+} \widehat{f}(\eta)\overline{\widehat{g}(\eta)}d\sigma(\eta)$$

$$= C_\Psi \langle \widehat{f}, \widehat{g} \rangle$$

$$= C_\Psi \langle f, g \rangle.$$

Theorem 19 *Let $\Psi \in L_\sigma^2(\mathbb{R}_+)$ and $f \in L_\sigma^2(\mathbb{R}_+)$. Then*

$$f(x) = \frac{1}{A_\Psi} \int \int (B_\Psi f)(a,b)\,\Psi(\frac{x-b}{a})\,\frac{d\sigma(a)d\sigma(b)}{a^{2\mu+1}}$$

in the L_σ^2-sense whenever $A_\Psi = \int_0^\infty t^{-2\mu-1}\,|\widehat{\Psi}(t)|^2\,d\sigma(t) < \infty.$

The proof is an immediate consequence of the previous theorem.

5.4 FRACTIONAL BESSEL WAVELETS

Definition 5.11 *The Hankel transform, also called the Bessel-Fourier transform, of a function $\varphi \in L^1(\mathbb{R}_+)$ is defined by*

$$(h_\mu\varphi)(y) = \widehat{\varphi}(y) = \int_0^\infty (xy)^{\frac{1}{2}} J_\mu(xy)\varphi(x)dx, \quad x \in \mathbb{R}_+, \ \mu \geq -\frac{1}{2}.$$

Its inverse formula is

$$(h_\mu^{-1}\widehat{\varphi}(y))(x) = \varphi(x) = \int_0^\infty (xy)^{\frac{1}{2}} J_\mu(xy)\widehat{\varphi}(y)dy, \quad y \in \mathbb{R},$$

where J_μ is the Bessel function of first kind and with order μ.

Definition 5.12 *The fractional Hankel transform, also called fractional Bessel-Fourier transform, of parameter θ on $L^1(\mathbb{R}^+)$ is defined by*

$$(h_\mu^\theta\varphi)(y) = \widehat{\varphi}_\mu^\theta(y) = \int_0^\infty K_\mu^\theta(x,y)\varphi(x)dx,$$

where K_μ^θ is the kernel

$$K_\mu^\theta(x,y) = \begin{cases} c_\mu^\theta e^{\frac{i}{2}(x^2+y^2)\cot\theta}(xy\cos\theta)^{\frac{1}{2}}J_\mu(xy\cos\theta), & \theta \neq qn\pi, \\ (xy)^{\frac{1}{2}}J_\mu(xy), & \theta = \frac{\pi}{2}, \\ \delta(x-y), & \theta = n\pi, \forall n \in \mathbb{Z} \end{cases}$$

and $\delta_s = 1$ for $s = 0$ and 0 elsewhere is the Dirac mass or the Kronecker symbol, and finally

$$c_\mu^\theta = \frac{exp\left[i(1+\mu)(\frac{\pi}{2}-\theta)\right]}{\sin\theta}, \quad \theta \neq n\pi.$$

Definition 5.13 Let $f, g \in L^1(\mathbb{R}_+)$. We define the fractional Hankel convolution by

$$(f \,\natural_\theta\, g)(x) = \int_0^{+\infty} f(y)\,(\tau_x^\theta g)(y)\,dy,$$

where the fractional Hankel translation $\tau_x^\theta g$ is given by

$$(\tau_x^\theta g)(y) = g^\theta(x,y) = e^{\frac{-i}{2}(x^2+y^2)\cot\theta}\int_0^\infty g(z)\,D_\mu^\theta(x,y,z)\,dz,$$

and

$$D_\mu^\theta(x,y,z) = \frac{2^{\mu-1}\,\nabla\,(x,y,z)^{2\mu-1}c_\mu^\theta e^{\frac{i}{2}(x^2+y^2+z^2)\cos\theta}}{(xyz)^{\mu-\frac{1}{2}}\sqrt{\pi}\,\Gamma(\mu+\frac{1}{2})},$$

where $\nabla(x,y,z)$ is the area of the triangle of vertices x,y,z.

Proposition 5.14

1. $D_\mu^\theta(x,y,z)$ is symmetric in x,y,z.

2. For $\xi = 0$, we obtain

$$\int_0^\infty \left|D_\mu^\theta(x,y,z)\right| z^{\mu+\frac{1}{2}}\,dz \leq \frac{(xy)^{\mu+\frac{1}{2}}}{2^\mu\,\Gamma(\mu+1)\,|\sin\theta|^{\mu+\frac{1}{2}}}.$$

Lemma 5.15 For $\varphi \in L^2(\mathbb{R})$, we have

$$\|x^{-\mu-\frac{1}{2}}\,(\tau_x^\theta\varphi)(y)\|_{L^2} \leq \frac{1}{|\sin\theta|^{\mu+\frac{1}{2}}\,2^\mu\,\Gamma(\mu+1)}\,\|\varphi\|_{L^2}.$$

Proof. We have

$$(\tau_x^\theta\varphi)(y) = \varphi^\theta(x,y) = e^{\frac{-i}{2}(x^2+y^2)\cot\theta}\int_0^\infty \varphi(z)\,D_\mu^\theta(x,y,z)\,dz,$$

Using Proposition 5.14. above, we obtain

$$|(\tau_x^\theta\varphi)(y)|$$

$$\leq \int_0^\infty \left|\varphi(z)\,D_\mu^\theta(x,y,z)\right|dz$$

$$\leq \int_0^\infty \left|\varphi(z)\,z^{\frac{-1}{2}(\mu+\frac{1}{2})}\,D_\mu^\theta(x,y,z)^{\frac{1}{2}}\,z^{\frac{1}{2}(\mu+\frac{1}{2})}\,D_\mu^\theta(x,y,z)^{\frac{1}{2}}\right|dz$$

$$\leq \left(\int_0^\infty z^{-(\mu+\frac{1}{2})}\,|\varphi(z)|^2\,\left|D_\mu^\theta(x,y,z)\right|dz\right)^{\frac{1}{2}}\left(\int_0^\infty z^{(\mu+\frac{1}{2})}\,\left|D_\mu^\theta(x,y,z)\right|dz\right)^{\frac{1}{2}}$$

$$\leq \left(\frac{(xy)^{\mu+\frac{1}{2}}}{2^\mu\,\Gamma(\mu+1)\,|\sin\theta|^{\mu+\frac{1}{2}}}\right)^{\frac{1}{2}}\left(\int_0^\infty z^{-(\mu+\frac{1}{2})}\,|\varphi(z)|^2\,\left|D_\mu^\theta(x,y,z)\right|dz\right)^{\frac{1}{2}}.$$

Hence,

$$\int_0^\infty \left| (\tau_x^\theta \varphi)(y) \right|^2 dy$$

$$\leq \frac{x^{\mu+\frac{1}{2}}}{2^\mu \, \Gamma(\mu+1) |\sin\theta|^{\mu+\frac{1}{2}}} \int_0^\infty z^{-(\mu+\frac{1}{2})} |\varphi(z)|^2 \, dz \int_0^\infty \left| D_\mu^\theta(x,y,z) \right| y^{\mu+\frac{1}{2}} dy$$

$$\leq \frac{x^{\mu+\frac{1}{2}}}{2^\mu \, \Gamma(\mu+1) |\sin\theta|^{\mu+\frac{1}{2}}} \int_0^\infty z^{-(\mu+\frac{1}{2})} |\varphi(z)|^2 \, \frac{(xz)^{\mu+\frac{1}{2}}}{2^\mu \, \Gamma(\mu+1) |\sin\theta|^{\mu+\frac{1}{2}}} \, dz$$

$$= \frac{x^{2(\mu+\frac{1}{2})}}{\left(2^\mu \, \Gamma(\mu+1) |\sin\theta|^{\mu+\frac{1}{2}} \right)^2} \int_0^\infty |\varphi(z)|^2 \, dz.$$

Consequently,

$$\| x^{-\mu-\frac{1}{2}} \, (\tau_x^\theta \varphi)(y) \|_{L^2} \leq \frac{1}{|\sin\theta|^{\mu+\frac{1}{2}} \, 2^\mu \, \Gamma(\mu+1)} \, \| \varphi \|_{L^2}.$$

Lemma 5.16 *If $\varphi \in L^2(\mathbb{R})$, then*

$$\int_0^\infty \left| (\tau_y^\theta \varphi)(x) \right|^2 dx \leq \frac{y^{2(\mu+\frac{1}{2})}}{\left(2^\mu \, \Gamma(\mu+1) |\sin\theta|^{\mu+\frac{1}{2}} \right)^2} \int_0^\infty |\varphi(z)|^2 \, dz.$$

Definition 5.17 *The function space $\mathcal{H}_\mu(\mathbb{R}_+)$ is composed of infinitely differentiable function φ on $\mathbb{R}_+$ with complex values, which satisfies*

$$\Gamma_{m,k}^\mu(\varphi) = \sup_{x\in\mathbb{R}_+} \left| x^m (x^{-1}D)^k [x^{-(\mu+\frac{1}{2})} \varphi(x)] \right| < \infty, \quad \forall \mu \in \mathbb{R}, m,k \in \mathbb{N}^*.$$

Definition 5.18 *The function space $\mathcal{H}_{\mu,\theta}(\mathbb{R}_+)$ is composed of class function C^∞, φ on $\mathbb{R}_+$ with complex values, which satisfies for $k,m \in \mathbb{N}^*$*

$$\Upsilon_{m,k}^\theta(\varphi) = \sup_{x\in\mathbb{R}_+} \left| x^m \Delta_{\mu,x}^k \varphi(x) \right| < \infty, \quad \forall\theta \neq n\pi, \, n \in \mathbb{Z},$$

where

$$\Delta_{\mu,k} = \left[\frac{d^2}{dx^2} + 2ix\cos\theta \frac{d}{dx} + (\frac{1-4\mu^2}{4x^2}) + i\cos\theta - x^2\cos^2\theta \right],$$

and

$$\Delta_{\mu,x}^k = x^{-2k} \sum_{r=0}^{2k} \left(\sum_{l=0}^{l=2k} a_l x^{2l} \right) \left(x^{-1} \frac{d}{dx} \right)^r,$$

where the constant a_l depends only on μ and θ.

Proposition 5.19 *Let $K_\mu^\theta(x,y)$ be the fractional Hankel transform kernel. Then, we have*

1. $\Delta_{\mu,x}^r K_\mu^\theta(x,y) = (-y^2 \cos^2 \theta)^r K_\mu^\theta(x,y), \quad r \in \mathbb{N}^*.$

2. $h_\mu^\theta \left((\Delta_{\mu,x}^*)^r \varphi(x) \right)(y) = (-y^2 \cos^2 \theta)^r h_\mu^\theta \varphi(y) \text{ and } \varphi \in \mathcal{H}_\mu(\mathbb{R}_+),$

where

$$\Delta_{\mu,x}^* = \left[\frac{d^2}{dx^2} - 2ix \cos\theta \frac{d}{dx} + \frac{1-4\mu^2}{4x^2} - i\cos\theta - x^2\cos^2\theta \right]$$

is the fractional Bessel operator of parameter θ.

Proposition 5.20 *Let $\varphi \in L^1(\mathbb{R}_+)$. Then, $\widehat{\varphi}_\mu^\theta$ satisfies the following properties*

1. $\widehat{\varphi}_\mu^\theta \in L^\infty(\mathbb{R}_+)$ *with* $\|\widehat{\varphi}_\mu^\theta\|_{L^\infty} \leq q A_{\mu,\theta} \|\varphi\|_{L^1}$, *where $A_{\mu,\theta}$ is a positive constant that depends on μ and θ.*

2. $\displaystyle\lim_{y \longrightarrow \pm\infty} \widehat{\varphi}_\mu^\theta(y) = 0.$

3. $\widehat{\varphi}_\mu^\theta$ *is continuous on $\mathbb{R}_+$.*

Proof.
1) It follows from Definition 5.12 that

$$\widehat{\varphi}_\mu^\theta(y) = \int_0^\infty K_\mu^\theta(x,y)\varphi(x)dx.$$

In addition, we have

$$\begin{aligned}
\|\widehat{\varphi}_\mu^\theta\|_{L^\infty} &= \sup ess_{y\in\mathbb{R}} |\widehat{\varphi}_\mu^\theta(y)|, \\[2mm]
&= \sup ess_{y\in\mathbb{R}} \left| \int_0^\infty K_\mu^\theta(x,y)\varphi(x)dx \right|, \\[2mm]
&\leq \sup ess_{y\in\mathbb{R}} \int_0^\infty |K_\mu^\theta(x,y)||\varphi(x)|dx, \\[2mm]
&\leq A_{\mu,\theta} \int_0^\infty |\varphi(x)|dx, \\[2mm]
&= A_{\mu,\theta} \|\varphi\|_{L^1} < \infty.
\end{aligned}$$

2) It follows from Proposition 5.19 and Assertion 2 for $r = 1$ that

$$h_\mu^\theta \left(\Delta_{\mu,x}^* \varphi(x) \right)(y) = (-y^2 \cos^2 \theta) \, h_\mu^\theta \varphi(y).$$

Hence,

$$h_\mu^\theta \varphi(y) = \widehat{\varphi}_\theta^\mu(y) = \frac{1}{-y^2 \cos^2 \theta} \, h_\mu^\theta \left(\Delta_{\mu,x}^* \varphi(x) \right)(y).$$

Consequently,

$$|\widehat{\varphi}_\theta^\mu(y)| = \frac{1}{|-y^2 \cos^2 \theta|} |h_\mu^\theta \left(\Delta_{\mu,x}^* \varphi(x) \right)(y)| \longrightarrow 0 \text{ as } y \longrightarrow \pm\infty.$$

3) Let $h > 0$. We shall show that

$$\|\widehat{\varphi}_\theta^\mu(y + h) - \widehat{\varphi}_\theta^\mu(y)\|_\infty \rightarrow 0 \text{ as } h \longrightarrow \pm 0.$$

Indeed,

$$\sup_{y \in \mathbb{R}} |\widehat{\varphi}_\theta^\mu(y + h) - \widehat{\varphi}_\theta^\mu(y)|$$

$$= \sup_{y \in \mathbb{R}} \left| \int_0^\infty K_\mu^\theta(x, y + h)\varphi(x)dx - \int_0^\infty K_\mu^{\text{-}\theta}(x, y)\varphi(x)dx \right|$$

$$= \sup_{y \in \mathbb{R}} \left| \int_0^\infty [K_\mu^\theta(x, y + h) - K_\mu^{\text{-}\theta}(x, y)] \, \varphi(x)dx \right|$$

$$\leq |c_\mu^\theta| \sup_{y \in \mathbb{R}} \left| \int_0^\infty e^{ih(\frac{h}{2}\cot\theta + y\cot\theta)}(x(y + h)\cos\theta)^{\frac{1}{2}} \right.$$

$$\left. J_\mu^\theta(x(y + h)\cos\theta) - (xy\cos\theta)^{\frac{1}{2}}J_\mu^\theta(xy\cos\theta) \right| |\varphi(x)| \, dx$$

$$\leq A_{\mu,\theta} \int_0^\infty |\varphi(x)|dx.$$

In addition, when $h \rightarrow 0$, we have

$$\left| (x\cos\theta)^{\frac{1}{2}} \left[e^{ih(\frac{h}{2}\cot\theta + y\cot\theta)}(y + h)^{\frac{1}{2}} J_\mu^\theta(x(y + h)\cos\theta) - y^{\frac{1}{2}}J_\mu^\theta(xy\cos\theta) \right] \right| \rightarrow 0.$$

Hence,

$$\sup_{y \in \mathbb{R}} |\widehat{\varphi}_\theta^\mu(y + h) - \widehat{\varphi}_\theta^\mu(y)| \rightarrow 0 \text{ as } h \rightarrow 0.$$

Proposition 5.21 *(Parseval formula) Let Φ and Ψ be the fractional Hankel transforms of φ and ψ, respectively, that is to say*

$$\Phi(y) = (h_\mu^\theta \varphi)(y) \text{ and } \Psi(y) = (h_\mu^\theta \psi)(y).$$

Then, we have

$$\int_0^\infty \varphi(x)\,\overline{\psi}(x)dx = \sin\theta \int_0^\infty \Phi(y)\,\overline{\Psi}(y)dy.$$

In particular,

$$\int_0^\infty |\varphi(x)|^2 dx = \sin\theta \int_0^\infty |\Phi(y)|^2 \, dy.$$

Proof. Observe firstly that

$$\psi(x) = \overline{(c_\mu^\theta)} \sin\theta \int_0^\infty e^{-\frac{i}{2}(x^2+y^2)\cot\theta}(xy\cos\theta)^{\frac{1}{2}} J_\mu^\theta(xy\cos\theta)(h_\mu^\theta \psi)(y)dy.$$

Denote next

$$E_\theta(x) = e^{-\frac{i}{2}(x^2+y^2)\cot\theta}(xy\cos\theta)^{\frac{1}{2}}.$$

We get

$$
\begin{aligned}
\langle \varphi, \psi \rangle &= c_\mu^\theta \sin\theta \int_0^\infty \overline{(h_\mu^\theta \psi)}(y) \left(\int_0^\infty E_\theta(x) J_\mu^\theta(xy\cos\theta)\varphi(x)dx \right) dy \\
&= \sin\theta \int_0^\infty (h_\mu^\theta \varphi)(y)\overline{(h_\mu^\theta \psi)}(y)dy \\
&= \sin\theta \int_0^\infty \Phi(y)\,\overline{\Psi}(y)dy.
\end{aligned}
$$

Hence, the essential part is proved. Now, for $\varphi = \psi$, we get the particular case.

Definition 5.22 *For $\theta \neq qn\pi$, $n \in \mathbb{Z}$, we define fractional mother wavelet by*

$$
\psi_{a,b,\theta}(x) = \frac{1}{\sqrt{a}}\, \psi\left(\frac{x-b}{a} \right)\, e^{\frac{-i}{2}(x^2-b^2)\cot\theta}.
$$

The continuous fractional Bessel wavelet transform (CFrBWT) is the generalization of the continuous Bessel wavelet transform (CBWT) of parameter θ.

Definition 5.23 *A fractional Bessel wavelet is a function $\psi \in L^2(\mathbb{R}_+)$ satisfying the admissibility condition*

$$
C_{\mu,\psi,\theta} = \int_0^\infty x^{-2\mu-2}|(h_\mu^\theta \psi)(x)|^2\, dx < \infty, \quad \forall \mu \geq -\frac{1}{2}.
$$

Definition 5.24 *By dilation and translation operations of $\psi \in L^2(\mathbb{R}_+)$, we obtain a fractional Bessel wavelet family $\psi_{a,b}^\theta(x)$ given by*

$$
\psi_{a,b}^\theta(x) = \frac{1}{\sqrt{a}}\, D_a \tau_b^\theta \psi(x) = \frac{1}{\sqrt{a}}\, e^{-\frac{i}{2}(\frac{b^2}{a^2}+\frac{x^2}{a^2})\cot\theta} \int_0^\infty \psi(z)\, D_\mu^\theta(\frac{b}{a},\frac{x}{a},z)\, dz.
$$

Lemma 5.25 *If $\psi \in L^2(\mathbb{R}_+)$, we have*

$$
\|\psi_{b,a}^\theta\|_{L^2} \leq \frac{b^{(\mu+\frac{1}{2})}\, a^{-(\mu+\frac{1}{2})}}{|\sin\theta|^{\mu+\frac{1}{2}}\, 2^\mu\, \Gamma(\mu+1)}\, \|\psi\|_{L^2}.
$$

Proof. Using Definitions 5.14 and 5.24 and the inequality of Cauchy-Schwarz, we obtain

$$\psi_{a,b}^{\theta}(x)|$$

$$= \left| \frac{1}{\sqrt{a}} \, e^{-\frac{i}{2}(\frac{b^2}{a^2} + \frac{x^2}{a^2}) \cot \theta} \int_0^{\infty} \psi(z) \, D_{\mu}^{\theta}(\frac{b}{a}, \frac{x}{a}, z) \, dz \right|$$

$$\leq \frac{1}{\sqrt{a}} \int_0^{\infty} |\psi(z)| \, |D_{\mu}^{\theta}(\frac{b}{a}, \frac{x}{a}, z)| \, dz$$

$$\leq \frac{1}{\sqrt{a}} \int_0^{\infty} |\psi(z)| \, |z^{-\frac{1}{2}(\mu+\frac{1}{2})}| \, |D_{\mu}^{\theta}(\frac{b}{a}, \frac{x}{a}, z)|^{\frac{1}{2}} \, |z^{\frac{1}{2}(\mu+\frac{1}{2})}| \, |D_{\mu}^{\theta}(\frac{b}{a}, \frac{x}{a}, z)|^{\frac{1}{2}} \, dz$$

$$\leq \frac{1}{\sqrt{a}} \left(\int_0^{\infty} |\psi(z)|^2 \, |z^{-(\mu+\frac{1}{2})}| \, |D_{\mu}^{\theta}(\frac{b}{a}, \frac{x}{a}, z)| \, dz \right)^{\frac{1}{2}}$$

$$\left(\int_0^{\infty} |D_{\mu}^{\theta}(\frac{b}{a}, \frac{x}{a}, z)| \, |z^{(\mu+\frac{1}{2})}| \, dz \right)^{\frac{1}{2}}$$

$$\leq \frac{1}{\sqrt{a}} \left(\frac{(bx)^{\mu+\frac{1}{2}}}{a^{2(\mu+\frac{1}{2})} \, 2^{\mu} \, \Gamma(\mu+1) |\sin \theta|^{\mu+\frac{1}{2}}} \right)^{\frac{1}{2}}$$

$$\left(\int_0^{\infty} |\psi(z)|^2 \, |z^{-(\mu+\frac{1}{2})}| \, |D_{\mu}^{\theta}(\frac{b}{a}, \frac{x}{a}, z)| \, dz \right)^{\frac{1}{2}}.$$

Denote next

$$h_{a,b}^{\theta,\mu} = \frac{b^{\mu+\frac{1}{2}}}{a^{2\mu+2} \, |\sin \theta|^{\mu+\frac{1}{2}} \, 2^{\mu} \Gamma(\mu+1)}.$$

We get

$$\int_0^{\infty} |\psi_{a,b}^{\theta}(x)|^2 \, dx$$

$$\leq h_{a,b}^{\theta,\mu} \int_0^{\infty} z^{-(\mu+\frac{1}{2})} \, |\psi(z)|^2 \, dz \int_0^{\infty} |D_{\mu}^{\theta}(\frac{b}{a}, \frac{x}{a}, z)| \, x^{\mu+\frac{1}{2}} \, dx$$

$$\leq (h_{a,b}^{\theta,\mu})^2 \int_0^{\infty} |\psi(z)|^2 \, dz.$$

Hence, the lemma is proved.

Definition 5.26 *We define the transform of the continuous fractional Bessel wavelet B_{ψ}^{θ} of $f \in L^2(\mathbb{R}_+)$ by*

$$(B_{\psi}^{\theta} f)(a, b) = \left\langle f, \psi_{b,a}^{\theta} \right\rangle.$$

Proposition 5.27 *Let $f, \psi \in L^2(\mathbb{R}_+)$. The continuous fractional Bessel wavelet transform B_{ψ}^{θ} of f satisfies*

$$(B_{\psi}^{\theta} f)(a, b) = \int_0^{\infty} E_{a,b}^{\theta,\mu}(x) \, J_{\mu}(bx \cos \theta) \, (h_{\mu}^{\theta} e^{\frac{-i}{2}(.)^2 \cot \theta} f)(x) \, \overline{(h_{\mu}^{\theta} \psi)}(ax) \, dx,$$

where

$$E_{a,b}^{\theta,\mu}(x) = a^{-\mu} \sin\theta\, \overline{c_\mu^\theta}\, e^{\frac{-i}{2}(\frac{1}{a^2}-1)(ax)^2 \cot\theta}\, x^{-\mu-\frac{1}{2}} (bx\cos\theta)^{\frac{1}{2}}.$$

Proof. We have

$$
\begin{aligned}
(B_\psi^\theta f)(a,b) &= \langle f, \psi_{b,a}^\theta \rangle \\
&= \int_0^\infty f(t)\overline{\psi_{b,a}^\theta}\, dt \\
&= \int_0^\infty f(t) \overline{\left(\frac{1}{\sqrt{a}}\, e^{-\frac{i}{2}(\frac{b^2}{a^2}+\frac{t^2}{a^2})\cot\theta} \int_0^\infty \psi(z)\, D_\mu^\theta(\frac{b}{a}, \frac{t}{a}, z)\, dz \right)}
\end{aligned}
$$

Denote next

$$\sigma_{a,\theta}(\xi, t) = e^{\frac{i}{2}(t^2+\frac{\xi^2}{a^2})\cot\theta}(t\frac{\xi}{a}\cos\theta)^{\frac{1}{2}} e^{-\frac{i}{2}t^2\cot\theta},$$

$$\sigma_{a,\theta}^{b,\mu}(\xi) = e^{-\frac{i}{2}(\frac{1}{a^2}-1)\xi^2\cot\theta}\xi^{(-\mu-\frac{1}{2})}(\frac{b}{a}\xi\cos\theta)^{\frac{1}{2}}$$

and

$$\widetilde{\sigma}_{a,\theta}^{b,\mu}(\xi) = e^{-\frac{i}{2}(\frac{1}{a^2}-1)\xi^2\cot\theta}\xi^{(-\mu-\frac{1}{2})}(\frac{b}{a}\xi\cos\theta)^{\frac{1}{2}}.$$

We obtain

$$
\begin{aligned}
&(B_\psi^\theta f)(a,b) \\
&= \frac{1}{c_\mu^\theta \sqrt{a}} \int_0^\infty \left(c_\mu^\theta \int_0^\infty \sigma_{a,\theta}(\xi, t)\, J_\mu^\theta(t\frac{\xi}{a}\cos\theta) f(t) dt \right) \\
&\qquad\qquad\qquad \times \sigma_{a,\theta}^{b,\mu}(\xi)\, J_\mu^\theta(\frac{b}{a}\xi\cos\theta)\overline{(h_\mu^\theta\psi(z))}(\xi) d\xi \\
&= \frac{1}{c_\mu^\theta \sqrt{a}} \int_0^\infty \widetilde{\sigma}_{a,\theta}^{b,\mu}(\xi) J_\mu^\theta(\frac{b}{a}\xi\cos\theta) \times (h_\mu^\theta e^{-\frac{i}{2}(.)^2\cot\theta} f)(\frac{\xi}{a})\overline{(h_\mu^\theta\psi(z))}(\xi) d\xi.
\end{aligned}
$$

For $x = \frac{\xi}{a}$, we obtain Proposition 5.27. In addition, we have

$$
\begin{aligned}
h_\mu^\theta\left(e^{-\frac{i}{2}b^2\cot\theta}(B_\psi^\theta f)(a,b) \right) &= a^{-\mu}\sin\theta\left(x^{-\mu-\frac{1}{2}} e^{\frac{i}{2}a^2 x^2\cot\theta} \right. \\
&\qquad\qquad \left. (h_\mu^\theta e^{-\frac{i}{2}(.)^2\cot\theta} f)(x))\overline{h_\mu^\theta\psi(ax)} \right).
\end{aligned}
$$

Theorem 20 *If ψ_1, ψ_2 are two wavelets and $(B_{\psi_1}f)(a,b)$, $(B_{\psi_2}g)(a,b)$ their respective continuous fractional Bessel wavelet transforms, then*

$$\int_0^\infty \int_0^\infty (B_{\psi_1}f)(a,b)\overline{(B_{\psi_2}g)}(a,b)\frac{db\,da}{a^2} = \sin^2\theta\, C_{\mu,\psi_1,\psi_2,\theta}^1\langle f, g\rangle,$$

where

$$C_{\mu,\psi_1,\psi_2,\theta}^1\langle f, g\rangle = \int_0^\infty a^{-2\mu-2}\, \overline{(h_\mu^\theta\psi_1)}(a)\,(h_\mu^\theta\psi_2)(a)\, da < \infty.$$

Proof. Denote

$$E_{a,b}^{\mu,\theta}(x) = a^{-\mu}\sin\theta\, e^{\frac{-i}{2}(\frac{1}{a^2}-1)(ax)^2\cot\theta}x^{-\mu-\frac{1}{2}}\,(bx\cos\theta)^{\frac{1}{2}},$$

$$E_{a,\mu}^{\theta}(x) = a^{-2\mu-2}\,\sin^3\theta\,\,e^{\frac{-i}{2}(1-a^2)x^2\cot\theta}\,x^{-\mu-\frac{1}{2}},$$

$$E_{b,\theta}(x) = e^{\frac{i}{2}(b^2+x^2)\cot\theta}(bx\cos\theta)^{\frac{1}{2}}$$

and

$$E_{a,\theta}(y) = e^{\frac{i}{2}(2-a^2)y^2\cot\theta}\,y^{-\mu-\frac{1}{2}}.$$

Applying Proposition 5.27, we get

$$\int_0^\infty\int_0^\infty (B_{\psi_1}f)(a,b)\,\overline{(B_{\psi_2}g)}(a,b)\,\frac{db\,da}{a^2}$$

$$= \int_0^\infty\int_0^\infty \left[\overline{c_\mu^\theta}\int_0^\infty E_{a,b}^{\mu,\theta}(x)\,J_\mu(bx\cos\theta)\,(h_\mu^\theta e^{\frac{-i}{2}(.)^2\cot\theta}f)(x)\,\overline{(h_\mu^\theta\psi_1)}(ax)\,dx\right]$$

$$\times \left[\overline{c_\mu^\theta}\int_0^\infty E_{a,b}^{\mu,\theta}(y)\,J_\mu(by\cos\theta)\,(h_\mu^\theta e^{\frac{-i}{2}(.)^2\cot\theta}g)(y)\,\overline{(h_\mu^\theta\psi_2)}(ay)\,dy\right]db\frac{da}{a_2}$$

$$= \int_0^\infty\int_0^\infty E_{a,\mu}^{\theta}(x)\,(h_\mu^\theta e^{\frac{-i}{2}(.)^2\cot\theta}f)(x)\,\overline{h_\mu^\theta\psi_1}(ax)$$

$$\times\; c_\mu^\theta\int_0^\infty E_{b,\theta}(x)J_\mu(bx\cos\theta)\times\overline{c_\mu^\theta}\int_0^\infty \overline{E_{b,\theta}(y)}J_\mu(by\cos\theta)$$

$$\times\; E_{a,\theta}^{\mu}(y)\,\overline{(h_\mu^\theta e^{\frac{-i}{2}(.)^2\cot\theta}g)}(y)(h_\mu^\theta\psi_2)(ay)dy\,db\,dx\,da$$

$$= \int_0^\infty\int_0^\infty E_{a,\mu}^{\theta}(x)\,(h_\mu^\theta e^{\frac{-i}{2}(.)^2\cot\theta}f)(x)\,\overline{h_\mu^\theta\psi_1}(ax)$$

$$\times\; c_\mu^\theta\int_0^\infty E_{b,\theta}(x)J_\mu(bx\cos\theta)(h_\mu^\theta)^{-1}(E_{a,\theta}^{\mu}(y)$$

$$\overline{,(h_\mu^\theta e^{\frac{-i}{2}(.)^2\cot\theta}g)}(y)(h_\mu^\theta\psi_2)(ay))(b)db\,dx\,da$$

$$= \sin^3\theta\int_0^\infty\int_0^\infty E_{a,\mu}^{\theta}(x)\,(h_\mu^\theta e^{\frac{-i}{2}(.)^2\cot\theta}f)(x)\,\overline{h_\mu^\theta\psi_1}(ax)$$

$$\times\; h_\mu^\theta(h_\mu^\theta)^{-1}(E_{a,\theta}^{\mu}(y)\,\overline{(h_\mu^\theta e^{\frac{-i}{2}(.)^2\cot\theta}g)}(y)(h_\mu^\theta\psi_2)(ay))(x)dx\,da$$

$$= \sin^3\theta\int_0^\infty\int_0^\infty a^{-2\mu-2}x^{-2\mu-1}(h_\mu^\theta e^{\frac{-i}{2}(.)^2\cot\theta}f)(x)\,\overline{h_\mu^\theta\psi_1}(ax)$$

$$\times\overline{(h_\mu^\theta e^{\frac{-i}{2}(.)^2\cot\theta}g)}(x)(h_\mu^\theta\psi_2)(ax)dx\,da$$

$$= \sin^3\theta\int_0^\infty (h_\mu^\theta e^{\frac{-i}{2}(.)^2\cot\theta}f)(x)\,\overline{(h_\mu^\theta e^{\frac{-i}{2}(.)^2\cot\theta}g)}(x)$$

$$\times\left(\int_0^\infty (ax)^{(-2\mu-2)}\overline{(h_\mu^\theta\psi_1)}(ax)\,(h_\mu^\theta\psi_2)(ax)x\,da\right)dx$$

$$= \sin^3\theta\,C_{\mu,\psi_1,\psi_2,\theta}\int_0^\infty (h_\mu^\theta f)(x)\overline{h_\mu^\theta}g(x)\,dx$$

$$= \sin^3\theta\,C_{\mu,\psi_1,\psi_2,\theta}\,\langle h_\mu^\theta f, h_\mu^\theta g\rangle$$

$$= \sin^2\theta\,C_{\mu,\psi_1,\psi_2,\theta}\,\langle f, g\rangle.$$

Proposition 5.28   *1.   For $\psi = \psi_1 = \psi_2$, we get*

$$\int_0^\infty \int_0^\infty (B_\Psi f)(a,b) \, \overline{(B_\psi g)}(a,b) \, \frac{db \, da}{a^2} = \sin^2 \theta \, C_{\mu,\psi,\theta}^1 \langle f, g \rangle.$$

2. For $f = g$ and $\psi = \psi_1 = \psi_2$, there holds that

$$\int_0^\infty \int_0^\infty |(B_\Psi f)(a,b)|^2 \, \frac{db \, da}{a^2} = \sin^2 \theta \, C_{\mu,\psi,\theta}^1 \, \|f\|^2.$$

Theorem 21 *Let $f \in L^2(\mathbb{R}_+)$, we have*

$$f(t) = \frac{1}{\sin^2 \theta \, C_{\mu,\psi,\theta}^1} \int_0^\infty \int_0^\infty (B_\psi^\theta f)(a,b) \, \psi_{b,a}^\theta(t) \, \frac{db \, da}{a^2}, \quad a > 0.$$

Proof. For $g \in L^2(\mathbb{R}_+)$,

$$
\begin{aligned}
\sin^2 \theta \, C_{\mu,\psi,\theta}^1 \langle f, g \rangle &= \int_0^\infty \int_0^\infty (B_\psi^\theta f)(a,b) \, \overline{(B_\psi^\theta g)}(a,b) \, \frac{db \, da}{a^2} \\
&= \int_0^\infty \int_0^\infty (B_\psi^\theta f)(a,b) \, \overline{\left(\int_0^\infty g(t) \, \overline{\psi_{b,a}^\theta} \, dt \right)} \, \frac{db \, da}{a^2} \\
&= \int_0^\infty \left[\int_0^\infty \int_0^\infty (B_\psi^\theta f)(a,b) \, \psi_{b,a}^\theta(t) \, \frac{db \, da}{a^2} \right] \overline{g(t)} \, dt \\
&= \left\langle \int_0^\infty \int_0^\infty (B_\psi^\theta f)(a,b) \, \psi_{b,a}^\theta(t) \, \frac{db \, da}{a^2}, g(t) \right\rangle.
\end{aligned}
$$

So,

$$\langle f, g \rangle = \frac{1}{\sin^2 \theta \, C_{\mu,\psi,\theta}^1} \left\langle \int_0^\infty \int_0^\infty (B_\Psi f)(a,b) \, \psi_{b,a}^\theta(t) \, \frac{db \, da}{a^2}, g(t) \right\rangle.$$

Theorem 22 *If $\psi \in L^2(\mathbb{R}_+)$, then*

$$\int_0^\infty \left[(B_\psi^\theta f)(a,b) \, \overline{(B_\psi^\theta g)(a,b)} \right] db = a^{-2\mu} \sin^2 \theta \, \langle F, G \rangle,$$

where

$$F(x) = e^{-\frac{i}{2}((2-a^2)x^2)\cot\theta} x^{-\mu-\frac{1}{2}} (h_\mu^\theta \, e^{-\frac{i}{2}((2-a^2)x^2)^2 \cot\theta} f)(x) \overline{(h_\mu^\theta)\psi}(ax)$$

and

$$G(x) = e^{-\frac{i}{2}((2-a^2)x^2)\cot\theta} x^{-\mu-\frac{1}{2}} (h_\mu^\theta \, e^{-\frac{i}{2}((2-a^2)x^2)^2 \cot\theta} \overline{g})(x) \overline{(h_\mu^\theta)\psi}(ax).$$

Proof. Denote firstly

$$\Gamma_{a,b}^\theta(\omega) = \frac{1}{\sqrt{a}} e^{-\frac{i}{2}(\frac{b^2}{a^2} + \frac{\omega^2}{a^2})\cot\theta}$$

and

$$\sigma_{a,\mu}^\theta(x) = e^{\frac{-i}{2}((2-a^2)x^2)\cot\theta} x^{-\mu-\frac{1}{2}}.$$

Using Propositions 5.27 and (20), we obtain

$$\int_0^\infty \left[(B_\psi^\theta f)(a,b) \, \overline{(B_\psi^\theta g)(a,b)} \right] db$$

$$= \int_0^\infty \langle f, \psi_{b,a}^\theta \rangle \, \overline{\langle g, \psi_{b,a}^\theta \rangle} \, db$$

$$= \int_0^\infty \left(\int_0^\infty f(\omega) \, \overline{\psi_{b,a}^\theta(\omega)} \, d\omega \right) \left(\int_0^\infty \overline{g}(\sigma) \, \psi_{b,a}^\theta(\sigma) \, d\sigma \right) db$$

$$= \int_0^\infty \left[\int_0^\infty f(\omega) \overline{\left(\Gamma_{a,b}^\theta(\omega) \int_0^\infty \psi(z) \, D_\mu^\theta(\frac{b}{a}, \frac{\omega}{a}, z) \, dz \right)} d\omega \right]$$

$$\times \left[\int_0^\infty \overline{g}(\sigma) \left(\Gamma_{a,b}^\theta(\sigma) \int_0^\infty \psi(z) \, D_\mu^\theta(\frac{b}{a}, \frac{\sigma}{a}, z) \, dz \right) d\sigma \right] db$$

$$= \sin^3 \theta \, a^{-2\mu} \int_0^\infty h_\mu^\theta(\sigma_{a,\mu}^\theta(x) \, (h_\mu^\theta e^{\frac{-i}{2}(.)^2 \cos \theta} f)(x) \, \overline{(h_\mu^\theta \psi)}(ax))(b)$$

$$\times \overline{h_\mu^\theta(\sigma_{a,\mu}^\theta(y) \, (h_\mu^\theta e^{\frac{-i}{2}(.)^2 \cot \theta} \overline{g})(y) \, \overline{(h_\mu^\theta \psi)}(ay))(b)} \, db$$

$$= \sin^3 \theta \, a^{-2\mu} \int_0^\infty (h_\mu^\theta F)(b) \, \overline{(h_\mu^\theta G)}(b) \, db$$

$$= \sin^3 \theta \, a^{-2\mu} \langle h_\mu^\theta F, h_\mu^\theta G \rangle$$

$$= \sin^2 \theta \, a^{-2\mu} \langle F, G \rangle.$$

Theorem 23 *Let $\psi \in L^2(\mathbb{R}_+)$ is a Bessel wavelet and f is an integrable function at the edges, then the convolution $(\psi *_\theta f)$ is a fractional Bessel wavelet given by*

$$(\psi *_\theta f)(x) = \int_0^\infty (\tau_x^\theta \psi)(y) \, y^{(-\mu+\frac{1}{2})} \, f(y) \, dy.$$

Proof. To show that the convolution $(\psi *_\theta f)$ is a fractional Bessel wavelet, it suffices to show that it satisfies the admissibility condition of Definition 5.23. Indeed,

$$\int_0^\infty |(\psi *_\theta f)(x)|^2 \, dx$$

$$= \int_0^\infty \left| \int_0^\infty (\tau_x^\theta \psi)(y) \, y^{(-\mu+\frac{1}{2})} \, f(y) \, dy \right|^2 dx$$

$$\leq \int_0^\infty \left(\left| \int_0^\infty (\tau_x^\theta \psi)(y) \, y^{(-\mu+\frac{1}{2})} \, f^{\frac{1}{2}}(y) \, \| \, dy \right| \, \left| f^{\frac{1}{2}}(y) \right| \, dy \right)^2 dx$$

$$\leq \int_0^\infty \left[\left(\int_0^\infty |f(y)| \, \left| (\tau_x^\theta \psi)(y) \, y^{(-\mu+\frac{1}{2})} \right|^2 dy \right)^{\frac{1}{2}} \left(\int_0^\infty |f(y)| \, dy \right)^{\frac{1}{2}} \right]^2 dx$$

$$= \left(\int_0^\infty |f(y)| \, dy \right) \left(\int_0^\infty \left(\int_0^\infty \left| (\tau_x^\theta \psi)(y) \, y^{(-\mu+\frac{1}{2})} \right|^2 dx \right) |f(y)| \, dy \right)$$

$$\leq \left(\int_0^\infty |f(y)| \, dy \right)^2 \left(\int_0^\infty \left| (\tau_y^\theta \psi)(x) \, y^{(-\mu+\frac{1}{2})} \right|^2 dx \right).$$

Using Lemma 5.25, we obtain

$$\|(\psi *_\theta f)(x)\|_{L^2} \leq \frac{1}{|\sin\theta|^{\mu+\frac{1}{2}} 2^\mu \, \Gamma(\mu+1)} \|\psi_{L^2} \|f\|_{L^1} < \infty.$$

Consequently, $(\psi *_\theta f) \in L^2(\mathbb{R}_+)$, which implies that $h_\mu^\theta(\psi *_\theta f)$ exists. Moreover,

$$\int_0^\infty x^{-2\mu-2} \left| h_\mu^\theta(\psi *_\theta f) \right|^2 dx$$

$$\leq \quad |\cos\theta| \int_0^\infty \left| x^{-3\mu-\frac{5}{2}} (h_\mu^\theta f)(x\cos\theta)\,(h_\mu^\theta\psi)(x\cos\theta) \right|^2 dx$$

$$\leq \quad |\cos\theta| \sup_x \left(x^{-\mu-\frac{1}{2}} |(h_\mu^\theta f)(x\cos\theta)|^2 \right)$$

$$\times \quad \int_0^\infty x^{-2\mu-2} |(h_\mu^\theta\psi)(x\cos\theta)|^2)\, dx$$

$$= \quad C'_{\mu,\psi,\theta} \sup_x \left(x^{-\mu-\frac{1}{2}} |(h_\mu^\theta f)(x\cos\theta)|^2 \right) < \infty.$$

So $(\psi *_\theta f)$ is a fractional Bessel wavelet.

Theorem 24 *If $f, \psi \in L^2(\mathbb{R}_+)$ and $(B_\psi^\theta f)(a,b)$ is the continuous fractional Bessel wavelet transform, then*

1. *The function $(a,b) \longmapsto (B_\psi^\theta f)(a,b)$ is continuous on $\mathbb{R}_+ \times \mathbb{R}_+$*

2. *The following holds*

$$\|(B_\psi^\theta f)(a,b)\|_{L^\infty} \leq \frac{b^{(\mu+\frac{1}{2})} a^{-(\mu+\frac{1}{2})}}{|\sin\theta|^{\mu+\frac{1}{2}} 2^\mu \, \Gamma(\mu+1)} \|f\|_{L^2} \|\psi\|_{L^2}.$$

Proof. 1) Let us show that $(B_\psi^\theta f)(a,b)$ is continuous on $\mathbb{R}_+ \times \mathbb{R}_+$. Let (b_0, a_0) be an arbitrary but fixed point. Using the Hölder's inequality, we obtain

$$\left| (B_\psi^\theta f)(a,b) - (B_\psi^\theta f)(b_0, a_0) \right|$$

$$= \left| \frac{1}{\sqrt{a}} \int_0^\infty f(t) \int_0^\infty \psi(z) \left[D_\mu^\theta(\frac{b}{a}, \frac{t}{a}, z) - D_\mu^\theta(\frac{b_0}{a_0}, \frac{t}{a_0}, z) \right] dt\, dz \right|$$

$$\leq \frac{1}{\sqrt{a}} \int_0^\infty \int_0^\infty \left| f(t)\psi(z) \left[D_\mu^\theta(\frac{b}{a}, \frac{t}{a}, z) - D_\mu^\theta(\frac{b_0}{a_0}, \frac{t}{a_0}, z) \right] \right| dt\, dz$$

$$\leq \frac{1}{\sqrt{a}} \left(\int_0^\infty t^{-\mu-\frac{1}{2}} |f(t)|^2 dt \int_0^\infty z^{\mu+\frac{1}{2}} \left[D_\mu^\theta(\frac{b}{a}, \frac{t}{a}, z) - D_\mu^\theta(\frac{b_0}{a_0}, \frac{t}{a_0}, z) \right] dz \right)^{\frac{1}{2}}$$

$$\times \left(\int_0^\infty z^{-\mu-\frac{1}{2}} |\psi(z)|^2 dz \int_0^\infty t^{\mu+\frac{1}{2}} \left[D_\mu^\theta(\frac{b}{a}, \frac{t}{a}, z) - D_\mu^\theta(\frac{b_0}{a_0}, \frac{t}{a_0}, z) \right] dt \right)^{\frac{1}{2}}.$$

Denote next $K_\mu^\theta = |\sin\theta|^{\mu+\frac{1}{2}} 2^\mu \Gamma(\mu+1)$. Using Proposition 5.14, we obtain

$$\int_0^\infty z^{\mu+\frac{1}{2}} \left[D_\mu^\theta(\frac{b}{a}, \frac{t}{a}, z) - D_\mu^\theta(\frac{b_0}{a_0}, \frac{t}{a_0}, z) \right] dz \leq \frac{t^{\mu+\frac{1}{2}}}{K_\mu^\theta} \left[\frac{b^{\mu+\frac{1}{2}}}{a^{2(\mu+\frac{1}{2})}} - \frac{b_0^{\mu+\frac{1}{2}}}{a_0^{2(\mu+\frac{1}{2})}} \right]$$

and similarly

$$\int_0^\infty t^{\mu+\frac{1}{2}} \left[D_\mu^\theta(\frac{b}{a}, \frac{t}{a}, z) - D_\mu^\theta(\frac{b_0}{a_0}, \frac{t}{a_0}, z) \right] dz \leq \frac{z^{\mu+\frac{1}{2}}}{K_\mu^\theta} \left[ab^{\mu+\frac{1}{2}} - a_0 b_0^{\mu+\frac{1}{2}} \right],$$

Moreover, from the continuity of $D_\mu^\theta(\frac{b}{a}, \frac{t}{a}, z)$ and the monotonic convergence theorem, we get

$$\lim_{b \longrightarrow b_0} \lim_{a \longrightarrow a_0} \left| (B_\psi^\theta f)(a, b) - (B_\psi^\theta f)(b_0, a_0) \right| = 0.$$

Hence, the continuity of $(B_\psi^\theta f)(a, b)$.

5.5 QUANTUM THEORY TOOLKIT

This section is devoted to the introduction of the quantum theory tools. Useful definitions, notations and properties of q-derivatives and q-integrals will be introduced. For $0 < q < 1$, denote

$$\mathbb{R}_q = \{\pm q^n, \ n \in \mathbb{Z}\}, \ \mathbb{R}_q^+ = \{q^n, \ n \in \mathbb{Z}\} \text{ and } \tilde{\mathbb{R}}_q^+ = \mathbb{R}_q^+ \bigcup \{0\}.$$

Definition 5.29 *[224] The q-shifted factorial are defined by*

$$(a, q)_0 = 1,$$

$$(a, q)_n = \prod_{k=0}^{n-1} (1 - aq^k) = (1-a)(1-aq)......(1-aq^{n-1}),$$

$$(a, q)_\infty = \lim_{n \longrightarrow +\infty} (a, q)_n = \prod_{k=0}^{+\infty} (1 - aq^k),$$

and the symbol of Pochhammer $(a)_n$ is given by

$$(a)_0 = 1, \ (a)_n = a(a+1)....(a+n-1), \quad n \in \mathbb{N}^*.$$

For all complex numbers $a_1, a_2, ..., a_r$, we define

$$(a_1, a_2, ..., a_k; q)_n = (a_1, q)_n (a_2, q)_n .... (a_k, q)_n.$$

Remark 5.30 *The q-shifted factorial is considered as the q-analogue of the Pochhammer symbol as it satisfies*

$$\lim_{q \longrightarrow 1^-} \frac{(q^a, q)_n}{(1-q)^n} = (a)_n.$$

Definition 5.31 *On $\widetilde{\mathbb{R}}_q^+$, the q-Jackson integrals from 0 to a and from 0 to $+\infty$ are defined, respectively, by (see [73], [213])*

$$\int_0^a f(x)d_q x = (1-q)\, a \sum_{n \geq 0} f(aq^n)\, q^n$$

and

$$\int_0^\infty f(x)d_q x = (1-q) \sum_{n \in \mathbb{N}} f(q^n)\, q^n$$

provided that the sums converge absolutely. On $[a,b]$, the integral is given by

$$\int_a^b f(x)d_q x = \int_0^b f(x)d_q x - \int_0^a f(x)d_q x.$$

This allows us to introduce next the functional space

$$\mathcal{L}_{q,p,v}(\widetilde{\mathbb{R}}_q^+) = \{f : \|f\|_{q,p,v} < \infty\},$$

where

$$\|f\|_{q,p,v} = \left[\int_0^\infty |f(x)|^p\, x^{2v+1}\, d_q x\right]^{\frac{1}{p}},$$

where $v > \dfrac{-1}{2}$ fixed. Let $C_q^0(\widetilde{\mathbb{R}}_q^+)$ be the space of functions defined on $\widetilde{\mathbb{R}}_q^+$, continuous in 0 and vanishing at $+\infty$, equipped with the induced topology of uniform convergence such that

$$\|f\|_{q,\infty} = \sup_{x \in \widetilde{\mathbb{R}}_q^+} |f(x)| < \infty.$$

Finally, $C_q^b(\widetilde{\mathbb{R}}_q^+)$ designates the space of functions that are continuous at 0 and bounded on $\widetilde{\mathbb{R}}_q^+$.

Definition 5.32 *The q-derivative of a function $f \in \mathcal{L}_{q,p,\alpha}(\widetilde{\mathbb{R}}_q^+)$ is defined by*

$$D_q f(x) = \begin{cases} \dfrac{f(x) - f(qx)}{(1-q)x}, & x \neq 0 \\ f'(0), & else, \end{cases}$$

provided that f is differentiable at 0.

Lemma 5.33 *[19] If f is differentiable, note that*

$$\lim_{q \longrightarrow 1} D_q f(x) = \frac{df(x)}{dx}.$$

The operator $D_q f$ is then the q-analogue of the classical derivative.

Proof. Let f be a differentiable function. Then

$$\lim_{h \longrightarrow 0} \frac{f(qx + h) - f(qx)}{h} = f'(qx).$$

For $h = x - qx$, when $q \longrightarrow 1$ we have $h \longrightarrow 0$. Hence,

$$
\begin{aligned}
\lim_{q \longrightarrow 1} D_q f(x) &= \lim_{h \longrightarrow 0} \frac{f(qx + h) - f(qx)}{h} \\
&= \lim_{q \longrightarrow 1} \frac{f(x) - f(qx)}{(1 - q)x} \\
&= \frac{df(x)}{dx}.
\end{aligned}
$$

Proposition 5.34 *[19] The q-derivative of a function is a linear operator. That is, for any functions f and g and any constants a and b, we have*

$$D_q(af + bg)(x) = aD_q f(x) + bD_q g(x).$$

Proof. For all $x \neq 0$, we have

$$
\begin{aligned}
D_q(af + bg)(x) &= \frac{(af + bg)(x) - (af + bg)(qx)}{(1 - q)\,x} \\
&= \frac{a\,f(x) + b\,g(x) - a\,f(qx) - b\,g(qx)}{(1 - q)\,x} \\
&= a\,\frac{f(x) - f(qx)}{(1 - q)\,x} + b\,\frac{g(x) - g(qx)}{(1 - q)\,x} \\
&= a\,D_q f(x) + b\,D_q g(x).
\end{aligned}
$$

Proposition 5.35 *[19] For $x \neq 0$, the q-derivative of product is given by*

$$D_q\left(f(x)g(x)\right) = f(qx)D_q g(x) + D_q f(x)g(x),$$

and for $g(x) \neq 0$, we have

$$D_q\left(\frac{f(x)}{g(x)}\right) = \frac{g(qx)D_q f(x) - f(qx)D_q g(x)}{g(qx)g(x)}.$$

Proof. For $x \neq 0$, we have

$$
\begin{aligned}
D_q(f\,g)(x) &= \frac{(f\,g)(x) - (f\,g)(qx)}{(1-q)\,x} \\[2mm]
&= \frac{f(x)\,g(x) - f(qx)\,g(x) + f(qx)\,g(x) - f(qx)\,g(qx)}{(1-q)\,x} \\[2mm]
&= \frac{f(x)\,g(x) - f(qx)\,g(x)}{(1-q)\,x} + \frac{f(qx)\,g(x) - f(qx)\,g(qx)}{(1-q)\,x} \\[2mm]
&= \frac{f(x) - f(qx)}{(1-q)\,x}\,g(x) + f(qx)\,\frac{g(x) - g(qx)}{(1-q)\,x} \\[2mm]
&= D_q f(x)\,g(x) + f(qx)\,D_q g(x).
\end{aligned}
$$

Proposition 5.36 *[19]*

1. *For any function f, we have*

$$
D_q \int_0^x f(t)d_q t = f(x).
$$

2. *The q-analogue of the rule of integration by parts is*

$$
\int_a^b f(x)\,D_q g(x)d_q x = [f(b)g(b) - f(a)g(a)] - \int_a^b g(qx)\,D_q f(x)d_q x,
$$

where the integration is understood in q-Jackson sense.

Proof. Assertion **1.** is somehow reproduced from [19] and is easy. We have

$$
\begin{aligned}
D_q \int_0^x f(t)d_q t &= D_q \left[(1-q)x \sum_{n=0}^{\infty} f(xq^n)q^n \right] \\[2mm]
&= \frac{(1-q)^2 x \sum_{n=0}^{\infty} \left[f(xq^n) - f(xq^{n+1}) \right] q^n}{(1-q)x} \\[2mm]
&= (1-q) \sum_{n=0}^{\infty} \left[f(xq^n) - f(xq^{n+1}) \right] q^n \\[2mm]
&= f(x).
\end{aligned}
$$

2. Using the definition of the q-integral, we have

$$
\int_a^b f(x)\,D_q g(x)d_q x = \int_0^b f(x)\,D_q g(x)d_q x - \int_0^a f(x)\,D_q g(x)d_q x.
$$

Applying again the definition of the q-integral, we get

$$
\int_0^b f(x)\,D_q g(x)d_q x = \sum_{n=0}^{\infty} f(bq^n)(g(bq^n) - g(bq^{n+1})).
$$

Similarly

$$\int_0^a f(x)\, D_q g(x) d_q x = \sum_{n=0}^{\infty} f(aq^n)(g(aq^n) - g(aq^{n+1})).$$

On the other hand, we get by using similar techniques

$$\int_0^b g(qx)\, D_q f(x) d_q x = \sum_{n=0}^{\infty} g(bq^{n+1})(f(bq^n) - f(bq^{n+1}))$$

and

$$\int_0^a g(qx)\, D_q f(x) d_q x = \sum_{n=0}^{\infty} g(aq^{n+1})(f(aq^n) - f(aq^{n+1})).$$

By regrouping all these equalities, we obtain

$$\int_a^b f(x)\, D_q g(x) d_q x + \int_a^b g(qx)\, D_q f(x) d_q x = \sum_{n=0}^{\infty} (I_n(b) - I_n(a) - K_n(b) + K_n(a),$$

where $I_n(x) = f(xq^n)(g(xq^n) - g(xq^{n+1}))$ and $K_n(x) = g(xq^{n+1})(f(xq^n) - f(xq^{n+1}))$. Next, observing that

$$\sum_{n=0}^{\infty} (I_n(b) - I_n(a) - K_n(b) + K_n(a) = f(b)g(b) - f(a)g(a),$$

we obtain

$$\int_a^b f(x)\, D_q g(x) d_q x + \int_a^b g(qx)\, D_q f(x) d_q x = [f(b)g(b) - f(a)g(a)].$$

So as Assertion 2.

Definition 5.37 *The q-Bessel operator is defined for all $f \in \mathcal{L}_{q,2,v}(\widetilde{\mathbb{R}}_q^+)$ by*

$$\Delta_{q,v} f(x) = \frac{f(q^{-1}x) - (1 + q^{2v})f(x) + q^{2v} f(qx)}{x^2}, \quad \forall x \neq 0.$$

The following relation is easy to show and constitutes an analogue of Stokes rule.

Lemma 5.38 *For all $f, g \in \mathcal{L}_{q,2,v}(\widetilde{\mathbb{R}}_q^+)$ such that $\Delta_{q,v} f, \Delta_{q,v} g \in \mathcal{L}_{q,2,v}(\widetilde{\mathbb{R}}_q^+)$, we have*

$$\int_0^{\infty} \Delta_{q,v} f(x)\, g(x)\, x^{2v+1}\, d_q x = \int_0^{\infty} f(x)\, \Delta_{q,v} g(x)\, x^{2v+1}\, d_q x.$$

5.6   SOME QUANTUM SPECIAL FUNCTIONS

We propose in this section to recall two basic functions that are applied almost everywhere in q-theory and its applications.

Definition 5.39 *The q-analogue of the classical exponential function, called q-exponential function, is defined by*

$$e_q(x) = \frac{1}{(1 - (1-q)x)_q^\infty}$$

and

$$E_q(x) = \sum_{k=0}^{\infty} q^{\frac{k(k-1)}{2}} \frac{x^k}{[k]_q!} = (1 + (1-q)x)_q^\infty,$$

where

$$[k]_q! = \frac{(q,q)_k}{(1-q)^k}.$$

Properties 5.40 *The q-exponential functions satisfy the following properties:*

1. $D_q e_q(x) = e_q x.$

2. $D_q E_q(x) = E_q(qx).$

3. $e_q(x) E_q(-x) = E_q(x) e_q(-x) = 1.$

Proof.

At the beginning of the twentieth century, Jackson introduced a q-analogue of the Gamma function.

Definition 5.41 *The q-Gamma Euler's function is defined by*

$$\Gamma_q(x) = \frac{(q,q)_\infty}{(q^x,q)_\infty}(1-q)^{1-x}, \ x \neq 0, -1, -2, ...$$

Properties 5.42 *The q-Gamma function satisfies the following properties:*

1. *The function Γ_q tends to the classical Euler function Γ when q tends to 1. That is, for all x,*
$$\lim_{q \to 1} \Gamma_q(x = \Gamma(x).$$

2. *For all x, we have*
$$\Gamma_q(x+1) = [x]_q \Gamma(x), \ \Gamma_q(1) = 1.$$

3. *Whenever q satisfies*
$$\frac{\log(1-q)}{\log(q)} \in \mathbb{Z},$$
we have the q-integral representation
$$\Gamma_q(x) = \int_0^\infty t^{x-1} E(-qt,q) d_q t.$$

4. *The Legendre duplication q-formula*
$$\Gamma_q(2x)\Gamma_{q^2}(\frac{1}{2}) = (1+q)^{2x-1} \Gamma_{q^2}(x)\Gamma_{q^2}(x + \frac{1}{2}).$$

Proof. Left to the reader.

We now introduce the q-Bessel analogue of the Bessel function.

Definition 5.43 *[296, 297] The generalized q-Bessel-type function is defined by*

$$j_v(x) = \frac{(x/2)^v}{(q;q)_v} \sum_{k=0}^{\infty} \frac{q^{\frac{3k}{2}(k+v)}}{(q^{v+1};a)_k} \frac{(x^2/4)^k}{(q;q)_k}.$$

The function j_v is a q-analogue of the ordinary modified Bessel function as it satisfies the limit relation

$$\lim_{q \to 1} j_v(x) = j_v(x),$$

where j_v is the ordinary modified Bessel function. If v is an integer, it satisfies the equality

$$j_v(x) = j_{-v}(x).$$

The function $j_v(x)$ is an even (odd) function if the parameter v is an even (odd) integer, since

$$j_{-v}(x) = (-1)^v j_v(x),$$

whenever $v \in \mathbb{N}$.

Definition 5.44 *[195] The q-Bessel function is defined by*

$$j_v(x, q^2) = \sum_{n \geq 0} (-1)^n \frac{q^{n(n+1)}}{(q^2, q^2)_n (q^2, q^2)_n} x^{2n}.$$

Proposition 5.45 *The q-Bessel function satisfies the following relations*

$$J_n(x, q) = (-1)^n q^{\frac{n}{2}} J_{-n}(q^{\frac{n}{2}}; q),$$

$$J_\alpha(q^{\frac{\beta}{2}}, q) = J_\beta(q^{\frac{\alpha}{2}}, q),$$

$$(1 - q^v + x^2) J_v(x, q) = x\{J_{v-1}(x, q) + J_{v+1}(x, q)\},$$

$$(1 - q) D_q J_v(x, q) = q^{(\frac{1-v}{2})} J_{v-1}(xq^{\frac{1}{2}}, q) - J_{v+1}(xq^{\frac{1}{2}}, q).$$

Proof. Left to the reader.

Definition 5.46 *[195] The normalized q-Bessel function is given by*

$$j_\alpha(x, q^2) = \sum_{n \geq 0} (-1)^n \frac{q^{n(n+1)}}{(q^{2\alpha+2}, q^2)_n (q^2, q^2)_n} x^{2n}.$$

The q-Bessel operator is defined as follows $(\alpha > -\frac{1}{2})$,

$$\Delta_{q,\alpha} f(x) = \frac{f(q^{-1}x) - (1 + q^{2\alpha}) f(x) + q^{2\alpha} f(qx)}{x^2}, \quad \forall x \neq 0.$$

Lemma 5.47 *Let* $a_n = (-1)^n \dfrac{q^{n(n+1)}}{(q^{2\alpha+2}, q^2)_n (q^2, q^2)_n}$ *and* $b_{2n} = \dfrac{1}{q^{2n}} - (1+q^{2\alpha}) + q^{2\alpha} q^{2n}$, $n \in \mathbb{N}$. *Then, for all* $n \in \mathbb{N}$, *we have* $b_{2n+2}\, a_{n+1} + a_n = 0$.

Proposition 5.48 *[213] For* $\lambda \in \mathbb{C}$, *the problem*

$$\Delta_{q,\alpha} u(x) = -\lambda^2\, u(x) \quad and \quad u(0) = 1,\ u'(0) = 0$$

has as unique solution: the function $x \mapsto j_\alpha(\lambda x, q^2)$.

Proof. We have

$$j_\alpha(\lambda x, q^2) = \sum_{n\geq 0} a_n\, \lambda^{2n}\, x^{2n}.$$

Consequently,

$$
\begin{aligned}
\Delta_{q,\alpha} j_\alpha(\lambda x, q^2)
&= \frac{\displaystyle\sum_{n\geq 0} a_n \frac{\lambda^{2n}}{q^{2n}} x^{2n} - (1+q^{2\alpha})\sum_{n\geq 0} a_n \lambda^{2n} x^{2n} + q^{2\alpha}\sum_{n\geq 0} a_n \lambda^{2n} q^{2n} x^{2n}}{x^2} \\[2mm]
&= \sum_{n\geq 0} a_n \lambda^{2n} x^{2n-2}\left[\frac{1}{q^{2n}} - (1+q^{2\alpha}) + q^{2\alpha} q^{2n}\right] \\[2mm]
&= \sum_{n\geq 0} a_n \lambda^{2n} b_{2n}\, x^{2n-2} \\[2mm]
&= a_0\, b_0\, x^{-2} + \sum_{n\geq 1} a_n \lambda^{2n} b_{2n}\, x^{2n-2} \\[2mm]
&= \sum_{n\geq 1} a_n \lambda^{2n} b_{2n}\, x^{2n-2}.
\end{aligned}
$$

Hence,

$$
\begin{aligned}
\Delta_{q,\alpha} j_\alpha(\lambda x, q^2) + \lambda^2\, j_\alpha(\lambda x, q^2)
&= \sum_{n\geq 1} a_n \lambda^{2n} b_{2n}\, x^{2n-2} + \lambda^2 \sum_{n\geq 0} a_n \lambda^{2n} x^{2n} \\[2mm]
&= \sum_{k\geq 0} a_{k+1} \lambda^{2k+2} x^{2k} + \sum_{n\geq 0} a_n \lambda^{2n+2} x^{2n} \\[2mm]
&= \sum_{n\geq 0} a_{n+1} \lambda^{2n+2} x^{2n} + \sum_{n\geq 0} a_n \lambda^{2n+2} x^{2n} \\[2mm]
&= \sum_{n\geq 0} \lambda^{2n+2} x^{2n} \left[b_{2n+2}\, a_{n+1} + a_n\right].
\end{aligned}
$$

Using Lemma 5.47, we obtain

$$\Delta_{q,\alpha} j_\alpha(\lambda x, q^2) + \lambda^2\, j_\alpha(\lambda x, q^2) = 0,$$

and

$$j_\alpha(0, q^2) = 1, \quad j'_\alpha(0, q^2) = 0.$$

The following relations are easy to show. The first is an analogue of Stokes rule. The second is an orthogonality relation for the normalized q-Bessel function.

Proposition 5.49 *It holds for all* $x, y \in \widetilde{\mathbb{R}}_q^+$ *that*

$$\int_0^\infty j_\alpha(xt, q^2) \, j_\alpha(yt, q^2) \, t^{2\alpha+1} \, d_q t = \frac{1}{c_{q,\alpha}^2} \, \delta_{q,\alpha}(x, y),$$

where

$$c_{q,\alpha} = \frac{1}{1-q} \frac{(q^{2\alpha+2}, q^2)_\infty}{(q^2, q^2)_\infty} \quad and \quad \delta_{q,\alpha}(x, y) = \frac{1}{(1-q) \, x^{2(\alpha+1)}} \delta_{x,y}.$$

Finally, the following preliminary result will be useful for the next.

Lemma 5.50 *Define the* (q, v)-*delta operator by* $\delta_{q,v}(x, y) = \dfrac{1}{(1-q) \, x^{2(|v|+1)}} \delta_{x,y}$. *It holds for all* $f \in \mathcal{L}_{q,2,v}(\widetilde{\mathbb{R}}_q^+)$ *and all* $t \in \widetilde{\mathbb{R}}_q^+$ *that*

$$f(t) = \int_0^\infty f(x) \, \delta_{q,v}(x, t) \, x^{2(|v|+1)} \, d_q x.$$

Proof. From the definition of the q-Jackson integral, we have

$$
\begin{aligned}
\int_0^\infty f(x) \delta_{q,v}(x, t) x^{2|v|+1} d_q x &= (1-q) \sum_{n=0}^\infty f(q^n) \delta_{q,v}(q^n, s) q^{n(2|v|+2)} \\
&= (1-q) f(q^k) \delta_{q,v}(q^k, t) q^{k(2|v|+2)} \\
&= f(q^k),
\end{aligned}
$$

where k is the unique integer such that $t = q^k$.

5.7 QUANTUM WAVELETS

In the present section, we propose to develop wavelet analysis based on the quantum Bessel functions. This class joins and extends the Bessel and q-Bessel wavelets to the q-theory framework.

Definition 5.51 *[213] The* q-*Bessel Fourier transform* $\mathcal{F}_{q,\alpha}$ *is defined by*

$$\mathcal{F}_{q,\alpha} f(x) = c_{q,\alpha} \int_0^\infty f(t) \, j_\alpha(xt, q^2) \, t^{2\alpha+1} \, d_q t,$$

where

$$c_{q,\alpha} = \frac{1}{1-q} \frac{(q^{2\alpha+2}, q^2)_\infty}{(q^2, q^2)_\infty}.$$

Definition 5.52 *The* q-*Bessel translation operator is defined by*

$$T_{q,x}^\alpha f(y) = c_{q,\alpha} \int_0^\infty \mathcal{F}_{q,\alpha} f(t) \, j_\alpha(xt, q^2) \, j_\alpha(yt, q^2) \, t^{2\alpha+1} \, d_q t.$$

Lemma 5.53 *The translation operator satisfies for all* $f \in \mathcal{L}_{q,2,\alpha}(\widetilde{\mathbb{R}}_q^+)$ *a Fourier invariance property ([1])*

$$\mathcal{F}_{q,\alpha}(T_{q,x}^\alpha f)(\lambda) = j_\alpha(\lambda x, q^2)\, \mathcal{F}_{q,\alpha} f(\lambda), \quad \forall \lambda, x \in \widetilde{\mathbb{R}}_q^+.$$

Proof. For $f \in \mathcal{L}_{q,2,\alpha}(\widetilde{\mathbb{R}}_q^+)$, we have

$$\mathcal{F}_{q,\alpha}(T_{q,x}^\alpha f)(\lambda) = c_{q,\alpha} \int_0^\infty (T_{q,x}^\alpha f)(t)\, j_\alpha(\lambda t, q^2)\, t^{2\alpha+1}\, d_q t$$

$$= c_{q,\alpha}^2 \int_0^\infty \int_0^\infty \left[\mathcal{F}_{q,\alpha} f(\mu)\, j_\alpha(x\mu, q^2)\, j_\alpha(t\mu, q^2)\, \mu^{2\alpha+1}\, d_q\mu \right] j_\alpha(\lambda t, q^2)\, t^{2\alpha+1}\, d_q t$$

$$= c_{q,\alpha}^2 \int_0^\infty \mathcal{F}_{q,\alpha} f(\lambda) \left[\int_0^\infty j_\alpha(x\mu, q^2)\, j_\alpha(t\mu, q^2)\, \mu^{2\alpha+1}\, d_q\mu \right] j_\alpha(\lambda t, q^2)\, t^{2\alpha+1}\, d_q t$$

$$= c_{q,\alpha}^2 \int_0^\infty \mathcal{F}_{q,\alpha} f(\lambda) \frac{1}{c_{q,\alpha}^2} \delta_{q,\alpha}(x, t) j_\alpha(\lambda t, q^2)\, t^{2\alpha+1}\, d_q t$$

$$= j_\alpha(\lambda x, q^2)\, \mathcal{F}_{q,\alpha} f(\lambda).$$

Theorem 25 *[1] The q-Bessel Fourier transform satisfies the following assertions:*

1. *For all* $f \in \mathcal{L}_{q,p,\alpha}(\mathbb{R}_q^+)$, $\mathcal{F}_{q,\alpha}^2 f(x) = f(x)$, $\quad \forall x \in \mathbb{R}_q^+$.

2. *For all* $f \in \mathcal{L}_{q,2,\alpha}(\mathbb{R}_q^+)$, $\|\mathcal{F}_{q,\alpha} f\|_{q,2,\alpha} = q^{2\alpha+1} \|f\|_{q,2,\alpha}$.

3. *For all* $f \in \mathcal{L}_{q,p,\alpha}(\mathbb{R}_q^+)$, $p \geq 1$, *we have* $\mathcal{F}_{q,\alpha} f \in \mathcal{L}_{q,\bar{p},v}(\mathbb{R}_q^+)$. *Furthermore, if* $1 \leq p \leq 2$, *then*

$$\|\mathcal{F}_{q,\alpha} f\|_{q,\bar{p},v} \leq B_{q,\alpha}^{\frac{2}{p}-1} \|f\|_{q,p,v},$$

where $B_{q,\alpha} = \dfrac{1}{1-q} \dfrac{(-q^2, q^2)_\infty (-q^{2\alpha+2}, q^2)_\infty}{(q^2, q^2)_\infty}$.

Proof. 1. If $f \in \mathcal{L}_{q,p,\alpha}(\mathbb{R}_q^+)$, then $\mathcal{F}_{q,\alpha} f$ exists, and we have

$$\mathcal{F}_{q,\alpha}^2 f = c_{q,\alpha} \int_0^{+\infty} \mathcal{F}_{q,\alpha} f(t)\, j_\alpha(xt, q^2)\, t^{2\alpha+1}\, d_q t$$

$$= \int_0^{+\infty} f(y) \left[c_{q,\alpha}^2 \int_0^\infty j_\alpha(xt, q^2)\, j_\alpha(yt, q^2)\, t^{2\alpha+1} d_q t \right] y^{2\alpha+1}\, d_q y$$

$$= \int_0^{+\infty} f(y)\, \delta_q(x, y)\, y^{2\alpha+1}\, d_q y = f(x).$$

The computations are justified by the Fubini's theorem: If $p > 1$, then we use the Hölder's inequality

$$\int_0^\infty |f(y)| \left[\int_0^\infty |j_\alpha(xt, q^2)\, j_\alpha(yt, q^2)| t^{2\alpha+1} d_q t \right] y^{2\alpha+1} d_q y$$

$$\leq \left[\int_0^\infty |f(y)|^p y^{2\alpha+1} d_q(y)\right]^{\frac{1}{p}} \left[\int_0^\infty |\sigma(y)|^{\bar{p}} y^{2\alpha+1} d_q(y)\right]^{\frac{1}{\bar{p}}}$$

and

$$\sigma(y) = \int_0^\infty |j_\alpha(xt, q^2) j_\alpha(yt, q^2)| t^{2\alpha+1} d_q t.$$

Then

$$\int_0^\infty \sigma(y)^{\bar{p}} y^{2\alpha+1} d_q(y) = \int_0^1 \sigma(y)^{\bar{p}} y^{2\alpha+1} d_q(y) + \int_1^\infty \sigma(y)^{\bar{p}} y^{2\alpha+1} d_q(y).$$

Notice that

$$\int_0^1 \sigma(y)^{\bar{p}} y^{2\alpha+1} d_q(y) \leq \|j_\alpha(\cdot, q^2)\|_{q,\infty}^{\bar{p}} \int_0^1 \left[\int_0^\infty |j_\alpha(xt, q^2)| t^{2\alpha+1} d_q t\right]^{\bar{p}} y^{2\alpha+1} d_q y$$

$$\leq \|j_\alpha(\cdot, q^2)\|_{q,\infty}^{\bar{p}} \|j_\alpha(\cdot, q^2)\|_{q,1,\alpha}^{\bar{p}} x^{-2(\alpha+1)\bar{p}} \left[y^{2\alpha+1} d_q y\right] < \infty$$

and

$$\int_1^\infty \sigma(y)^{\bar{p}} y^{2\alpha+1} d_q(y) \leq \|j_\alpha(\cdot, q^2)\|_{q,\infty}^{\bar{p}} \|j_\alpha(\cdot, q^2)\|_{q,1,\alpha}^{\bar{p}} \int_1^\infty \frac{y^{2\alpha+1}}{y^{2(\alpha+1)\bar{p}}} d_q y$$

$$\leq \|j_\alpha(\cdot, q^2)\|_{q,\infty}^{\bar{p}} \|j_\alpha(\cdot, q^2)\|_{q,1,v}^{\bar{p}} \int_1^\infty \frac{1}{y^{2(\alpha+1)(\bar{p}-1)+1}} d_q y < \infty.$$

For $p = 1$, we get

$$\int_0^\infty |f(y)| \left[\int_0^\infty |j_\alpha(xt, q^2) j_\alpha(yt, q^2)| t^{2\alpha+1} d_q t\right] y^{2\alpha+1} d_q y$$

$$\leq \|f\|_{q,1,\alpha} \|j_\alpha(\cdot, q^2)\|_{q,\infty} \|j_\alpha(\cdot, q^2)\|_{q,1,\alpha} \frac{1}{x^{2(\alpha+1)}}.$$

2. We introduce the function Ψ_x as follows

$$\Psi_x(t) = c_{q,\alpha} j_\alpha(tx, q^2).$$

The inner product $<;>$ in the Hilbert space $\mathcal{L}_{q,2,\alpha}(\mathbb{R})$ is defined for $f, g \in \mathcal{L}_{q,2,\alpha}(\mathbb{R})$ by

$$\langle f, g \rangle = \int_0^\infty f(t) g(t) t^{2\alpha+1} d_q t.$$

Using the orthogonality relation of $j_\alpha(\cdot, q^2)$, we obtain $x \neq y \implies \langle \Psi_x, \Psi_y \rangle = 0$ and $\|\Psi_x\|_{q,2,\alpha}^2 = \frac{1}{1-q} x^{-2(\alpha+1)}$. Next, $\mathcal{F}_{q,\alpha} f(x) = \langle f, \Psi_x \rangle$ and $f \in \mathcal{L}_{q,2,\alpha} \implies \mathcal{F}_{q,\alpha}^2 f = f$. Then

$$\langle f, \Psi_x \rangle = 0, \ \forall x \in \mathbb{R}_q^+ \implies \mathcal{F}_{q,\alpha} f(x) = 0, \ \forall x \in \mathbb{R}_q^+ \implies f = 0.$$

Hence, $\{\Psi_x, x \in \mathbb{R}_q^+\}$ form an orthogonal basis of the Hilbert space $\mathcal{L}_{q,2,\alpha}$ and we have $\overline{\{\Psi_x, x \in \mathbb{R}_q^+\}} = \mathcal{L}_{q,2,\alpha}$. Now, $f \in \mathcal{L}_{q,2,\alpha}$ yields that

$$f = \sum_{x \in \mathbb{R}_q^+} \frac{1}{\|\Psi_x\|_{q,2,\alpha}^2} \langle f, \Psi_x \rangle \, \Psi_x.$$

Consequently,

$$\|f\|_{q,2,\alpha}^2 = \sum_{x \in \mathbb{R}_q^+} \frac{1}{\|\Psi_x\|_{q,2,\alpha}^2} \langle f, \Psi_x \rangle^2 = (1-q) \sum_{x \in \mathbb{R}_q^+} x^{2(\alpha+1)} \mathcal{F}_{q,\alpha}^2 f(x)^2 = \|\mathcal{F}_{q,\alpha} f\|_{q,2,\alpha}^2.$$

3. This is an immediate consequence of the Assertion 1, the Riesz-Thorin theorem and the inversion formula.

The following proposition summarizes some results about q-Bessel translation operator.

Proposition 5.54 *[213] The following assertions hold.*

1. *For any function $f \in \mathcal{L}_{q,2,\alpha}(\widetilde{\mathbb{R}}_q^+)$, we have*

$$T_{q,x}^\alpha f(y) = T_{q,y}^\alpha f(x) \quad \text{and} \quad T_{q,x}^\alpha f(0) = f(x).$$

2. *For $f, g \in \mathcal{L}_{q,p,\alpha}(\widetilde{\mathbb{R}}_q)$, we have*

$$\int_0^\infty T_{q,x}^\alpha(f)(y) \, g(y) \, y^{2\alpha+1} \, d_q y = \int_0^\infty f(y) \, T_{q,x}^\alpha g(y) \, y^{2\alpha+1} \, d_q y, \forall y \in \widetilde{\mathbb{R}}_q^+$$

and $T_{q,x}^\alpha j_\alpha(ty, q^2) = j_\alpha(tx, q^2) \, j_\alpha(ty, q^2), \ \forall t, x, y \in \widetilde{\mathbb{R}}_q^+$.

3. *For $\Psi \in \mathcal{L}_{q,2,\alpha}(\widetilde{\mathbb{R}}_q^+)$, we have*

$$\|T_{q,x}^\alpha \Psi\|_{q,2,\alpha} \le \frac{1}{(q,q^2)_\infty^2} \|\Psi\|_{q,2,\alpha}.$$

Proof. 1. For $f \in \mathcal{L}_{q,2,\alpha}(\widetilde{\mathbb{R}}_q^+)$, we have

$$
\begin{aligned}
T_{q,x}^\alpha f(y) &= c_{q,\alpha} \int_0^\infty \mathcal{F}_{q,\alpha} f(t) \, j_\alpha(xt, q^2) \, j_\alpha(yt, q^2) \, t^{2\alpha+1} \, d_q t \\[2mm]
&= c_{q,\alpha} \int_0^\infty \mathcal{F}_{q,\alpha} f(t) \, j_\alpha(yt, q^2) \, j_\alpha(xt, q^2) \, t^{2\alpha+1} \, d_q t \\[2mm]
&= T_{q,y}^\alpha f(x).
\end{aligned}
$$

Similarly, we have

$$
\begin{aligned}
T_{q,x}^\alpha f(0) &= c_{q,\alpha} \int_0^\infty \mathcal{F}_{q,\alpha} f(t) \, j_\alpha(xt, q^2) \, j_\alpha(0, q^2) \, t^{2\alpha+1} \, d_q t \\[2mm]
&= c_{q,\alpha} \int_0^\infty \mathcal{F}_{q,\alpha} f(t) \, j_\alpha(xt, q^2) \, t^{2\alpha+1} \, d_q t \\[2mm]
&= \mathcal{F}_{q,\alpha}^2 f(x) = f(x).
\end{aligned}
$$

2. Denote for simplicity $d_q(st) = d_q s d_q t$ and $d_q(sty) = d_q s d_q t d_q y$. Denote also for $k \geq 1$; $\widetilde{\mathbb{R}}_+^{q,k} = \widetilde{\mathbb{R}}_+^q \times \widetilde{\mathbb{R}}_+^q \times \cdots \times \widetilde{\mathbb{R}}_+^q$. We have for $f, g \in \mathcal{L}_{q,2,\alpha}(\widetilde{\mathbb{R}}_q^+)$,

$$\int_0^\infty T_{q,x}^\alpha f(y) g(y) d_q y = c_{q,\alpha}^2 \int_{\widetilde{\mathbb{R}}_+^{q,3}} f(s) g(y) j_\alpha(ts, q^2) j_\alpha(tx, q^2) j_\alpha(ty, q^2)(tsy)^{2\alpha+1} d_q(sty)$$

$$= c_{q,\alpha}^2 \int_{\widetilde{\mathbb{R}}_+^{q,3}} g(y) j_\alpha(ty, q^2) y^{2\alpha+1} d_q y f(s) j_\alpha(ts, q^2) j_\alpha(tx, q^2)(ts)^{2\alpha+1} d_q(st)$$

$$= c_{q,\alpha} \int_{\widetilde{\mathbb{R}}_+^{q,2}} \mathcal{F}_{q,\alpha} g(t) f(s) j_\alpha(ts, q^2) j_\alpha(tx, q^2)(ts)^{2\alpha+1} d_q t d_q s$$

$$= \int_{\widetilde{\mathbb{R}}_+^q} T_{q,x}^\alpha(g)(s) f(s) s^{2\alpha+1} d_q s.$$

The remaining part of Assertion 2 is an easy consequence of Lemma 5.53 and Theorem 25, Assertion 1.

3. Denote

$$\widetilde{\mathcal{K}}_{q,v}(x, y, t, s) = \mathcal{K}_{q,v}(x, t, s) \mathcal{K}_{q,v}(y, t, s) \quad \text{and} \quad \mathcal{Q}_{q,v} f(t, s) = \mathcal{F}_{q,v} f(t) \overline{\mathcal{F}_{q,v} f(s)}.$$

We have

$$\|T_{q,x}^v f\|_{q,2,v}^2 = c_{q,v}^2 \int_0^\infty \int_0^\infty \int_0^\infty \mathcal{Q}_{q,v} f(t, s) \widetilde{\mathcal{K}}_{q,v}(x, y, t, s)(tsy)^{2|v|+1} d_q t d_q s d_q y$$

$$= c_{q,v}^2 \int_0^\infty \int_0^\infty \mathcal{Q}_{q,v} f(t, s) \int_0^\infty \widetilde{\mathcal{K}}_{q,v}(x, y, t, s) y^{2|v|+1} d_q y (ts)^{2|v|+1} d_q t d_q s$$

$$= \int_0^\infty \int_0^\infty \mathcal{Q}_{q,v} f(t, s) \delta_{q,v}(t, s) \mathcal{K}_{q,v}(x, t, s)(ts)^{2|v|+1} d_q t d_q s$$

$$= \int_0^\infty \mathcal{F}_{q,v} f(t) \widetilde{j}_{q,v}(xt, q^2) t^{2|v|+1} \int_0^\infty \overline{\mathcal{F}_{q,v} f(s)} \delta_{q,v}(t, s) \overline{\widetilde{j}_{q,v}(xs, q^2)} s^{2|v|+1} d_q s d_q t$$

$$= \int_0^\infty |\mathcal{F}_{q,v} f(t)|^2 |\widetilde{j}_{q,v}(xt, q^2)|^2 t^{2|v|+1} d_q t.$$

The second and the fourth equalities are simple applications of Fubini's rule. The third and the fifth ones are applications of the second and the first assertions in Lemma 5.50, respectively. Next, observing that

$$|\widetilde{j}_{q,v}(xt, q^2)| \leq \frac{1}{(q, q^2)_\infty^2},$$

we get

$$\int_0^\infty |\mathcal{F}_{q,v} f(t)|^2 |\widetilde{j}_{q,v}(xt, q^2)|^2 t^{2|v|+1} d_q t \leq \frac{1}{(q, q^2)_\infty^4} \int_0^\infty |\mathcal{F}_{q,v} f(t)|^2 t^{2|v|+1} d_q t$$

$$= \frac{1}{(q, q^2)_\infty^4} \|\mathcal{F}_{q,v} f\|_{q,2,v}^2.$$

Definition 5.55 *[213]A q-Bessel wavelet is an even function* $\Psi \in \mathcal{L}_{q,2,\alpha}(\mathcal{R}_q^+)$ *satisfying the following admissibility condition:*

$$C_{\alpha,\Psi} = \int_0^\infty |\mathcal{F}_{q,\alpha}\Psi(a)|^2 \frac{d_q a}{a} < \infty.$$

The continuous q-Bessel wavelet transform of a function $f \in \mathcal{L}_{q,2,\alpha}(\widetilde{\mathbb{R}}_q^+)$ *is defined by*

$$C_{q,\Psi}^\alpha(f)(a,b) = c_{q,\alpha} \int_0^\infty f(x)\,\overline{\Psi_{(a,b)}^\alpha}(x)\,x^{2\alpha+1}\,d_q x,\ \forall a \in \mathbb{R}_q^+,\ \forall b \in \widetilde{\mathbb{R}}_q^+,$$

where

$$\Psi_{(a,b)}^\alpha(x) = \sqrt{a}\mathcal{T}_{q,b}^\alpha(\Psi_a);\ \forall a,b \in \mathbb{R}_q^+,$$

and

$$\Psi_a(x) = \frac{1}{a^{2\alpha+2}}\,\Psi(\frac{x}{a}).$$

Proposition 5.56 *[73] Let* Ψ *be a q-Bessel wavelet in* $\mathcal{L}_{q,2,\alpha}(\mathbb{R}_q^+)$. *Then, the function* $F : (a,b) \mapsto \Psi_{(a,b)}^\alpha$ *is continuous on* $\mathbb{R}_q^+ \times \widetilde{\mathbb{R}}_q^+$.

Proof. It is clear that F is a mapping from $\mathbb{R}_q^+ \times \widetilde{\mathbb{R}}_q^+$ into $\mathcal{L}_{q,2,\alpha}(\mathbb{R}_q^+)$ and it is continuous at all $(a,b) \in \mathbb{R}_q^+ \times \widetilde{\mathbb{R}}_q^+$. Now, fix $a \in \mathbb{R}_q^+$. For $b \in \widetilde{\mathbb{R}}_q^+$, we have

$$\|F(a,b) - F(a,0)\|_{q,2,\alpha}^2$$

$$= \|\mathcal{T}_{q,b}^\alpha(\Psi_a) - \Psi_a\|_{q,2,\alpha}^2$$

$$= q^{-4\alpha-2}\|\mathcal{F}_{q,\alpha}(\mathcal{T}_{q,b}^\alpha(\Psi_a) - \Psi_a)\|_{q,2,\alpha}^2$$

$$= q^{-4\alpha-2}\int_0^\infty \left|1 - j_\alpha(xb,q^2)\right|^2 |\mathcal{F}_{q,\alpha}(\Psi_a)|^2\,(x)x^{2\alpha+1}d_q x.$$

However, for all $x \in \mathbb{R}_q^+$ and $b \in \widetilde{\mathbb{R}}_q^+$, we have

$$\left|1 - j_\alpha(xb,q^2)\right|^2 |\mathcal{F}_{q,\alpha}(\Psi_a)|^2\,(x) \le (1 + \frac{1}{(q,q^2)_\infty^2})^2\,|\mathcal{F}_{q,\alpha}(\Psi_a)|^2\,(x),$$

and $\mathcal{F}_{q,\alpha}(\Psi_a) \in \mathcal{L}_{q,2,\alpha}(\mathbb{R}_q^+)$. So, the Lebesgue theorem leads to

$$\lim_{b \longrightarrow 0} \|F(a,b) - F(a,0)\|_{q,2,\alpha} = 0.$$

Then, for all open neighborhood V of $F(a,0)$ in $\mathcal{L}_{q,2,\alpha}(\mathbb{R}_q^+)$, there exists an open neighborhood U of 0 in $\widetilde{\mathbb{R}}_q^+$ such that

$$\forall b \in U, F(a,b) \in V.$$

Thus, $\{a\} \times U$ is an open neighborhood of $(a,0)$ in $\mathbb{R}_q^+ \times \widetilde{\mathbb{R}}_q^+$ and $F(\{a\} \times U) \in V$, which proves the continuity of F at $(a,0)$.

The following result is a variant of Parseval-Plancherel rules for the case of q-Bessel wavelet transforms.

Theorem 26 *[213] Let Ψ be a q-Bessel wavelet in $\mathcal{L}_{q,2,\alpha}(\mathbb{R}_q^+)$. Then,*

1. ; $\forall g \in \mathcal{L}_{q,2,\alpha}(\mathbb{R}_q^+)$,

$$\frac{1}{C_{\alpha,\Psi}} \int_0^\infty \int_0^\infty |C_{q,\Psi}^\alpha(g)(a,b)|^2 \, b^{2\alpha+1} \frac{d_q a \, d_q b}{a^2} = \|g\|_{q,2,\alpha}^2,$$

2. $\forall f_1, f_2 \in \mathcal{L}_{q,2,\alpha}(\mathbb{R}_q^+)$,

$$\int_0^\infty f_1(x) \overline{f}_2(x) \, x^{2\alpha+1} \, d_q x = \frac{1}{C_{\alpha,\Psi}} \int_0^\infty \int_0^\infty C_{q,\Psi}^\alpha(f_1)(a,b)$$

$$\overline{C_{q,\Psi}^\alpha}(f_2)(a,b) \, b^{2\alpha+1} \frac{d_q a \, d_q b}{a^2}.$$

Proof. 1. By using Fubini's theorem, and Assertion 3 in Theorem 25, we obtain

$$q^{4\alpha+2} \int_0^\infty \int_0^\infty |c_{q,\Psi}^\alpha(g)(a,b)|^2 \, b^{2\alpha+1} \frac{d_q a \, d_q b}{a^2}$$

$$= q^{4\alpha+2} \int_0^\infty \left(\int_0^\infty |\mathcal{F}_{q,\alpha}(g)(x)|^2 |\mathcal{F}_{q,\alpha}(\overline{\Psi}_a)|^2(x) x^{2\alpha+1} d_q x \right) \frac{d_q a}{a}$$

$$= \int_0^\infty |\mathcal{F}_{q,\alpha}(g)(x)|^2 \left(|\mathcal{F}_{q,\alpha}(\Psi)(ax)|^2 \frac{d_q a}{a} \right) x^{2\alpha+1} d_q x$$

$$= C_{\alpha,\Psi} \int_0^\infty |\mathcal{F}_{q,\alpha}(g)(x)|^2 x^{2\alpha+1} d_q x$$

$$= C_{\alpha,\Psi} q^{4\alpha+2} \|g\|_{q,2,\alpha}^2.$$

Assertion 2 is a direct consequence of Assertion 1.

Theorem 27 *Let Ψ is a q-Bessel wavelet in $\mathcal{L}_{q,2,\alpha}(\mathbb{R}_q^+)$. Then, for all $f \in \mathcal{L}_{q,2,\alpha}(\mathbb{R}_q^+)$, we have*

$$f(x) = \frac{c_{q,\alpha}}{C_{\alpha,\Psi}} \int_0^\infty \int_0^\infty C_{q,\Psi}^\alpha(f)(a,b) \, \Psi_{(a,b)}^\alpha(x) \, b^{2\alpha+1} \frac{d_q a \, d_q b}{a^2}; \; \forall x \in \mathbb{R}_q^+.$$

Proof. For $x \in \mathbb{R}_q^+$, we have $h = \delta_{q,\alpha}$ belonging to $\mathcal{L}_{q,2,\alpha}(\mathbb{R}_q^+)$. On the other hand, according to Theorem 26, the definition of $C_{q,\Psi}^\alpha$ and the definition of the q-Jackson integral, we have

$$(1-q)x^{2\alpha+2}f(x)$$

$$= \int_0^\infty f(t)\overline{h}(t)t^{2\alpha+1}d_q t$$

$$= \frac{1}{C_{\alpha,\Psi}} \int_0^\infty \int_0^\infty C_{q,\Psi}^\alpha(f)(a,b)\overline{C_{q,\Psi}^\alpha}(h)(a,b)b^{2\alpha+1}\frac{d_q a d_q b}{a^2}$$

$$= \frac{c_{q,\alpha}}{C_{\alpha,\Psi}} \int_0^\infty \int_0^\infty C_{q,\Psi}^\alpha(f)(a,b) \left(\int_0^\infty \overline{h}(t)\Psi_{(a,b)}^\alpha(x)\, t^{2\alpha+1}d_q t \right) b^{2\alpha+1}\frac{d_q a\, d_q b}{a^2}$$

$$= (1-q)x^{2\alpha+1}\frac{c_{q,\alpha}}{C_{\alpha,\Psi}} \int_0^\infty \int_0^\infty C_{q,\Psi}^\alpha(f)(a,b)\, \Psi_{(a,b)}^\alpha(x)\, b^{2\alpha+1}\frac{d_q a\, d_q b}{a^2}.$$

Thus,

$$f(x) = \frac{c_{q,\alpha}}{C_{\alpha,\Psi}} \int_0^\infty \int_0^\infty C_{q,\Psi}^\alpha(f)(a,b)\, \Psi_{(a,b)}^\alpha(x)\, b^{2\alpha+1}\frac{d_q a\, d_q b}{a^2},$$

which completes the proof.

5.8 EXERCISES

Exercise 1.

Let $n \in \mathbb{N}$ and f, g be two functions n-times q-differentiable. Show the q-Leibniz derivation rule

$$D_q^n(fg)(x) = \sum_{k=0}^n C_{k,q}^n D_q^k(f)(xq^{n-k})D_q^l[n-k](g)(x),$$

where

$$C_{k,q}^n = \frac{(q;q)_n}{(q;q)_k(q;q)_{n-k}}$$

and where for $l \in \mathbb{N}$, $D_q^l = D_q \circ D_q \circ \cdots \circ D_q$ is the composition of the q-derivative operator D_q l-times.

Exercise 2.

Show that for all real numbers $x, y > 0$, we have

$$\int_0^\infty sJ_v(xs)J_v(ys)ds = \frac{\delta(x-y)}{x}$$

where $\delta(.)$ is the Kronecker symbol.

Exercise 3.

Denote

$$d\sigma(x) = \frac{x^{2\mu}}{2^{\mu-\frac{1}{2}}\,\Gamma(\mu+\frac{1}{2})}\chi_{[0,\infty[}(x)\,dx$$

and

$$j_\mu(x) = 2^{\mu-\frac{1}{2}}\,\Gamma(\mu+\frac{1}{2})\,x^{\frac{1}{2}-\mu}\,J_{\mu-\frac{1}{2}}(x),$$

where $J_{\mu-\frac{1}{2}}(x)$ is the Bessel function of order $v = \mu - \frac{1}{2}$. Denote next

$$D(x,y,z) = \int_0^\infty j_\mu(xt)\,j_\mu(yt)\,j_\mu(zt)\,d\sigma(t).$$

Show that

$$\int_0^\infty D(x,y,z)\,d\sigma(z) = 1.$$

Exercise 4.

Show that

$$h_\mu^\theta\left[(\Delta_{\mu,x}^*)^r\delta(x-c)\right](y) = (-y^2\cos^2\theta)^r\,K_\mu^\theta(c,y),\quad x,c\in\mathbb{R}_+.$$

Exercise 5.

Develop a proof for Proposition 5.14.

Exercise 6.

Develop a proof for Lemma 5.16.

Exercise 7.

Develop a proof for Lemma 5.47

Exercise 8.

Develop a proof for Proposition 5.49

Exercise 9.

Consider the q-Bessel function

$$j_\alpha(x,q^2) = \sum_{n\geq0}(-1)^n\frac{q^{n(n+1)}}{(q^{2\alpha+2},q^2)_n\,(q^2,q^2)_n}\,x^{2n}$$

and the q-Bessel operator defined for $\alpha > -\dfrac{1}{2}$ by

$$\Delta_{q,\alpha}f(x) = \frac{f(q^{-1}x) - (1+q^{2\alpha})f(x) + q^{2\alpha}f(qx)}{x^2},\qquad \forall x\neq 0.$$

Show that for all $\lambda \in \mathbb{C}$,

$$\Delta_{q,\alpha} u(x) = -\lambda^2 \, u(x) \quad \text{and} \quad u(0) = 1, \ u'(0) = 0$$

has as unique solution: the function $u(x) = j_\alpha(\lambda x, q^2)$.

Exercise 10.

With the same notations as in Exercise 9, denote the q-Bessel Fourier transform by $\mathcal{F}_{q,\alpha}$, defined as

$$\mathcal{F}_{q,\alpha} f(x) = c_{q,\alpha} \int_0^\infty f(t) \, j_\alpha(xt, q^2) \, t^{2\alpha+1} \, d_q t,$$

where $c_{q,\alpha}$ is an appropriate constant. Define also a q-Bessel translation operator by

$$T_{q,x}^\alpha f(y) = c_{q,\alpha} \int_0^\infty \mathcal{F}_{q,\alpha} f(t) \, j_\alpha(xt, q^2) \, j_\alpha(yt, q^2) \, t^{2\alpha+1} \, d_q t.$$

1) Show that for all $f \in \mathcal{L}_{q,2,\alpha}(\tilde{\mathbb{R}}_q^+)$,

$$\mathcal{F}_{q,\alpha}(T_{q,x}^\alpha f)(\lambda) = j_\alpha(\lambda x, q^2) \, \mathcal{F}_{q,\alpha} f(\lambda), \quad \forall \lambda, x \in \tilde{\mathbb{R}}_q^+.$$

2) A q-Bessel wavelet is an even function $\Psi \in \mathcal{L}_{q,2,\alpha}(\tilde{\mathbb{R}}_q^+)$ satisfying

$$C_{\alpha,\Psi} = \int_0^\infty |\mathcal{F}_{q,\alpha}\Psi(a)|^2 \, \frac{d_q a}{a} < \infty.$$

The continuous q-Bessel wavelet transform of a function $f \in \mathcal{L}_{q,2,\alpha}(\tilde{\mathbb{R}}_q^+)$ is defined by

$$C_{q,\Psi}^\alpha(f)(a,b) = c_{q,\alpha} \int_0^\infty f(x) \, \overline{\Psi_{(a,b)}^\alpha}(x) \, x^{2\alpha+1} \, d_q x,$$

where $\Psi_{(a,b)}^\alpha(x) = \sqrt{a} T_{q,b}^\alpha(\Psi_a)$ and $\Psi_a(x) = \dfrac{1}{a^{2\alpha+2}} \, \Psi(\dfrac{x}{a})$, $\forall a \in \mathbb{R}_q^+$, $\forall x, \in \tilde{\mathbb{R}}_q^+$.

Show that the function $F : (a,b) \mapsto \Psi_{(a,b)}^\alpha$ is continuous on $\mathbb{R}_q^+ \times \tilde{\mathbb{R}}_q^+$.

3) Prove that $\forall g \in \mathcal{L}_{q,2,\alpha}(\tilde{\mathbb{R}}_q^+)$,

$$\frac{1}{C_{\alpha,\Psi}} \int_0^\infty \int_0^\infty |C_{q,\Psi}^\alpha(g)(a,b)|^2 \, b^{2\alpha+1} \frac{d_q a \, d_q b}{a^2} = \|g\|_{q,2,\alpha}^2.$$

4) Prove that $\forall f_1, f_2 \in \mathcal{L}_{q,2,\alpha}(\tilde{\mathbb{R}}_q^+)$,

$$\int_0^\infty f_1(x) \overline{f}_2(x) x^{2\alpha+1} d_q x = \frac{1}{C_{\alpha,\Psi}} \int_0^\infty \int_0^\infty C_{q,\Psi}^\alpha(f_1, f_2)(a,b) b^{2\alpha+1} d\mu_q(a,b),$$

where $C_{q,\Psi}^\alpha(f_1, f_2)(a,b) = C_{q,\Psi}^\alpha(f_1)(a,b)\overline{C_{q,\Psi}^\alpha}(f_2)(a,b)$ and $d\mu_q(a,b) = \dfrac{d_q a \, d_q b}{a^2}$.

5) Prove that for all $f \in \mathcal{L}_{q,2,\alpha}(\tilde{\mathbb{R}}_q^+)$,

$$f(x) = \frac{c_{q,\alpha}}{C_{\alpha,\Psi}} \int_0^\infty \int_0^\infty C_{q,\Psi}^\alpha(f)(a,b) \, \Psi_{(a,b)}^\alpha(x) \, b^{2\alpha+1} \frac{d_q a \, d_q b}{a^2}; \ \forall x \in \tilde{\mathbb{R}}_q^+.$$

CHAPTER 6

Wavelets in statistics

6.1   INTRODUCTION

In Statistical modeling, especially of financial series, the familiar and most known models are gaussian and log-normal ones. The gaussian models are based on the price increment hypothesis, while the log-normal ones are based on the log-price increment hypothesis. One can then suggest that the statistics of price variations is gaussian. However, such a hypothesis is not sufficient as it does not estimate the importance of rare and intensive increments. Therefore, it is necessary to go back to the multi-scaling aspects of the problem by studying the stock exchange of the different financial products with the suitable concepts such as scale low and self-similar statistical aspects. Indeed, one can observe in practice that some return distributions have heavy tails than the gaussians. This fact has been noticed firstly by B. Mandelbrot when studying the cotton price. One can also remark that the tail indices of the distribution seem to increase with the time intervals. These facts and some others such as non-autocorrelation of some stock market models, clustering volatility and time scaling behaviors allow researches to think about some more general models to describe or to study financial series. There are many models that can be stated as classical ones such as brownian motion, fractional brownian motion, martingales and semi-martingales. For more details and backgrounds on these subjects and for a comprehensive literature on wavelet analysis, wavelet filter, time series, self-similarity, price models, volatility and related topics, we have listed an exhaustive updated list of references.

Time series constitute a very delicate area of study due to the specific characteristics. Most of the time-varying series are non-linear, in particular, the financial and economic series, which present an intellectual challenge. Their behavior seems to change dramatically, and uncertainty is always present. The financial data available are always approximate. Increasingly, artificial intelligence techniques such as fuzzy systems, neural networks and genetic and hybrid algorithms have been used to successfully model complex fundamental relationships of non-linear time series. Such models, referred to as universal estimators, are theoretically capable of uniformly approximating any true continuous function over a compact set to any degree of accuracy. Consider the case of fuzzy systems, defined as techniques for reasoning uncertainty, and based on the theory of fuzzy sets developed by Zadeh (1973). Such systems are characterized by linguistic interpretation. In addition, in the case of non-stationary financial series, the relationship between dependent and other independent variables is tainted with ambiguity. Introducing fuzzy logic into regression methods reduces the degree of uncertainty. It thus appears that the concept of vagueness covers indifferently the notions of knowledge badly specified, badly described, imperfect, partial, incomplete or approximate.

Other than uncertainty and imprecision, most of the actual processes in the financial markets consist of complex combinations of sub-processes or components, which operate at different frequencies. On the other hand, the examination of a time series allows us, in general, to recognize three types of components: a trend, a seasonal component and a random variation. Other evolutionary characteristics, such as shocks, can also be observed but are less closely linked to the structure of the series. It is therefore useful to separate these components for two reasons. The first is to answer common sense questions such as that of the growth or the general decrease of the phenomenon observed. Extracting the trend and analyzing it will answer this question. It is also interesting to highlight the possible presence of a periodic variation, thanks to the analysis of the seasonal component. The second of these reasons is to rid the phenomenon of its tendency and its periodic variations to more easily observe the random phenomenon. Methods based on multi-scale wavelet transformation provide powerful analysis tools that decompose time series data into coefficients related to time and a specific frequency band. Wavelets are considered capable of isolating fundamental low-frequency dynamics from non-stationary time series, and are robust to the presence of noise. The interpretation of the complex financial time series devices is therefore made easy by first applying the wavelet transformation and later interpreting each sub-series individually.

6.2 WAVELET ANALYSIS OF TIME SERIES

6.2.1 Wavelet time series decomposition

Wavelet analysis allows the representation of time series into species relative to the time and frequency information known as time-frequency decomposition. It consists in decomposing a series in different frequency components with a scale-adapted resolution and thus permits observation and analysis of data at different scales. A time series $X(t)$ is projected onto W_j, yielding a component $DX_j(t)$ given by

$$DX_j(t) = \sum_k d_{j,k} \psi_{j,k}(t), \tag{6.1}$$

where the $d_{j,k}$ are the detail coefficients of the series $X(t)$. The following decomposition is proved for $j \in \mathbb{Z}$:

$$X(t) = \sum_j DX_j(t) = \sum_{j \leq J} DX_j(t) + \sum_{j > J} DX_j(t). \tag{6.2}$$

The component $AX_J(t) = \sum_{j \leq J} DX_j(t)$ is called the approximation of $X(t)$ at the level J and reflects the trend or the global shape of the series in the space V_J. Therefore, it may be expressed as

$$AX_J(t) = \sum_k a_{J,k}\varphi_{J,k}(t), \tag{6.3}$$

where the $a_{J,k}$ are the approximation coefficients of the series. As a result, we obtain the wavelet decomposition of $X(t)$ at the truncation level J:

$$X(t) = AX_J(t) + \sum_{j>J} DX_j(t). \tag{6.4}$$

It is composed of one part reflecting the global behavior or the trend of the series and a second part reflecting the higher frequency oscillations or the fine scale deviations of the series near its trend.

In practice, we cannot obviously compute the complete set of coefficients. We thus fix a maximal level of decomposition J and consider the decomposition

$$X_J(t) = AX_{J_0}(t) + \sum_{J_0<j\leq J} DX_j(t) \tag{6.5}$$

for $J_0 \leq J$ arbitrary. There is no theoretical method for the exact choice of the parameters J_0 and J. However, the minimal parameter J_0 does not have an important effect on the total decomposition and is usually chosen to be 0. But, the choice of J is always critical. One selects J related to the error estimates. In finance, economics, management and generally actuarial sciences, compared to classical theories wavelet analysis is still less used, although it provides good results and needs to be more developed. The literature is growing rapidly. See [20, 21, 24, 25, 48, 55, 56, 59, 64, 169, 166, 166, 198, 207, 225, 226, 227, 255, 256, 257, 319, 370, 371, 372, 373].

For a better understanding of wavelet analysis in the context of time series modeling, several factors have to be taken into consideration ([319]) such as the choice of wavelet filter, the boundary conditions and the size of the sample.

The problem with the small-width wavelet filter is that it can sometimes introduce unwanted things into the analysis results , such as unreal slices and fine sharks. The larger wavelet filters can be better aligned with the characteristics of the time series. However, with the increase in width, (i) several coefficients will be excessively influenced by the boundary conditions, (ii) there is a decrease in the degree of localization of the transform coefficients in discrete wavelets and (iii) there is an increase in computational burden; thus, we have to look for the smallest L that allows us to have a reasonable result afterwards.

The transformation into discrete wavelets uses circular filtering. This means that the time series are treated as a proportion of a periodic sequence of period N. For financial time series, this is a bit difficult in the sense that there is a rare evidence to support this hypothesis. Furthermore, there is a large discontinuity between the first and the last observation. The extension that deals with the circular influences of the discrete wavelet transformation coefficients and the corresponding multi-resolution analysis are quantified in [319]. Note that Haar wavelets produce coefficients that are independent of the circularity assumption. A short path has been proposed for an analysis with the hypothesis of circularity. The authors in [319] suggest indicating exactly on the DWT graphics and the corresponding multi-resolution analysis the areas that are affected by the borders. However, they argue that the influence of circularity can be negligible, in particular, in the case of an insignificant contradiction between the start and the end of the series. Hence, the most marked areas are often fairly conservative measures of the influence of circularity. To reduce the impact of circularity, we can reflect the time series around these latter points. The resulting series of width $2N$ have the same mean and the same variance as the original series. This method can eliminate the effects due to a serious inconsistency between the first and the last values ([225]).

Transformation into ordinary DWT discrete wavelets requires N observations, multiple of 2, while transformation into partial discrete wavelets requires size N multiple of 2^{J_0}. However, it is too rare in reality to have a database of dyadic width or even an integer that is a multiple of 2^{J_0}. There are several methods adopted to deal with this kind of problems. The most obvious is to truncate (swap) the series to integers roughly multiple of 2^{J_0}, but we then need to choose the most reasonable value of J_0. The simplest alternative is to familiarize yourself with Fourier analysis and therefore dilute the series with zero or the sample mean. Note that having the series diluted with the sample average does not change the sample average compared to the original series.

6.2.2 The wavelet decomposition sample

It is more widely recommended to introduce the transformation into discrete wavelets through a simple matrix operation: let X be the vector of the observations of dyadic width (for example, $N = 2^j$). The column vector for the width N of the discrete wavelet coefficients w is obtained by $w = WX$, where $W \in M(N \times N)$ is a normal orthogonal matrix that defines the DWT matrix. We adopt the construction presented in [225] during a spatial limitation.

The matrix W is composed of the filtered wavelet coefficients and the filtered scale coefficients. The resulting structure w and the matrix W can be seen through the sub-vectors $w_1, w_2, w_3, \ldots\ldots, w_j, v_j$ and the sub-matrices $W_1, W_2, \ldots\ldots\ldots W_j, V_j$, respectively:

$$
w = \begin{pmatrix} w_1 \\ w_2 \\ \vdots \\ w_j \\ v_j \end{pmatrix} \quad \text{and} \quad W = \begin{pmatrix} W_1 \\ W_2 \\ \vdots \\ W_j \\ V_j \end{pmatrix},
$$

where w_j is a column vector with size $N/2^j$ composed of associated details' coefficients to the scale $\lambda_j = 2^{j-1}$ and v_J is the column vector of size $N/2^J$ composed of scaling coefficients associated with the scale λ_J. Hence, $W_j \in M(N/2^j \times N)$ and $V_j \in M(N/2^J \times N)$.

To illustrate the structure of the matrix W, consider a width filter $L = 2$ and a width signal $N = 8$. The matrix $W_i \in M(4 \times 8)$ is therefore

$$
W_1 = \begin{pmatrix} h_1 & h_0 & 0 & 0 & 0 & 0 & 0 & 0 \\ 0 & 0 & h_1 & h_0 & 0 & 0 & 0 & 0 \\ 0 & 0 & 0 & 0 & h_1 & h_0 & 0 & 0 \\ 0 & 0 & 0 & 0 & 0 & 0 & h_1 & h_0 \end{pmatrix}.
$$

The h_i's are the wavelet filter coefficients.

The number of affected coefficients increases with the level j and the width L. Precisely, Percival and Walden in [319] showed that the number of details' coefficients is estimated by $L_j = [(L-2)(1-2^{-j})]$, where for a real number x the notation $[x]$ stands for the smallest integer that is greater than or equal to x. (See also [225, 226, 227].)

An additive decomposition of the time series can be obtained by using the discrete wavelet transform. Firstly, according to the definition of the j-level, we have $d_j = W_j^T w_j$. For dyadic width wavelet series $N = 2^J$, the final wavelet detail $d_{J+1} = V_J^T v_J$ is equal to the sample mean of the observations. Then, we define the j-level of wavelet regularity as

$$S_j = \sum_{k=j+1}^{J+1} d_k \quad \text{for} \quad j = 0, \dots, J,$$

where S_{J+1} is the zero vector. Unlike the wavelet detail d_j, the scaling part S_j becomes more regular when increasing the level. Moreover, X is approximated by its projection

$$X_J = D_j + S_j = \sum_{k=1}^{j} d_k + \sum_{k=j+1}^{J+1} d_k.$$

6.3   WAVELET VARIANCE AND COVARIANCE

One of the important characteristics of the discrete wavelet transform is their ability to decompose the sample variance of time series into bases of scales-by-scales. Indeed,

$$||X||^2 = X^T X = (Ww)^T Ww = w^T W^T Ww = w^T w = ||w||^2.$$

Otherwise,

$$||X||^2 = \sum_{t=0}^{N-1} x_i^2 = \sum_{j=1}^{J} \sum_{t=0}^{N/2^j - 1} \omega_{j,t}^2 + v_{j,0}^2 = ||w||^2.$$

Given that the structure of the wavelet coefficients is split by means of the scales, we get

$$||X||^2 = \sum_{j=1}^{J} ||w_j||^2 + ||v_J||^2.$$

$||w_j||^2$ is the energy of X due to the scale λ_j and $||v_j||^2$ is the information due to the change of the scales less than λ_j. Since W and V are orthonormal matrices, we may have $d_j^T d_j = w_j^T w_j$ for $1 \le j \le J$ and $S_j^T S_j = v_j^T v_j$. So, alternately

$$||X||^2 = \sum_{j=1}^{J} (||D_j||^2 + ||S_j||^2).$$

In a parallel study, the case of a locally stationary long memory process has been investigated by applying the maximum overlap discrete wavelet transform (MODWT). It is shown that the wavelet coefficients obtained by MODWT form a zero mean locally stationary process at each level. However, the variance is a non-constant time function. Let X_t, $0 < t \leq T$, be the stochastic process and for a level j, $\widetilde{\omega}_{j,t}$ be the process formed by the MODWT coefficients at a level j. The variance satisfies

$$V(\widetilde{\omega}_{j,t}) = V^2(\lambda_j) \longrightarrow \sigma^2 2^{j[2d-1]} \text{ as } j \longrightarrow \infty.$$

Consequently, we get the approximation

$$\log V^2(\lambda_j) \simeq \log \sigma^2 + jj[2d-1] \log 2 \text{ as } j \longrightarrow \infty. \tag{6.6}$$

d is called the fractional differentiating parameter. It holds, in fact, from simulation studies that the median of $\widehat{d}$ may induce an estimator of the real value of the fractional differentiating parameter with a negative bias at the boundaries. In the special case of stationary global ARFIMA model, the MSE due to $\widehat{d}$ increases. This is essentially due to the inefficiency of the information used to build the local estimator. Recall that ARFIMA processes are usually characterized by distributions with sudden shifts in long memory parameters to mimic local stationarity.

The wavelet covariance is simply the covariance between the wavelet coefficients at a scale or a level j. Let $X_t = (X_t^1, X_t^2)$, $t > 0$, be a bivariate stochastic process associated with the univariate spectra $S_1(f)$ and $S_2(f)$, respectively. Let, for $i = 1, 2$, $W_{j,t}^i = (w_{j,1,t}^i, w_{j,2,t}^i, ; w_{j,K_j,t}^i)$ be the vector of wavelet coefficients of X_t^i at the scale j. The wavelet covariance of X_t at the scale j is defined by

$$\gamma_{X_t}(j) = \frac{1}{2\lambda_j} cov(W_{j,t}^1 W_{j,t}^2).$$

Whenever the wavelet filter length is sufficient to eliminate any non-stationary characteristic in the process X_t, we may define the wavelet covariance in a simple way as

$$\gamma_{X_t}(j) = \frac{1}{2\lambda_j} \mathbb{E}(W_{j,t}^1, W_{j,t}^2)$$

provided that we have a zero mean of the wavelet coefficients.

The wavelet covariance decomposes the covariance of a bivariate process into a series of covariances as follows

$$cov(X_t^1, X_t^2) = \sum_j \gamma_{X_t}(j).$$

The same decomposition remains valid in the case of MODWT.

We now explain the construction of wavelet estimator of the covariance. Let $X_t = (X_t^1, X_t^2)$, $t = 0, 1, \ldots, N-1$, be a realization of the bivariate process X_t. Consider the MODWT at a level $J < \log_2 T$ for each component X_t^i, $i = 1, 2$. We obtain J matrices $\widetilde{W}_j = (\widetilde{W}_{j,0}, \widetilde{W}_{j,1}, \ldots, \widetilde{W}_{j,N-1})$ whose columns are the vectors $\widetilde{W}_{j,k} = (\widetilde{\omega}_{j,k}^1, \widetilde{\omega}_{j,k}^2)$, where $0 \leq j \leq J$ and $0 \leq k \leq N/2^j - 1$ issued from the MODWT wavelet coefficients, provided with the vector of the MODWT scale coefficients $\widetilde{V}_J = (\widetilde{V}_{J,0}, \widetilde{V}_{J,1}, \ldots, \widetilde{V}_{J,N-1})$ whose columns are the vectors $\widetilde{V}_{J,k} = (\widetilde{v}_{j,k}^1, \widetilde{v}_{j,k}^2)$, where $0 \leq$

$k \leq N/2^J - 1$ issued from the MODWT scaling coefficients. The unbiased MODWT covariance estimator is given by

$$\tilde{\gamma}_{X_t}(j) = \frac{1}{\widetilde{N}_j} \sum_t \tilde{\omega}_{j,t}^1 \tilde{\omega}_{j,t}^2,$$

where $\widetilde{N}_j = N - L_j + 1$.

This estimator is characterized essentially by the exclusion of the coefficients affected by the boundaries. A biased estimator for the covariance may therefore be obtained by including these coefficients. Gencay et. al ([225]) proved that the variance of the estimator $\gamma_{X_t}(j)$ obtained by DWT depends on the offset between the two components of the bivariate process.

However, the estimator based on the MODWT is invariant relative to such shifting. This may be more suitable since, in practice, the shifting between time series is generally inadequate. Besides, a study of confidence intervals for covariance estimators has been already studied in details in [225, 226, 227].

To resume, for estimating the variance and the covariance of the statistical (time) series and introducing the analogues for the components due to the wavelet decomposition, we split here also the variance of the series into sub-variances relative to scales or the levels j, which will be called the variance of the series at the scale j ([319]). Let V_X^2 be the variance of $X(t)$ and $V_X^2(j)$ be the variance of the projection at the scale or the level j. We have

$$V_X^2 = \sum_j v_X^2(j). \tag{6.7}$$

This permits us to focus on the sub-variances at a level j instead of considering the whole series. This, in turn, facilitates the analysis of the fluctuations and the dynamics of the series. Let L be the length of the wavelet support, $N_j = [N/2^j]$ the number of wavelet coefficients at the level j and $L_j = [2^{-j}(2^j - 1)(L - 2)]$ the number of boundary wavelet coefficients at the level j. The variance at the level j is estimated as

$$\tilde{V}_X^2 = \frac{1}{2^j(N_j - L_j)} \sum_{k=L_j-1}^{N_j-1} d_{j,k}^2. \tag{6.8}$$

Similarly, we have an analogue formulation for the covariance at the level j for a couple of series (X, Y) as

$$\tilde{V}_{XY}^2 = \frac{1}{2^j(N_j - L_j)} \sum_{k=L_j-1}^{N_j-1} d_{j,k}^X d_{j,k}^Y. \tag{6.9}$$

Sometimes, we may apply the so-called partial DWT, which consists of a direct generalization of the transform into a discrete wavelet. It offers more flexibility due to the choices of the scale beyond which wavelet analysis in large individual scales is no longer of real interest. The practical benefit of this flexibility is that we no longer need to have a sample size of dyadic width. It suffices therefore that the sample size is a multiple of 2^{J_0}.

Variance based on the wavelet method is also called wavelet spectrum. This spectrum can be interesting for several reasons ([319]).

- The variance decomposition is used if the phenomenon contains variances above the rank of the different scales.

- The wavelet variance (to be defined later) is closely related to the concept of spectral density function (SDF).

- The wavelet variance is a usable substitute for the variance of a process with infinite variance.

6.4 WAVELET DECIMATED AND STATIONARY TRANSFORMS

6.4.1 Decimated wavelet transform

The discrete decimated wavelet transform is a generalization of the discrete wavelet transform. Let $(s_i)_{0 \leq i \leq J-1}$ be a binary sequence. We introduce the decimation operators D_{s_j} and R_{s_j}, $j \in \mathbb{N}$, which act on the recursive decomposition algorithm. To be more precise, let us explain the procedure. Let $(Sx)_j = x_{j+1}$ be the shift operator. The operators D_{s_j} and R_{s_j} will be defined recursively as follows:

- $D_1 = D_0 S$ and $R_1 = S^{-1} R_0$.

- $S D_0 = D_0 S^2$.

- S commutes with the filters G and H.

Let $r \in \mathbb{N}$ be such that its binary representation is $\varepsilon_0, \varepsilon_1, ..., \varepsilon_{J-1}$. The sequences of coefficients C_j and d_j obtained by means of the discrete decimated transform are, in fact, shifted versions of those obtained by the discrete wavelet transform applied to the sequence $S^r x$. Indeed, let t_1 and t_2 be two integers with respective binary representation $s_0, s_1, \ldots, s_{j-1}$ and $s_j, s_{j+1}, \ldots, s_J$. From the discrete wavelet transform, one obtains

$$d^j = D_0 G (D_0 H)^{J-j-1} c^J.$$

Consequently,

$$
\begin{aligned}
d^j &= D_{s_j} G D_{s_{j+1}} H D_{s_{j+2}} H ... D_{s_{J-1}} H c^J \\
&= D_0 S^{s_j} G D_0 S^{s_{j+1}} H D_0 S^{s_{j+2}} H ... D_0 S^{s_{J-1}} H . c^J \qquad (6.10) \\
&= D_0 G (D_0 H)^{J-j-1} S^{t_2} c^J.
\end{aligned}
$$

By applying the operator S^{t_1}, we obtain

$$
\begin{aligned}
S^{t_1} d^j &= S^{s_1} D_0 G (D_0 H)^{J-j-1} S^{t_2} c^J \\
&= D_0 G (D_0 H)^{J-j-1} S^r c^J.
\end{aligned}
$$

Hence, the sequence d^j shifted from s^1 is the jth sequence of the detail coefficients applied to the original sequence shifted from r. Moreover, the C^j's can be obtained by replacing $D_0 G (D_0 H)^{J-j-1}$ by $(D_0 H)^{J-j}$.

6.4.2 Wavelet stationary transform

The stationary wavelet transform is a modification of the discrete wavelet transform algorithm. Indeed, it consists of choosing the filters H and G in order to generate some new sequences with the same length as the original sequences and to eliminate the decimation obtained from the discrete wavelet transform algorithm. Let Y be an operator alternating a given sequence with zeros such that

$$(Yx)_{2j} = x_j \quad \text{and} \quad (Yx)_{2j+1} = 0, \quad \forall j.$$

Let $H^{[r]}$ and $G^{[r]}$ be the filters with respective weights $Z^r h$ and $Z^r g$. By inserting the zeros into each adjacent pairs of the filter $H^{[r-1]}$, we obtain the filter $H^{[r]}$ as follows

$$H^{[r]}_{2^r j} = h_j \quad \text{and} \quad h^{[r]}_k = 0.$$

In the case where k is not a multiple of 2^r, one proceeds similarly for the filter $G^{[r]}$. Since the filters $H^{[r]}$ and $G^{[r]}$ commute with S, one obtains

$$D_0 H^{[r]} = H D_0^r \quad \text{and} \quad D_0^r G^{[r]} = G D_0^r.$$

Let a^J be the original data sequence. The discrete wavelet transform algorithm is generated by the recursive relations

$$a^{j-1} = H^{[J-j]} a^j \quad \text{and} \quad b^{j-1} = G^{[J-j]} a^j.$$

6.5 WAVELET DENSITY ESTIMATION

In this part, we propose to review different types of wavelet density estimations such as nonparametric wavelet density estimators, linear estimators and generally conventional estimators. We will also show the differences especially in the representation of discontinuities and local oscillations. Properties as well as comparisons and multivariate extensions will be considered.

6.5.1 Orthogonal series for density estimation

For a realization $(x_1, x_2, ..., x_n)$ of a random variable X of unknown density f, an automatic approximation is the empirical density

$$f_\varepsilon(x) = \frac{1}{n} \sum_{j=1}^{n} \delta(x - x_i).$$

It is an unbiased estimator of f since $\mathbb{E}(f_\varepsilon(x)) = \delta * f = f$. However, for an absolutely continuous distribution, the estimator f_ε is not always the efficient choice as it constitutes indeed an irregular function. For this reason, different estimators are proposed in order to obtain a regular functional estimator.

Centsov proposed some types of density estimators known as orthogonal series density estimators or projection estimators (see [154]). The idea was based on the fact that a square integrable density could be represented by a convergent orthogonal series such as

$$f = \sum_j a_j \psi_j, \tag{6.11}$$

where $(\psi_j)_j$ is a complete orthogonal system in $L^2(\mathbb{R})$. The coefficients a_j are given by

$$a_j = <f, \psi_j> = \int_{\mathbb{R}} f(x)\psi_j(x)dx = \mathbb{E}(\psi_j(X)).$$

Given a sample $(x_1, x_2, ..., x_n)$, from an unknown distribution f, an empirical counterpart of a_j should be

$$\widehat{a}_j = \frac{1}{n} \sum_{i=1}^{n} \psi_j(x_i). \tag{6.12}$$

The density f will be estimated by

$$\widehat{f} = \sum_j \widehat{a}_j \psi_j. \tag{6.13}$$

The nomination as projection estimator comes from the derivative f_ε of the empirical distribution function,

$$F_\varepsilon(x) = \frac{1}{n} \sum_{i=1}^{n} H(x - x_i),$$

where H is the Heaviside function projected on the space $H_j = \overline{<\psi_j, j>}$. The estimator (6.13) is also called linear as it is a linear function of the associated empirical measure F_ε. We have indeed

$$\widehat{f} = \sum_j <\psi_j, \frac{dF_\varepsilon}{dx}> \psi_j.$$

For an infinite set of indices j, the estimator (6.13) cannot be well defined. The variance may be infinite and thus it cannot be consistent in the integrated quadratic error sense. Moreover, whenever the ψ_j are wavelet functions and thus the series is a wavelet one, the coefficients a_j satisfy

$$\lim_{j \to +\infty} \widehat{a}_j = 0.$$

In practice, it consists of selecting a high number of empirical coefficients $\widehat{a}_j$ and shrinking them appropriately. Indeed, it is shown that the wavelet density estimator can reach the optimal mean integrated squared error rate for multiple families of smooth densities (see [236], [280], [281]). To illustrate this procedure, we will discuss through the orthogonal wavelet bases the construction of non-parametric estimators of densities. A density f will be written in a wavelet representation form

$$f = \sum_{j,k \in \mathbb{Z}} d_{j,k}\psi_{j,k} = \sum_{k \in \mathbb{Z}} a_{j_0,k}\varphi_{j_0,k} + \sum_{j \geq j_0} \sum_{k \in \mathbb{Z}} d_{j,k}\psi_{j,k}, \tag{6.14}$$

where we have
$$d_{j,k} = \mathbb{E}(\psi_{j,k}(X)) \quad \text{and} \quad a_{j,k} = \mathbb{E}(\varphi_{j,k}(X)).$$

Therefore, the wavelet estimator of f will be obtained by truncation of its wavelet representation and replacing the coefficients $a_{j,k}$ and $d_{j,k}$ by the respective means

$$\widehat{a}_{j,k} = \frac{1}{n} \sum_{i=1}^{n} \varphi_{j,k}(x_i) \tag{6.15}$$

and

$$\widehat{d}_{j,k} = \frac{1}{n} \sum_{i=1}^{n} \psi_{j,k}(x_i). \tag{6.16}$$

However, we obtain here unbiased estimators. Moreover, these wavelet estimators are orthogonal series.

A general class of nonparametric density estimators has been introduced by Centsov ([154]). Different orthogonal systems in $L^2(\mathbb{R})$ have been applied to obtain orthogonal series estimators such as Fourier and Haar system in the interval $[0,1]$.

A major drawback that can be noticed in these cases is that the series constructed do not allow us to describe the local properties of the density. This is due basically to the fact that the orthogonal systems constructed are not well localized in time-frequency. Such a property is perfectly satisfied by wavelets which are characterized by good localizations in time and in frequency. This makes wavelet estimators more capable to detect local characteristics of the estimated densities. Compared to conventional estimators, the wavelet ones permit an asymptotic control of the convergence of estimators and to outpoint the statistical properties.

6.5.2 δ-series estimators of density

The class of δ-series estimators has been known since the 1970s as follows.

Definition 6.1 *Let $I \subset \mathbb{R}$ be an open interval. A sequence $(\delta_n)_n$ of bounded $\mathcal{C}^\infty$ functions on $I \times I$ is said to be a δ-sequence of I if for all $x \in D$ and for all $\mathcal{C}^\infty$-function φ defined on I, we have*

$$\lim_{m \to +\infty} \int_D \delta_m(x, y) \varphi(y) dy = \varphi(x).$$

The δ-series estimator of a density f based on a sample $(x_1, x_2, ..., x_n)$ is given by

$$\widehat{f}_{n,m} = \frac{1}{n} \sum_{i=1}^{n} \delta_m(X_i, x).$$

We may consider the standard histogram as an example of δ-series estimator of density. It is given by

$$\delta_m(x, y) = \frac{1}{m} \sum_{i=1}^{m} \chi_i(x) \chi_i(y),$$

where χ_i is the characteristic function of the interval $]\frac{i-1}{m}, \frac{i}{m}[$.

Otherwise, let $(Y_1, Y_2, ..., Y_m)$ be a vector of random variables with zero means and infinite variances. Let g_m be the density of the mean value random variable

$$\overline{Y} = \frac{1}{m} \sum_{i=1}^{m} Y_i.$$

The function $\delta_m(x, y) = g_m(x - y)$ defines a δ-density. We cite, for example, the gaussian

$$\delta_m(x, y) = \sqrt{\frac{m}{2\pi}} \exp\left(-\frac{m}{2}(x - y)^2\right).$$

Remark that the last sequence is positive, which guarantees the uniform convergence in Definition 6.1. Besides, the estimators are also positive. However, convergence, in terms of mean square integration and/or mean integrated square error, is not optimal. The convergence rate for the histogram is bounded by $n^{-4/5}$ and does not respect the regularity of the density.

6.5.3 Linear estimators

A first idea to build linear estimators is obtained by truncating at a level J the wavelet series of the density

$$f = \sum_k a_{J,k}\varphi_{J,k} + \sum_{l \geq J} \sum_k d_{l,k}\psi_{l,k}$$

to get

$$\widehat{f}_J = \sum_k \widehat{a}_{J,k}\varphi_{J,k}. \tag{6.17}$$

For an orthogonal multiresolution analysis $(V_j)_j$, this estimator is an orthogonal projection of f on the subspace V_j. The performance of the estimator (6.17) depends on the choice of the level J, which adjusts the exchange between the bias and the variance of the estimator. In the case of Haar wavelet, the estimator (6.17) is a usual histogram whose cells are centered at the dyadic points $k2^{-j}$, $k \in \mathbb{Z}$, with constant width 2^{-j}. Recall that, in this case, the scale function is $\varphi = \chi_{[0,1[}$. The coefficients $\widehat{a}_{j,k}$ will be given explicitly by

$$\widehat{a}_{j,k} = \frac{\sqrt{2^j}}{n} card\left(\left\{n; \ X_n \in [\frac{k}{2^j}, \frac{k+1}{2^j}[\right\}\right).$$

The wavelet estimator becomes

$$\widehat{f}_j(x) = \frac{2^j}{n} \sum_k N_k(x), \tag{6.18}$$

where $N_{j,k}$ is the number of sample's values in the interval $I_k = [\frac{k}{2^j}, \frac{k+1}{2^j}[$.

Linear wavelet estimators have been the subject of several studies such as [14],[15], [236],[280], [281],[359]. Convergence results of the mean square error for the wavelet estimators in the case of continuous densities have been proven. It has also been shown that the rate of convergence is fast for regular functions. These findings are not known for previous estimators of orthogonal series.

A wavelet estimator can be written in terms of a convolution kernel

$$\widehat{f}_j(x) = \frac{1}{n} \sum_{i=1}^{n} q_j(X_i, x), \tag{6.19}$$

where q_j is the kernel defined by

$$q_j(t, x) = \sum_k \varphi_{j,k}(t)\varphi_{j,k}(x).$$

It is shown that $(q_j)_j$ is a δ-sequence. The bias is expressed by $\mathbb{E}(\widehat{f}_j(X) - f(X))$. As a result, we obtain for the above estimator

$$\mathbb{E}(\widehat{f}_j(X) - f(X)) = \int_{\mathbb{R}} q_j(t, x)f(x)dx - f(X).$$

It is therefore a nonzero quantity and the estimator is biased. This is because in general the density f does not belong to V_j. However, uniformly (when f is defined and continuous on a compact interval) the estimator $\widehat{f}$ is asymptotically without bias. The quadratic error brought by this estimator satisfies

$$\mathbb{E}[\widehat{f}_j(X) - f(X)]^2 \leq \frac{1}{n} \int_{\mathbb{R}} q_j(t, x)f(x)dx \leq \frac{2^j}{n}.$$

This inequality is important in the classification of linear wavelet estimators in the δ-series sense.

Another similar class of wavelet estimators has been introduced by Walter et al. in [359] (see also [280]). The rate of convergence of the wavelet estimators for some densities in Besov spaces has been characterized.

Wu in [366] proved that we may obtain multi-scale wavelet estimators by convex combinations of linear estimators. Let Y^* and Z^* be disjoint subsamples of a whole sample with respective sizes N_1 and N_2. Let also M be a fixed integer. The M-linear multi-scale wavelet estimator is defined by

$$\widehat{f}(x, t) = \sum_{m=1}^{M} \lambda_m(y^*, t)\widehat{f}_{N_2}^{(m)}(z^*, t), \tag{6.20}$$

where $\lambda_m(y^*, t)$ are positive random variables satisfying

$$\sum_{m=1}^{M} \lambda_m(y^*, t) = 1.$$

$\widehat{f}_{N_2}^{(m)}(z^*, t)$ is the linear wavelet estimator associated with the z^* samples. It turns out that these estimators are unbiased and symptomatic and verify a property of normality.

6.5.4   Donoho estimator

In this part, we propose to introduce the thresholding estimators due to Donoho and his collaborators as well as their main properties. We will also discuss the role of wavelets in the adaptability of these estimators. The wavelet coefficients thresholding allows estimators for an automatic adaptation of local properties of densities.

Donoho et al in [203] proposed a non-linear thresholding estimator by truncating the wavelet series issued from both the approximation and the detail coefficients and (6.16). The method consists of considering the estimator

$$\widehat{f} = \sum_k \widehat{a}_{j,k}\varphi_{j,k} + \sum_{l \geq j}\sum_k \widetilde{d}_{l,k}\psi_{l,k}, \tag{6.21}$$

where $\widetilde{d}_{l,k}$ is the threshold version of $\widehat{d}_{l,k}$. This estimator can be seen as a coarse approximation of f. The linear part of the estimator describes the low frequency. The non-linear part is adapted to local fluctuations such as discontinuities and high-frequency oscillations of the density. The goal is to exclude from the series the terms that are poorly estimated in a feasible way giving a large number of $\widehat{d}_{l,k}$ coefficients and keep a reduced number of coefficients taking into account the discontinuity.

There are two types of thresholding. The first type is called hard thresholding in which we consider for a threshold parameter $\delta > 0$,

$$\widetilde{d}_{j,k} = \widehat{d}_{j,k}\chi_{\{|\widehat{d}_{j,k}|>\delta\}},$$

which means that we put zero as all the coefficients absolutely non-crossing the threshold δ.

The second type is called soft thresholding and is based on the choice

$$\widetilde{d}_{j,k} = sign\left(\widehat{d}_{j,k}\right)\left(|\widehat{d}_{j,k}| - \delta\right)_+,$$

where

$$\left(|\widehat{d}_{j,k}| - \delta\right)_+ = \begin{cases} |\widehat{d}_{j,k}| - \delta & if \quad |\widehat{d}_{j,k}| \geq \delta \\ 0 & , \quad else \end{cases}$$

The exchange between bias and variance is adjusted by the parameters j and δ. For more details on these types, the readers may refer to [199], [200], [201], [202] and [203].

6.5.5   Hall-Patil estimator

Hall and Patil in [233],[234] proposed a class of wavelet thresholding estimators by the use of a second parameter of smoothing. Instead of the scale function φ and the wavelet ψ, they considered some modified versions

$$\varphi^p_{j,k}(x) = p^j\varphi(p^{2j}x - k) \qquad and \qquad \psi^p_{j,k}(x) = p^j\psi(p^{2j}x - k), \tag{6.22}$$

where $p > 0$ to get an orthogonal family that leads to a generalized formulation in Fourier series. The proposed wavelet thresholding estimator is given by

$$\widehat{f} = \sum_k \widehat{a}_{j,k}\varphi^p_{j,k} + \sum_{l=0}^{q-1}\sum_k \widehat{d}_{j,k}\omega\left(\frac{\widehat{d}_{j,k}}{\lambda}\right)\psi^p_{j,k}, \tag{6.23}$$

where the coefficients $\widehat{a}_{j,k}$ and $\widehat{d}_{j,k}$ are

$$\widehat{a}_{j,k} = \frac{1}{n}\sum_{i=1}^{n}\varphi_{j,k}^{p}(x_i) \qquad \text{and} \qquad \widehat{d}_{j,k} = \frac{1}{n}\sum_{i=1}^{n}\psi_{j,k}^{p}(x_i).$$

The integer parameter q gives the order of truncation. The parameter λ defines the threshold. The function ω is positive and known as the weight and is strongly related to threshold. A simple example is $\omega = \chi_{[1,+\infty]}$, which allows obtaining a hard threshold. The thresholding wavelet estimators proposed by Donoho et al may be obtained by taking $p = \sqrt{2}$.

Hall and Patil proved that a theoretical advantage of more flexibility in choosing the parameter p is that it allows to avoid log factors that may affect convergence. They noticed the analogy of the linear part estimators proposed with kernel estimators and they have developed a formula for variance, bias and mean square error. An asymptotic formula for the mean integrated quadratic error is provided for the wavelet hresholding estimators (6.21).

Unlike kernel methods, wavelet estimators have an excellent rate of convergence when the unknown density is only piecewise continuous. Wavelet estimators have been shown to be also extremely robust against oversmoothing. We notice finally that wavelet estimators are essentially consistent if the number of jumps multiplied by the jump size is smaller than the size of the sample.

6.5.6 Positive density estimators

With the exception of Haar-based estimators, these later may take negative values in the distribution tails or in the missing data parts of the density support. This constitutes a persistent problem in non-parametric estimations.

One solution consists of truncating the estimator in its positive part and then re-normalizing the truncated estimator. Otherwise, we may estimate some transform of the density f such as $\log(f)$ or $\sqrt{f}$ and then return to f.

The use of the transform $\sqrt{f}$ was introduced by Good and Gaskins in 1971 in the context of finalization of the maximum likelihood estimator to satisfy the constraint of non-negativity. This approach presents theoretical problems related to the multiplicity of solutions. The transformation $\log(f)$ is used in a Bayesian approach by Silverman and others.

Linear and thresholding estimators are expressed by

$$\widehat{a}_{j,k} = \frac{1}{n}\sum_{i=1}^{n}\frac{\varphi_{j,k}(x_i)}{\sqrt{\widehat{f}_n(x_i)}} \qquad \text{and} \qquad \widehat{d}_{j,k} = \frac{1}{n}\sum_{i=1}^{n}\frac{\psi_{j,k}(X_i)}{\sqrt{\widehat{f}_n(X_i)}}, \tag{6.24}$$

where $\widehat{f}_n$ is a suitable estimator of f. As an example, we may choose as in [353]

$$\widehat{f}_n(x_i) = card\left(\{x_j;\ x_j \in]x_i - r, x_i + r[\}\right)$$

or in the multivariate case

$$\widehat{f}_n(x_i) = card\left(\left\{x_j;\ x_j \in]x_i - r, x_i + r[^{d}\right\}\right).$$

We may also consider

$$\widehat{f}_n(x_{n,i}) = \frac{C}{n(x_{n,i+1} - x_{n,i})},$$

where $x_{n,1} \le x_{n,2} \le \dots \le x_{n,n}$ are ranked in some statistical order and where C is a suitable constant.

6.6 WAVELET THRESHOLDING

When applying estimators of orthogonal series, a crucial problem is the choice of the truncation used. This is the number of terms included in the estimator. Truncation is essential not only in the practice but also for theoretical reasons. Indeed, the estimator is not generally well defined for an infinite indices as in this case the quadratic error $\|\widehat{f} - f\|_2$ may not converge to zero. Standard practices are to lengthen the estimator by multiplying the empirical coefficients with symmetrical weights, which return them towards the origin, or more simply to select a finite number of coefficients, which results, however, generally to a minimal bias. Perhaps, a large variance may also be achieved by introducing more terms. In general, a limited number of terms can lead to smooth estimators, losing important features of the density, while large estimators of a number mistreated may contain false artifacts. As an extreme case, the estimator may be close to the sum of delta functions centered on observations.

For wavelet estimators, a number of choices of parameters is required. For example, the linear estimator defined in (6.17) depends only on the truncation and therefore on the level of decomposition j. Donoho thresholding estimators require the choice of a lower scale j_0 and a higher scale j_1. Hall-Patil estimator (6.21) depends on the truncation parameter q and the regularity parameter p, which is the bandwidth equivalent in the kernel estimate. For the threshold estimators, the choice of the threshold is essential.

Generally, it turns out that the problem of choosing the scaling parameters is more important in the linear case when using Hall-Patil estimators and thresholding estimators.

6.6.1 Linear case

Walter in [359] proposed the choice of the set of the truncation level for the estimator (6.17) to be fixed by minimizing the square of error

$$\mathbb{E}\left(\|\widehat{f} - f\|_2 \right)^2 = \mathbb{E}\left(\|\widehat{f} - \mathbb{E}(\widehat{f})\|_2 \right)^2 + \mathbb{E}\left(\|\mathbb{E}(\widehat{f}) - f\|_2 \right)^2,$$

where $\mathbb{E}\left(\|\widehat{f} - \mathbb{E}(\widehat{f})\|_2 \right)^2$ is the stochastic error due to random observations and $\mathbb{E}\left(\|\mathbb{E}(\widehat{f}) - f\|_2 \right)^2$ is the bias error of a method used. Let e_j be the mean integrated square error at the level j. We have

$$e_j - e_{j-1} = \frac{1}{n} \int 2^j q(2^j x, 2^j x) f(x) dx - \frac{n+1}{n} \sum_k d_{j-1,k}^2. \qquad (6.25)$$

A possible estimate of the above integral is given by

$$\int 2^j q(2^j x, 2^j x) f(x) f(x) dx \sim \sum_{i=1}^{n} 2^j q(2^j x_i, 2^j x_i).$$

In the case of Haar wavelet, for example, we get

$$e_j - e_{j-1} = \frac{2^{j-1}}{n} - \frac{n+1}{n} \sum_k d_{j-1,k}^2. \tag{6.26}$$

Walter suggested for a given scale, which should be chosen not finer than that required to put each observation in a different interval, application of the empirical wavelet coefficients for estimating increments for the next grosser balance until the error increases by a large amount. The last scale reached will be the scale of interest. Other possible procedures have been also proposed in [359] for the special case of Haar wavelet estimators.

Scott in [338] has shown that the histogram estimators are consistent relatively to the mean integrated square error. Moreover, such error may be minimized asymptotically by the choice of an optimal width. The optimal bandwidth depends on the norm of f; therefore, the approximate values will be chosen by practical proposals. Scott noticed that the bandwidth approximation is $h \sim 3.5 \frac{s}{\sqrt[3]{n}}$, where s is the standard deviation of the sample, which is suitable for gaussian samples and leads to over-smoothing. Thus, when using Haar linear estimator, a starting point is to reach the scale of interest, which will be

$$j = \frac{1}{3} \frac{\log n}{\log 2} - \frac{\log \sigma}{\log 2} - 2,$$

where σ is a parameter to be estimated from the standard deviation of the sample.

Vannuci method ([352, 353]) and Vidakovic ([355]) proposed to apply a functional transform known as Fisher transform, defined by

$$F(f) = \int_{\mathbb{R}} \frac{(f'(x))^2}{f(x)} dx. \tag{6.27}$$

This function is applied to measure the rigidity of a function by discriminating it by itself when it is subject to a small translation. Fisher transform has been introduced since the 1970s in the context of maximum likelihood estimation. Next, by using a wavelet representation of the derivative from the linear estimator, an estimate of Fisher transform is obtained as

$$\widehat{f}(x) = \left(\sum_k \widehat{a}_{J,k} \varphi_{J,k}(x) \right)^2. \tag{6.28}$$

This estimate depends on the parameter of the truncation J and it provides rigidity in the measurement of coefficients of the wavelet estimator. The authors in ([352, 353, 355]) also provided rigorous optimal value for compactly supported densities and thus have proposed the choice of the value J to be such that the wavelet estimator yields an estimated Fisher transform close to its optimum.

6.6.2 General case

Consider the threshold wavelet estimators due to Hall-Patil (6.23). The parameters j_0, j_1 and δ need to be chosen. Donoho et al [199, 200, 201, 202, 203] have proposed a choice of j_0 corresponding to the regularity of the function f. In addition, they proposed a scale that depends on the threshold value $\delta_j = C\sqrt{\dfrac{j}{n}}$. Delyon and Juditky in [188, 189], when studying the wavelet threshold estimators (6.23), proposed the choice $j_1 = [\log_2(n) - \log_2(\log(n))]$.

In the same context of the scaling parameter choice, we can consider the wavelet scalogram for the density f as

$$S_f(j) = \sum_k |< f, \psi_{j,k} >|^2 , \tag{6.29}$$

which describes the distribution of the total energy of f to a scale j. It has been shown that the scalogram behaves well for small values of j and decreases in exponential form for j large enough. Therefore, whenever the sample is contained in an interval $[a, b]$, by choosing the well-known Lorentz curve relative to the energy of empirical wavelet coefficients as smoothing parameter, a major part of the energy of the density is shown to be concentrated in a reduced number of large coefficients. Hall and Patil [232, 233, 234] provided necessary and sufficient conditions for the threshold form that the optimal rate should reach in the case of continuous or piecewise smooth functions.

We provide hereafter a simple example for the choice of the parameters p and q. Let $p = p_j = 2^j$ and q be such that

$$q_0 - C_1 \leq q \leq C_2 \log_2(n)$$

with

$$\frac{1}{2r+1} \leq C_1, C_2 \leq 1 \quad \text{and} \quad q_0 = \left[\frac{\log(n)}{(2r+1)\log(2)} \right].$$

The real number r designates the maximum regularity of the density f. Three possibilities may be applied

- $\delta_j = 0$ for $0 \leq j \leq q_1$ and $\delta_j = C\sqrt{\dfrac{j - q_1}{n}}$ for $j > q_1$.

- $\delta_j = 0$ for $0 \leq j \leq q_1$ and $\delta_j = C\sqrt{\dfrac{\log(n)}{n}}$ for $j > q_1$.

- $\delta_j = C\sqrt{\log(n)}$ for all j.

The first two rules produce uniform rates of convergence of the optimal mean square error for all classes of densities, which are r-times differentiable. However, the third one produces a smoother estimator with quadratic bias participation in the increase of the mean integrated square error compared to the variance. In the general case $p_j = p^{2j}$, both the first two thresholding rules are kept but with a constant $C > \sqrt{2\log 2 \max(f)}$. Parameters that allow smoothing are C and p_{q_1}, which ensure the exchange between the quadratic bias and the variance for the first rule, and only p_{q_1} in the second.

6.6.3 Local thresholding

In general, the threshold value that guarantees a behavior that is asymptotically optimal is given by

$$\delta = C\sqrt{\frac{\log(n)}{n}}.$$

(See [236].) Besides, other types of local thresholds have been also proposed where the density function depends on the scale or the level j. These thresholding procedures are obtained for (see [201])

$$\delta = \delta_j = C\sqrt{\frac{j - j_0}{n}} \qquad \text{and} \qquad \delta = \delta_j = C\sqrt{\frac{j}{n}}.$$

A third local thresholding subclass corresponds to a choice depending simultaneously on both j and k and is expressed as

$$\delta = \delta_{j,k} = \sqrt{2\sigma_{j,k}^2[\psi]\log(M_j)},$$

where

- $\sigma_{j,k}^2[\psi]$ denotes the variance of the empirical wavelet coefficients $\widehat{d}_{j,k}$.

- M_j is the number of non-zero coefficients at the scale or the level j.

We notice here that the variance is in general unknown and it is therefore replaced by its empirical counterpart. This leads to a random threshold.

6.6.4 Global thresholding

Global thresholding procedures affect the whole grid of detail coefficients of a level j as

$$\widehat{f}(x) = \sum_k \widehat{a}_{j_0,k}\varphi_{j_0,k}(x) + \sum_{j=j_0}^{j_1} \eta_j\left(\sum_k \widehat{d}_{j,k}\psi_{j,k}(x)\right), \qquad (6.30)$$

where η_j is a nonlinear transformation. In [281], the authors considered the global thresholding technique to estimate the probability densities. They proposed the model η_j as follows.

- $\eta_j(u) = \eta_j^H(u) = u\chi_{\{S_j(u)>2^j n^{-p/2}\}}$ for a hard global thresholding.

- $\eta_j(u) = \eta_j^H(u) = u\dfrac{S_j(u) - 2^j n^{-p/2}}{S_j(u)}$ for a soft global thresholding.

$S_j(u)$ is a statistic that depends on u and the sample $x_1, x_2, ..., x_n$. For example, in the case where u is an even integer smaller than the size n of the sample, $S_j(u)$ is given by

$$S_j(u) = \frac{1}{C_n^u} \sum_{i_1 \neq i_2 \neq ... \neq i_p} \sum_k \psi_{j,k}(x_{i_1})\psi_{j,k}(x_{i_2})...\psi_{j,k}(x_{i_n}).$$

Compared to the local estimator technique, to determine a global thresholding estimator, we focus only on empirical observations. Notice also that whenever u or the size n of the sample increases, the S_j statistic requires more and more computational efforts.

6.6.5   Block thresholding

Block thresholding is an intermediate procedure between global thresholding and local thresholding. This technique recommends to retain or delete blocks of wavelet coefficients specifically chosen at each level j. The block thresholding procedure has been initially introduced in [232] as follows.

- We divide the whole grid of indices into different blocks of length $l = l(u)$ such that
$$B_k = \{m; \ (k-1)l + 1 \le m \le kl\}.$$

- Next we put $b_{j,k} = \dfrac{1}{l} \displaystyle\sum_{m \in B_k} d_{j,k}^2.$

- We consider the estimator of $b_{j,k}$ defined by
$$\widehat{b}_{j,k} = \frac{1}{l} \sum_{m \in B_k} \widehat{d}_{j,k}^2.$$

- The block threshold estimator of f will be

$$\widehat{f}(x) = \sum_{k} \widehat{a}_{j_0,k} \varphi_{j_0,k}(x) + \sum_{j=j_0}^{j_1} \sum_{k} \left(\sum_{m \in B_k} \widehat{d}_{j,k} \psi_{j,m}(x) \right) \chi_{\widehat{b}_{j,k} > cn^{-1}}, \qquad (6.31)$$

where c is an appropriately chosen constant.

The authors in [232] have shown that the asymptotic properties of the block threshold estimator are better than those of local thresholding techniques. Unlike the global threshold estimator that we can extract from the observations $x_1, x_2, ..., x_n$, the block threshold estimator also depends on the choice of the constant c, which is not explicated theoretically.

6.6.6   Sequences thresholding

In order to proceed to the study of the properties of wavelet estimators, an interesting model has been discussed and shown to be statistically performant and is known as sequence space model (SSM).

Consider the estimators $\widehat{a}_{j_0,k}$ and $\widehat{d}_{j,k}$ defined in (6.15) and (6.16). We write

$$\widehat{a}_{j_0,k} = a_{j_0,k} + \sigma_{j_0,k}(\varphi)\zeta_{j_0,k} \qquad \text{and} \qquad \widehat{d}_{j,k} = d_{j,k} + \sigma_{j,k}(\psi)\xi_{j,k}, \qquad (6.32)$$

where

- $\widehat{a}_{j_0,k}$ and $\widehat{d}_{j,k}$ are the real wavelet coefficients of the function f.

- $\zeta_{j_0,k}$ and $\xi_{j,k}$ are random variables with zero expectation and variance 1.

- $\sigma_{j_0,k}(\varphi)$ and $\sigma_{j,k}(\psi)$ denote the scale factors.

To get a finite number of coefficients, compactly supported φ and ψ are considered. Let M be the number of non-zero coefficients $\widehat{d}_{j,k}$. Without loss of the generality, we may assume further that $1 \leq k \leq M$. The $\xi_{j,k}$ are asymptotically gaussian and are approximately uncorrelated. An asymptotic approximation of the SSM is

$$Z_k = \theta_k + \sigma_k \xi_k, \tag{6.33}$$

where

- The Z_k plays the role of $\widehat{d}_{j,k}$.

- The θ_k are unknown parameters.

- The ξ_k are random variables i.i.d. with normal central reduced distribution.

This is to estimate the unknown vector θ given the vector of gaussian observations Z. For gaussian white noise, the errors $\xi_{j,k}$ follow the normal central reduced distribution and $\sigma_{j,k}(\psi) = \varepsilon$. Thus, the SSM would be in the form

$$Z_k = \theta_k + \varepsilon \xi_k \qquad \text{with} \quad \xi_k \sim \mathcal{N}(0,1).$$

6.7 APPLICATION TO WAVELET DENSITY ESTIMATIONS

In this section, we propose to conduct some illustrative examples for wavelet estimators of densities. We will consider the most used densities especially in financial applications such as normal law, Kurtotic law and Claw. We first discuss some practical problems. The first is related to the fixation of the translation parameter k especially that some densities are not compactly supported. Such parameter permits to guarantee a finite number of wavelet coefficients for holding an estimator.

Consider, for example, the Daubechies wavelets DbN, $N \in \mathbb{N}$. These are compactly supported with support of the interval $[0, 2N - 1]$ for the scale function φ. On an interval $[a, b]$, the sample size is estimated by the values of the parameter k such that

$$[a2^j] - 2N + 1 \leq k \leq [b2^j].$$

Similarly, the support of the DbN mother wavelet ψ is $[1 - N, N]$, which yields for k and $[a, b]$ the estimations

$$[a2^j] - N \leq k \leq [b2^j] + N - 1.$$

The data applied for the graphical illustrations are simulated observations of mixed density laws using the $\boldsymbol{R}$ software. One of the causes and/or motivation for this choice is the fact that mixed densities are relatively complex compared to conventional densities. The mixed densities actually present discontinuities, which justify the application of wavelets in their estimates.

In practice, the choice of the mother/father wavelet is of great importance to build estimators. Indeed, each wavelet and each density present have a degree of regularity well defined. A suitable choice seems to apply an analyzing wavelet with a regularity order that is large and close to that of the density to be estimated.

Besides, it is always noticed that wavelets allow very efficient approximation despite the large differences of the regularity orders. This is one of the advantages of wavelet analysis and one of the strong causes of its success. We point out at the end that the simulations are using Matlab software.

6.7.1 Gaussian law estimation

In this section, we start by applying wavelet estimators to the most used and the most famous density consisting of the well-known gaussian density. We consider the standard normal reduced and centered one, denoted usually by $\mathcal{N}(0, 1)$. Recall that the density f is expressed as

$$f(x) = \frac{1}{\sqrt{2\pi}} \exp\left(-\frac{x^2}{2}\right).$$

Figure 6.1 illustrates such density.

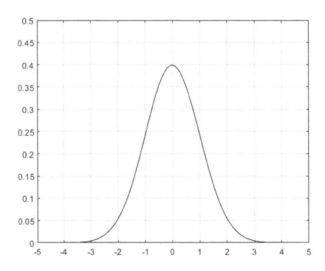

Figure 6.1 The normal reduced and centered density $\mathcal{N}(0, 1)$.

We used a sample of 1024 size. A soft thresholding was tested at resolution level $J = 5$. The method gave an optimal threshold value 0.27. We thus provide next the wavelet estimator for the gaussian law at the same level $J = 5$. Figure 6.2 illustrates the wavelet estimator and its closeness to the original density.

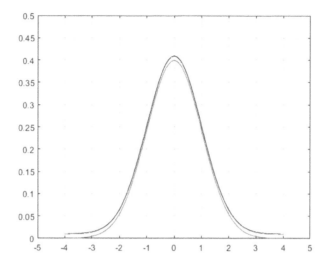

Figure 6.2 The gaussian density and its wavelet estimator.

6.7.2   Claw density wavelet estimators

Here, a sample of 4096 randomly distributed observations has been generated to construct a wavelet estimator for the Claw density. Recall that the Claw density is a mixture of normal-gaussian densities as

$$Claw = 0.5\mathcal{N}(0,1) + 0.1\sum_{i=0}^{4}\mathcal{N}(\frac{i}{2}-1, 0.1).$$

Daubechies $Db4$ wavelet has been applied at the resolution level $J = 4$. An optimal threshold value equal to 0.16 has been obtained. Figures 6.3 and 6.4 illustrate graphically the findings.

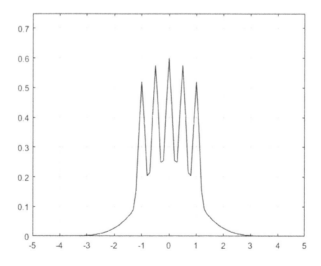

Figure 6.3 Claw density.

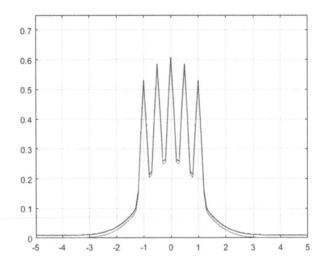

Figure 6.4 Claw density wavelet estimator at the level $J = 4$.

6.8 EXERCISES

Exercise 1.

Prove that

a. $D_1 = D_0 S$.

b. $R_1 = S^{-1} R_0$.

c. $SD_0 = D_0 S^2$.

d. $SH = HS$.

e. $SG = GS$.

Exercise 2.

Let X_t be a time series with size N. Let for j, k, $A_{j,k}$ be the k-th element of the j-level smooth approximation A_j of X_t. Show that

$$\frac{1}{N} \sum_{k=0}^{N-1} \left(A_{j,k} - \overline{A_j} \right)^2 \leq \frac{1}{N} \|A_j\|^2 - \overline{X}^2.$$

Exercise 3.

Let X_t be a time series with $\overline{X} = 0$ and $W = DWTX$ its discrete wavelet transform. Is it true that $\overline{W} = 0$?

Exercise 4.

Let for $T > 0$, $h : \mathbb{R} \to \mathbb{R}$ be such that $h(t) = 0$, $\forall t \notin [0, T]$. Denote next

$$H(t) = \sum_{n \in \mathbb{Z}} h(t - nT).$$

1) Compute the Fourier transform of H.
2) Let now g be a window function and define the windowed transform of H relatively to g by

$$SH(x, \xi) = \int_{\mathbb{R}} H(t)g(t - u)e^{-i\xi t}dt.$$

a) Express SH by means of $\widehat{H}$ and $\widehat{g}$.
b) Estimate the maximal size of the support of $\widehat{g}$ in such a way SH presents bands.

Exercise 5.

Consider the same assumptions as in Exercise 4 and let now g be the gaussian

$$g(x) = \frac{1}{\sigma\sqrt{2\pi}} \exp\left(-\frac{x^2}{2\sigma^2}\right).$$

1) Compute the Fourier transform of g.
2) What would be the best order for σ?
3) What should be the characteristic size of the support of $\widehat{g}$?

Exercise 6.

Consider the same assumptions as in Exercise 4 and let ψ be an analyzing wavelet.
1) Compute the wavelet transform $W_{\mathcal{L}}$ of $\mathcal{L}(x) = \exp(i\alpha x)$, $\alpha > 0$ fixed.
2) Compute the wavelet transform W_H of H.
3) Assume for the next that $\widehat{\psi}$ is negligible outside an interval $[1 - a, 1 + a]$ for $0 < a < 1$ small enough. Show that we may observe bands on the graph of $|W_f|$.
4) Estimate the number of bands by means of a.

Exercise 7.

Let $f \in L2(\mathbb{R})$ and denote $\sigma = \|f\|_2$. Denote also

$$u = \frac{1}{\sigma^2} \int_{\mathbb{R}} t|f(t)|^2 dt \quad \text{and} \quad \xi = \frac{1}{2\pi\sigma^2} \int_{\mathbb{R}} w|\widehat{f}(w)|^2 dw.$$

$$\sigma_t^2 = \frac{1}{\sigma^2} \int_{\mathbb{R}} (t - u)^2|f(t)|^2 dt \quad \text{and} \quad \sigma_\xi^2 = \frac{1}{2\pi\sigma^2} \int_{\mathbb{R}} (w - \xi)^2|\widehat{f}(w)|^2 dw.$$

1) Prove that $4\sigma_t^2\sigma_\xi^2 \geq 1$.
2) For any $A > 0$ provide f such that $\sigma_t > A$ and $\sigma_\xi > A$.

Exercise 8.

Consider the statistical series

$$S = [2\ 4\ 8\ 12\ 14\ 0\ 2\ 1],$$

corresponding to the initial data with the Haar wavelet transform. Let also H and G be the Haar low and high pass filters.

1) Show that the application of Haar wavelet transform on S once yields the series $S_1 = [3\ 10\ 7\ 1.5]$ as approximation coefficients and $D_1 = [-1\ -2\ 7\ 0.5]$ as details at the level 1.

2) Express S_1 and D_1 by means of S, H and G.

3) Deduce the decomposition of S at the level 3.

Exercise 9.

1) Recall the construction of Haar scaling and wavelet functions on $\mathbb{R}^2$.

2) Provide the corresponding filters H_2 and G_2.

3) Consider next the image

$$I = \begin{bmatrix} 0 & 250 & 25 & 50 & 200 & 0 & 0 & 0 \\ 50 & 50 & 50 & 25 & 50 & 0 & 25 & 0 \\ 25 & 50 & 0 & 250 & 0 & 50 & 0 & 0 \\ 75 & 200 & 200 & 0 & 250 & 250 & 0 & 200 \\ 250 & 25 & 250 & 200 & 0 & 75 & 25 & 25 \\ 50 & 0 & 50 & 0 & 0 & 75 & 250 & 250 \\ 250 & 250 & 25 & 250 & 50 & 25 & 50 & 50 \\ 250 & 200 & 50 & 50 & 50 & 25 & 0 & 200 \end{bmatrix}.$$

Provide a decomposition at the level 5 of the image I.

4) Apply soft and hard thresholding procedures on the decomposition.

5) Compute the errors due to the three decompositions.

Exercise 10.

Show that a mixture of hard- and soft-thresholding rules

$$\widehat{\theta} = (|d| - \lambda)_+ 1_{(|d| \leq 1.5\lambda)} + |d| 1_{(|d| > 1.5\lambda)}$$

is a minimizing solution to the variational problem

$$\frac{1}{2} \sum_{i=1}^{n} (d_i - \theta_i)^2 + \lambda \sum_{i=1}^{n} \min(|\theta_i|, \lambda).$$

CHAPTER 7

Wavelets for partial differential equations

7.1 INTRODUCTION

One of the most important and known mathematical models used to describe nature are partial differential equations (PDEs). Many methods and tools have been developed to resolve and/or to understand PDEs in both theory and applied fields. In this chapter, we propose to discuss the contribution of wavelet theory in such a topic and to re-develop some wavelet-based methods in the solutions of PDEs.

The idea is generally not complicated and it consists essentially of developing the unknown solution of the PDE into its eventual wavelet series decomposition (and its derivatives included in the PDE), and next using the concept of wavelet basis, we generally obtain algebraic and/or matrices/vector equations on the wavelet coefficients that should be resolved.

As we know, the wavelet series of a function is either infinite or finite approximation at a given level. To introduce in an easy way the use of wavelets in PDEs, let us discuss it for the simple real case. Let $(V_j)_j$ be a multiresolution analysis of $L^2(\mathbb{R})$ with its pair (φ, ψ) of scaling and wavelet functions and consider the simple first-order differential equation

$$y' + y = f(x), \tag{7.1}$$

where, as usual, y is the unknown function assumed to be continuously differentiable and f is a continuous function of the one real variable x. The idea in the majority of studies dealing with wavelet solutions of PDE consists of, as for the other functional basis, considering a formal wavelet series decomposition of y as

$$y(x) = \sum a_k(y)\varphi_k(x) + \sum d_{jk}(y)\psi_{jk}(x), \tag{7.2}$$

which by differentiation yields

$$y'(x) = \sum a_k(y)\varphi'_k(x) + \sum d_{jk}(y)\psi'_{jk}(x). \tag{7.3}$$

The first problem arising in the last equation (7.3) is about the differentiability of the scaling and the wavelet functions φ and ψ such as the oldest examples of Haar.

However, we may overcome such problem by considering and/or applying somehow more regular wavelets. A second problem is the fact that the derivative φ' of the scaling function φ is not a scaling function; similarly, the derivative ψ' of the wavelet ψ is not a scaling function. The next step thus consists of developing these derivatives themselves into wavelet series such as

$$\varphi'_k(x) = \sum_l \Gamma^l_k \varphi_l + \sum_{i,l} \Gamma^{il}_k \psi_{il} \tag{7.4}$$

for the derivative of φ and

$$\psi'_{jk}(x) = \sum_l \widetilde{\Gamma}^l_{jk} \varphi_l + \sum_{i,l} \widetilde{\Gamma}^{il}_{jk} \psi_{il} \tag{7.5}$$

for the derivative of ψ.

Next, by substituting (7.4) and (7.5) in (7.3) and by developing the right-hand term f in (7.3) by means of its wavelet series, we obtain generally an algebraic equation of the form $AY = F$, where A is a matrix depending on the coefficients Γ^l_{jk}, Γ^{il}_{jk}, $\widetilde{\Gamma}^l_{jk}$ and $\widetilde{\Gamma}^{il}_{jk}$.

Another way that has been applied in solving PDEs using wavelets was introduced by Cattani et al [368], which avoids the problem of being highly smooth for the wavelet. Instead of developing the unknown function y, we develop its higher derivative included in the PDEs. For the example (7.1), we put

$$y'(x) = \sum \alpha_k(y)\varphi_k(x) + \sum \beta_{jk}(y)\psi_{jk}(x) \tag{7.6}$$

and integrate along the interval of x. This avoids the problem of the derivatives of the scaling and wavelet functions.

In this chapter, we propose to discuss with some details the use of wavelets in approximating solutions of PDEs. This is based on some adapted multiresolution analyses on the intervals that are the most suitable to approximate the solutions of PDEs and more suitable to compute the so-called connection coefficients.

Indeed, a compactly supported wavelet, for example, consists of a compactly supported scaling function φ, which induces the wavelet function ψ to be also compactly supported. In this case, the sequence $\{h_k = \sqrt{2} \int_{\mathbb{R}} \varphi(x)\varphi(2x-k)dx\}_{k \in \mathbb{N}}$ contains only a finite number of non-zero terms, and the corresponding wavelet ψ is a finite linear combination of compactly supported functions. Compactly supported wavelets have been the object of many studies such as B-splines wavelets, which are explicitly expressed. We recall hereafter some example due to Daubechies. It was based on a scale function φ supported on $[0, 2N - 1]$, where $N \in \mathbb{N}^*$. The two-scale relation will be

$$\varphi(x) = \sum_{k=0}^{2N-1} h_k \varphi(2x - k).$$

The h_k's are called the filter coefficients of the multi-scale analysis. The associated wavelet will be

$$\psi(x) = \sum_{k=0}^{2N-1} (-1)^k h_{2N-1-k} \varphi(2x - k).$$

For more details on Daubechies wavelets and their filter coefficients values, the readers may refer to [177].

7.2 WAVELET COLLOCATION METHOD

To illustrate the method easily, we will proceed by applying it on an evolutive problem of the form

$$u_t + u_{xx} = f(u) \tag{7.7}$$

on $\Omega_T = [0, T] \times [0, 1]$, where $u = u(t, x)$, u_t is the first derivative of u relatively to the time variable t, u_{xx} is the second derivative of u relatively to the space variable x and f is a generally nonlinear suitable one real variable function.

Let $(e_1, e_2, ..., e_L)$ be the basis of $V_J^{[0,1]}$ defined in the previous section. Here, $L = 2^j + 1$ designs the dimension of $V_J^{[0,1]}$. The method consists in finding a solution u of the form

$$u(x, t) = \sum_{k \in I_J} u_k(t) \bar{\varphi}_{J,k}(x) = \sum_{k=1}^{L} C_k(t) e_k(x).$$

We next apply a finite difference discretization in time by the following representation

$$\sum_{k=1}^{L} \frac{C_k^{n+1} - C_k^n}{\Delta t} e_k(x) = \sum_{k=1}^{L} \frac{C_k^{n+1} + C_k^n}{2} e_k''(x) + f(u^n(x)).$$

Next we consider a subdivision $\{x_i\}_{i \in I}$ of $[0, 1]$ composed of dyadic grid where the I is an index set with $card(I) = dim V_J^{[0,1]}$. The above equality yields, for all $i \in I$,

$$\sum_{k=1}^{L} C_k^{n+1} \left(e_k(x_i) - \frac{\Delta t}{2} e_k''(x_i) \right) = \sum_{k=1}^{L} C_k^n \left(e_k(x_i) + \frac{\Delta t}{2} e_k''(x_i) \right) + \Delta t f(u^n(x_i)),$$

where $u^n(x_i) = u(x_i, n\Delta t)$. The last equality can be expressed in a matrix form

$$A(\Delta t) C^{n+1} = A(-\Delta t) C^n + \Delta t F^n, \tag{7.8}$$

where $A_{ij}(\Delta t) = e_j(x_i) - \frac{\Delta t}{2} e_j''(x_i)$, C^n and F^n are the following vectors

$$C^n = {}^t (C_1^n, ..., C_L^n) \text{ and } F^n = {}^t (f(u^n(x_1)), ..., f(u^n(x_L))).$$

A simplified algorithm to compute the solution is the following.

Algorithm 1: A simple wavelet collocation algorithm.

- $u_0 \longmapsto P_J(u_0) \in V_J^{[0,1]}$
- Stock the sequence of wavelet coefficients of u_0; $\left(C_k^0 \right)_k$

$$\Longrightarrow u_0(x_i) = \sum_{k=0}^{2^j} C_k^0 \bar{\varphi}_{J,k}(x_i).$$

- For $0 \leq i \leq 2^j$, evaluate $\bar{\varphi}_{J,k}(x_i)$ and $\bar{\varphi}_{J,k}''(x_i)$.
- Resolve the system (7.8)

7.3 WAVELET GALERKIN APPROACH

To illustrate the Galerkin approach, here also we will apply it on a simple form of one-dimensional equation on the interval $[0, 1]$. Thus, we deal with a method of resolution of some partial differential equations of the form

$$\mathcal{L}u = f \quad \text{with} \quad u(0), \ u(1) \text{ given,} \tag{7.9}$$

where $\mathcal{L}$ is a differential operator and f some given function on $[0, 1]$. For $j \in Z$ fixed, the equation above is equivalent in the approximation space V_j to the following integral equality

$$\int_0^1 \mathcal{L}u(x)\varphi_{j,k}(x)dx = \int_0^1 f(x)\varphi_{j,k}(x)dx.$$

In all that follows, we set

$$f_{j,k} = \int_0^1 f(x)\varphi_{j,k}(x)dx \quad \text{and} \quad L_{j,k} = \int_0^1 \mathcal{L}u(x)\varphi_{j,k}(x)dx.$$

To compute the $f_{j,k}$ by collocation method, consider

$$\tilde{f}(x) = \begin{cases} f(x) & if \quad x \in [0, 1] \\ f(0) & if \quad x \le 0. \end{cases}$$

Let θ be the primitive of φ vanishing at 0. Recall that $Support(\varphi) = [0, 2N - 1]$ and $Support(\varphi_{j,k}) = [\dfrac{k}{2^g}, \dfrac{k + 2N - 1}{2^j}]$. This gives the following result.

Proposition 7.1 *Denote*

$$C_{j,k}(\tilde{f}) = \int_{supp\varphi_{j,k}} \tilde{f}(x)\varphi_{j,k}(x)dx.$$

We have

$$f_{j,k} = \begin{cases} C_{j,k}(\tilde{f}) - 2^{-j/2}\theta(-k)f(0) & if \quad -2N + 2 \le k \le -1 \\ C_{j,k}(\tilde{f}) & if \quad 0 \le k \le 2^j - 2N + 1 \end{cases}$$

Proof 7.1 *First Case :*$-2N + 2 \le k \le -1$. *We have* $0 \in supp\varphi_{j,k}$, *which is not included in* $[0, 1]$. *Then*

$$\begin{aligned} f_{j,k} &= \int_0^{2^{-j}(k+2N-1)} f(x)\varphi_{j,k}(x)dx \\ &= \int_{k2^{-j}}^{2^{-j}(k+2N-1)} \tilde{f}(x)\varphi_{j,k}(x)dx - \int_{k2^{-j}}^0 \tilde{f}(x)\varphi_{j,k}(x)dx \\ &= C_{j,k}(\tilde{f}) - \int_{k2^{-j}}^0 \tilde{f}(x)\varphi_{j,k}(x)dx. \end{aligned}$$

It results that

$$\begin{aligned} C_{j,k}(\tilde{f}) &= \int_{k2^{-j}}^0 f(0)\varphi_{j,k}(x)dx + f_{j,k} \\ &= f(0)2^{j/2}\int_{k2^{-j}}^0 \varphi(2^j x - k)dx + f_{j,k} \\ &= f(0)2^{-j/2}\int_0^{-k} \varphi(x)dx + f_{j,k} \\ &= f(0)2^{-j/2}\theta(-k) + f_{j,k}. \end{aligned}$$

The second case can be checked by similar techniques.

Remark 7.2 *For all $k \leq -1$, we have $C_{j,k}(\tilde{f}) = f(0)2^{-j/2}$.*
Consider now a differentiable function f at 0 and let

$$\tilde{f}(x) = \begin{cases} f(x) & if \quad x \in [0,1] \\ f(0) + f'(0)x & if \quad x \leq 0. \end{cases}$$

We obtained the following result.

Proposition 7.3 *For all $k; -2N+2 \leq k \leq -1$, we have*

$$C_{j,k}(\tilde{f}) = f(0)2^{-j/2}\theta(-k) + f'(0)2^{-3j/2}(k + M_1(\varphi)) + f_{j,k}.$$

Proof.

$$\begin{aligned}
C_{j,k}(\tilde{f}) &= \int_{k2^{-j}}^{0} \Big(f(0) + f'(0)x \Big)\varphi_{j,k}(x)dx + f_{j,k} \\
&= f(0)2^{j/2} \int_{k2^{-j}}^{0} \varphi(2^j x - k)dx + f'(0)2^{j/2} \int_{k2^{-j}}^{0} x\varphi(2^j x - k)dx + f_{j,k} \\
&= f(0)2^{-j/2}\theta(-k) + f'(0)2^{-3j/2}\int_{0}^{-k} \varphi(x)dx + f_{j,k} \\
&= f(0)2^{-j/2}\theta(-k) + f_{j,k}. \\
&+ f(0)2^{-j/2}\theta(-k) + f_{j,k} + f'(0)2^{-3j/2}\Big[\int_{0}^{-k} x\varphi(x)dx + k\theta(-k) \Big].
\end{aligned}$$

Let $I(k) = \int_{0}^{-k} x\varphi(x)dx$. By means of an integration by parts, we obtain

$$\begin{aligned}
I(k) &= -k\theta(-k) - \int_{0}^{-k} \int_{0}^{x} \varphi(t)dtdx \\
&= -k\theta(-k) - \int_{\mathbb{R}} \int_{\mathbb{R}} \chi_{[0,x]}(t)\chi_{[0,-k]}(x)\varphi(t)dtdx.
\end{aligned}$$

Using Fubini's theorem, one can obtain

$$\begin{aligned}
I(k) &= -k\theta(-k) - \int_{\mathbb{R}} \varphi(t) \int_{\mathbb{R}} \chi_{[0,x]}(t)\chi_{[0,-k]}(x)dxdt \\
&= -k\theta(-k) - \int_{\mathbb{R}} \varphi(t) \int_{t}^{-k} dxdt \\
&= -k\theta(-k) + \int_{\mathbb{R}} \varphi(t) \int_{\mathbb{R}} \varphi(t)dt \\
&= -k\theta(-k) + kM_0(\varphi) + M_1(\varphi),
\end{aligned}$$

where $M_l(\varphi)$ designs the lth order moment of φ.
We will now compute the $C_{j,k}(\tilde{f})$ by collocation. For this aim, consider the set $\Lambda = \{x_l = \frac{l}{2^j}; l = -2N+2, -2N+3, ..., 2^j\}$. Remark that for every $x_l \in \Lambda$,

$$\begin{aligned}
\tilde{f}(x_l) &= \sum_{k=-2N+2}^{2^j} C_{j,k}(\tilde{f})\varphi_{j,k}(x_l) \\
&= \sum_{k=-2N+2}^{2^j} C_{j,k}(\tilde{f})2^{j/2}\varphi(l - k).
\end{aligned}$$

Proposition 7.4 *For $l = 2N + 2, ..., 2^j$, we have*

$$C_{j,l-1}(\tilde{f}) = \frac{2^{-j/2} f(x_l) - \sum\limits_{k=-2N+2+l}^{l-2} C_{j,k}(\tilde{f})\varphi(l-k)}{\varphi(1)},$$

where $\tilde{f}$ is any extension of f on $]-\infty, 1]$.

Proof. Since $Support(\varphi) = [0, 2N - 1]$, it holds that $\varphi(l - k) = 0$ for $l \leq k$ and $l \geq 2N - 1 + k$. It follows that

$$\tilde{f}(x_l) = \sum_{k=-2N+2}^{2^j} 2^{j/2} C_{j,k}(\tilde{f})\varphi(l-k)$$

$$= \sum_{k=-2N+2+l}^{l-1} 2^{j/2} C_{j,k}(\tilde{f})\varphi(l-k).$$

It results that

$$C_{j,l-1}(\tilde{f}) = \frac{2^{-j/2} f(x_l) - \sum\limits_{k=-2N+2+l}^{l-2} C_{j,k}(\tilde{f})\varphi(l-k)}{\varphi(1)}.$$

We now propose to evaluate the $L_{j,k}$. For n and k integers and x real number, denote

$$\Gamma_k^n(x) = \int_0^x \varphi^{(n)}(t-k)\varphi(t)dt.$$

Without loss of generality, we can consider the operator $\mathcal{L} = \dfrac{\partial^n}{\partial x^n}$. For $j, p \in \mathbb{Z}$, denote

$$L_{p,j} = \int_0^1 \mathcal{L}u(x)\varphi_{p,j}(x)dx.$$

Observing that

$$u = \sum_{k=2-2N}^{2^p-1} u_{p,k}\varphi_{p,k},$$

we obtain

$$L_{p,j} = 2^{(n+1)p} \sum_{k=2-2N}^{2^p-1} u_{p,k} \int_0^1 \varphi^{(n)}(2^p x - k)\varphi(2^p x - j)dx$$

$$= 2^{np} \sum_{k=2-2N}^{2^p-1} u_{p,k} \int_{1-j}^{2^p-j} \varphi^{(n)}(x+j-k)\varphi(x)dx$$

$$= 2^{np} \sum_{k=2-2N}^{2^p-1} u_{p,k} \left[\Gamma_{k-j}^n(2^p - j) - \Gamma_{k-j}^n(-j)\right]$$

$$= 2^{np} \sum_{k=2-2N}^{2^p-1} u_{p,k} a_{j,k}^n(p),$$

where we have denoted by $a^n_{j,k}(p) = \Gamma^n_{k-j}(2^p - j) - \Gamma^n_{k-j}(-j)$. This yields a matrix representation system

$$2^{np} \boldsymbol{A}^{(n)}_p \mathcal{U}_p = L_p,$$

where we have

$$
\begin{aligned}
\boldsymbol{A}^{(n)}_p &= \left(a^n_{j,k}(p) \right)_{2-2N \le j,k \le 2^p - 1} \\
\mathcal{U}_p &= T_{(u_{p,2-2N}, \ldots, u_{p,2^p-1})} \\
L_p &= T_{(L_{p,2-2N}, \ldots, L_{p,2^p-1})}.
\end{aligned}
$$

We propose next to set some basic properties of $\Gamma_{j,k}$. We firstly provide some properties.

Proposition 7.5 *The following assertions hold.*

1. $\Gamma^n_k(x) = 0$ *for* $|k| \ge 2N - 1$.

2. $\Gamma^n_k(x) = 0$ *for* $|k| \ge 2N - 1$.

3. $\Gamma^n_k(x) = 0$ *for* $x \ge 0$ *or* $x \le k$.

4. $\Gamma^n_{-k}(2N - 1) = (-1)^n \Gamma^n_k(2N - 1)$ *for* $x + k \ge 2N - 1$ *and* $k \ge 0$.

The properties (1), (2) and (3) are obtained from the fact that the support of ϕ is $[0, 2N - 1]$. However, (4) is a sequence of n integrations by parts. Now observing that $Support(\phi) = [0, 2N - 1]$, we can write

$$
\begin{aligned}
\Gamma^n_{-k}(x) &= \int_0^x \phi^{(n)}(t + k)\phi(t)\,dt \\
&= \int_{-k}^{2N-1-k} \phi^{(n)}(t + k)\phi(t)\,dt \\
&= \Gamma^n_{-k}(2N - 1).
\end{aligned}
$$

Now by applying (4), we obtain

$$\Gamma^n_{-k}(x) = (-1)^n \Gamma(2N - 1).$$

Hence, we have proved property (5).

Remark 7.6 *For* $1 \le x \le 2N - 2, -2N + 3 \le k \le x - 1$ *and* $x - k \le 2N - 2$, *the number of* $\Gamma^n_k(x)$ *to compute is* $\mathfrak{N} = (2N - 2)^2$.

We are now able to evaluate $\Gamma^n_k(2N - 1)$. The two-scale relation yields

$$\Gamma^n_k(x) = 2^n \sum_{i=0}^{2N-1} \sum_{j=0}^{2N-1} a_i a_j \int_0^x \phi^{(n)}(2t - 2k - i)\phi(2t - j)\,dt$$

$$\Gamma^n_k(x) = 2^n \sum_{i=0}^{2N-1} \sum_{j=0}^{2N-1} a_i a_j \Gamma^n_{2k+i-j}(2x - j). \tag{7.10}$$

It results that

$$
\begin{aligned}
\Gamma_k^n(2N-1) &= 2^{n-1} \sum_{i=0}^{2N-1} \sum_{j=0}^{2N-1} a_i a_j \Gamma_{2k+i-j}^n(4N-2-j) \\
&= 2^{n-1} \sum_{i=0}^{2N-1} \sum_{j=0}^{2N-1} a_i a_j \Gamma_{2k+i-j}^n(2N-1).
\end{aligned}
$$

By replacing k by $1, 2, ..., 2N-2$ in the preceding formula, one obtains the matrix representation

$$
\Gamma(2N-1) = D.\Gamma(2N-1),
$$

where $\Gamma(2N-1)$ is given by

$$
\Gamma(2N-1) =^T \left(\Gamma_1^n(2N-1), ..., \Gamma_{2N-2}^n(2N-1) \right)
$$

and D is the matrix

$$
D = [d_{l,m}]_{1 \leq l,m \leq 2N-2}
$$

with

$$
d_{l,m} = 2^{n-1} \left(\sum_{\substack{0 \leq i,j \leq 2N-1 \\ 2l+i-j=m}} a_i a_j + (-1)^n \sum_{\substack{0 \leq i,j \leq 2N-1 \\ 2l+i-j=-m}} a_i a_j \right).
$$

$\Gamma(2N-1)$ is an eigenvector of D relatively to the eigenvalue 1. Since the choice of such a vector is multiple, we demand that it satisfies the condition $\int_{-\infty}^{+\infty} x^k \psi(x) dx = 0$, $k = 0, ..., N-1$, so that each polynomial x^s can be expressed by means of $(\phi(.-l))_{l \in Z}$. Let for $n = 0, 1, ..., N-1$,

$$
x^n = \sum_{l=-\infty}^{+\infty} a_{l,n} \phi(x-l).
$$

The coefficients $a_{l,n}$ are given by

$$
\begin{aligned}
a_{l,n} &= \int_{-\infty}^{+\infty} t^n \phi(t-l) dt \\
&= \sum_{j=0}^n C_n^j l^{n-j} \int_{-\infty}^{+\infty} t^n \phi(t) dt \\
&= \sum_{j=0}^n C_n^j l^{n-j} M_j(\phi),
\end{aligned}
$$

where $M_j(\phi) = \int_{-\infty}^{+\infty} x^j \phi(x) dx$. Now the nth derivative of the above relation implies that

$$
\sum_{l=-\infty}^{+\infty} \Gamma_l^n(2N-1) \sum_{j=0}^n C_n^j l^{n-j} M_j(\phi) = n!.
$$

Since $\Gamma_l^n(2N-1)=0$ for $|l| \geq 2N-1$ and $\Gamma_0^n(2N-1)=0$, it results that

$$\sum_{l=1}^{l=2N-2} \Gamma_l^n(2N-1) \sum_{j=0}^{n} C_n^j l^{n-j} M_j(\phi) = \frac{n!}{2}.$$

Suppose that $v \neq 0$ is an eigenvector of the matrix D associated with the eigenvalue 1, $v =^T (v_0, ..., v_{2N-2})$. Then, there exists $\alpha \neq 0$ such that $v = \alpha \Gamma(2N-1)$. Therefore,

$$\alpha = 2\Big(\sum_{l=1}^{2N-2} l^n v_l \Big) \frac{1}{n!}.$$

Thus, we obtain

$$\Gamma(2N-1) =^T (\frac{v_0}{\alpha}, ..., \frac{v_{2N-2}}{\alpha}).$$

7.4   REDUCTION OF THE CONNECTION COEFFICIENTS NUMBER

Our purpose in this section is to reduce the number of Γ_k^n to compute. Let us prove firstly the following result.

Lemma 7.7 *For all x and $k \geq 0$, we have*

$$\Gamma_{-k}^n(x) = (-1)^n \Gamma_k^n(x+k) + \alpha_{n,k}(x),$$

where $\alpha_{n,k}$ satisfies

$$\alpha_{n,k}(x) = \sum_{i=0}^{n-1} (-1)^i \phi^{n-1-i}(x+k)\phi^{(i)}(x).$$

Proof.

$$\begin{aligned}
\Gamma_{-k}^n(x) &= \int_0^x \phi^{(n)}(t+k)\phi(t)dt \\
&= \phi^{(n-1)}(x+k)\phi(x) - \int_0^x \phi^{(n)}(t+k)\phi'(t)dt.
\end{aligned}$$

Iterating this relation by parts, we obtain

$$\Gamma_{-k}^n(x) = \sum_{i=0}^{n-1} (-1)^i \phi^{(n-1-i)}(x+k)\phi^{(i)} + (-1)^n \int_0^x \phi(k+t)\phi^{(n)}(t)dt.$$

Let $z = t + k$. We get

$$\begin{aligned}
\int_0^x \phi(t+k)\phi^{(n)}(t) &= \int_{kx}^{x+k} \phi(z)\phi^{(n)}(z-k)dz \\
&= \int_0^x \phi(z)\phi^{(n)}(z-k)dz \\
&= \Gamma_k^n(x+k).
\end{aligned}$$

Hence, it results

$$\Gamma_{-k}^n = \alpha_{n,k}(x) + (-1)^n \Gamma_k^n(x+k).$$

For $k = 0$ and n odd, the above relation becomes

$$\Gamma_0^n(x) = \alpha_{n,0}(\alpha) - \Gamma_0^n(x).$$

As a consequence,

$$\Gamma_0^n(x) = \frac{1}{2} \sum_{i=0}^{n} (-1)^i \phi^{(n-1-i)}(x) \phi^{(i)}(x).$$

The method requires to compute, for every k and x such that $0 \leq k \leq 2N - 3$ and $1 \leq x \leq 2N - 2$, the quantities Γ^n_{-k} and $\Gamma_k^n(x)$ for $k + 1 \leq x \leq 2N - 2$. Property (4) permits us not to compute $\Gamma^n_{-k}(x)$ for every x satisfying $x + k \leq 2N - 1$. This gives a number of $\mathbf{N} = (2N - 2)^2$ integrals that would be computed. Due to the above lemma, we have no need to compute all $\Gamma_k^n(x)$ for positive values of k. This reduces the number to $\mathbf{N}_r = (N - 1)(2N - 1)$.

As an example, we obtained the following correspondences.

$$n \; even \begin{cases} N &= 3 \rightarrow \mathbf{N} = 16 & \mathbf{N}_r = 10 \\ N &= 5 \rightarrow \mathbf{N} = 64 & \mathbf{N}_r = 36 \\ N &= 6 \rightarrow \mathbf{N} = 100 & \mathbf{N}_r = 55 \end{cases}$$

$$n \; odd \begin{cases} N &= 3 \rightarrow \mathbf{N} = 16 & \mathbf{N}_r = 6 \\ N &= 5 \rightarrow \mathbf{N} = 64 & \mathbf{N}_r = 28 \\ N &= 6 \rightarrow \mathbf{N} = 100 & \mathbf{N}_r = 45 \\ N &= 9 \rightarrow \mathbf{N} = 256 & \mathbf{N}_r = 119 \end{cases}$$

Next, we proceed to the computation of $\Gamma^n_{-k}(x)$. To do it, we start by the values $0 \leq k \leq 2N - 3$. It follows from (7.10) that

$$\Gamma^n_{-k}(x) = 2^{n-1} \sum_{i=0}^{2N-1} \sum_{j=0}^{2N-1} a_i a_j \Gamma^n_{-2k+i-j}(2x - j),$$

where $2N - 3 \geq k \geq 0$ and $1 \leq x \leq 2N - 2 - k$. It follows that for all j satisfying $0 \leq j \leq 2x - 1$

$$\Gamma^n_{-k}(x) = 2^{n-1} \sum_{i=0}^{2N-1} \sum_{j=0}^{h_1} a_i a_j \Gamma^n_{-2k+i-j}(2x - j),$$

where $h_1 = min(2N - 1, 2x - 1)$. Now similar arguments allow that

$$\Gamma^n_{-k}(x) = 2^{n-1} \sum_{i=0}^{h_2} \sum_{j=0}^{h_1} a_i a_j \Gamma^n_{-2k+i-j}(2x - j),$$

with $h_2 = min(2(x + k) - 1, 2N - 1)$. This last formula can be formulated in a matrix form as

$$\Gamma = 2^{n-1} \mathbf{A} \Gamma + \beta \Leftrightarrow (I - 2^{n-1} \mathbf{A}) \Gamma = \beta,$$

where I is the identity matrix and $\Gamma =^T (U_1, U_2, ..., U_{2N-2})$. U_s are vectors defined in the following from

$$U_s =^T (\Gamma^n_{-2N+2+s}(s), \Gamma^n_{-2N+2+s}(s - 1), ..., \Gamma^n_{-2N+2+s}(1)).$$

Let $d = 2N - 1$. $\mathbf{A}_p^{(n)}$ is an $M_{2^p+2N-2}(\mathbb{R})$ band matrix of $4N - 3$ diagonals expressed in the following form

$$\mathbf{A}_p^{(n)} = \left(\begin{pmatrix} \begin{pmatrix} A_0 - A_1 \\ T_1 \\ 0 \end{pmatrix} & \begin{pmatrix} T_2 & 0 & 0 \\ D & & 0 \end{pmatrix} & \begin{pmatrix} 0 \\ 0 \end{pmatrix} \\ \begin{pmatrix} 0 \\ 0 \end{pmatrix} & (T_3) & \begin{pmatrix} T - 4 \\ A_1 \end{pmatrix} \end{pmatrix} \right)$$

where the matrices A_i, T_i and D are

$$A_0 = \begin{pmatrix} \Gamma_0(d) & \Gamma_1(d) & \cdots & \Gamma_{d-2}(d) \\ \Gamma_{-1}(d) & \Gamma_0(d) & \cdots & \Gamma_{d-3}(d) \\ \vdots & & & \\ \Gamma_{-d+2}(d) & \Gamma_{-d+1}(d) & \cdots & \Gamma_0(d) \end{pmatrix}$$

$$A_1 = \begin{pmatrix} \Gamma_0(d-1) & \Gamma_1(d-1) & \cdots & \Gamma_{d-2}(d-1) \\ \Gamma_{-1}(d-2) & \Gamma_0(d-2) & \cdots & \Gamma_{d-3}(d-2) \\ \vdots & & & \\ \Gamma_{-d+2}(1) & \Gamma_{-d+1}(1) & \cdots & \Gamma_0(1) \end{pmatrix}$$

$$T_1 = \begin{pmatrix} \Gamma_{-d+1}(d) & \Gamma_{-d+2}(d) & \cdots & \Gamma_{-1}(d) \\ 0 & \Gamma_{-d+1}(d) & \cdots & \Gamma_{-2}(d) \\ \vdots & & & \\ 0 & 0 & \cdots & \Gamma_{-d+1}(d) \end{pmatrix}$$

$$T_2 = \begin{pmatrix} \Gamma_{d-1}(d) & 0 & \cdots & 0 \\ \Gamma_{d-2}(d) & \Gamma_{d-1}(d) & \cdots & 0 \\ \vdots & & & \\ \Gamma_{-d+3}(d) & \cdots & \Gamma_{d-1}(d) & 0 \\ \Gamma_{-d+2}(d) & \Gamma_{-d+3}(d) & \cdots & \Gamma_{d-1}(d) \end{pmatrix}$$

$$T_3 = \begin{pmatrix} (X) & (Y) \\ (0) & (Z) \end{pmatrix}$$

$$X = \begin{pmatrix} \Gamma_{-d+1}(d) & \Gamma_{-d+2}(d) & \cdots & \Gamma_0(d) \\ 0 & \Gamma_{-d+1}(d) & \cdots & \\ \vdots & & & \\ 0 & & \cdots & 0 \;\; \Gamma_{-d+1}(d) \end{pmatrix}$$

$$Y = \begin{pmatrix} \Gamma_1(d) & \cdots & \cdots & \Gamma_{d-2}(d) \\ \Gamma_0(d) & \cdots & \cdots & \Gamma_{d-3}(d) \\ \vdots & & & \\ 0 & 0 & \cdots & \Gamma_{-d+1}(d) \\ \Gamma_{-d+2}(d) & \cdots & \cdots & \Gamma_0(d) \end{pmatrix}$$

$$Z = \begin{pmatrix} \Gamma_{-d+1}(d-1) & & .. & & .. & \Gamma_{-1}(d-1) \\ 0 & & \Gamma_{-d+1}(d-2) & .. & & \Gamma_{-2}(d-2) \\ \vdots & & & & & \\ 0 & & & .. & & 0 & \Gamma_{-2N+2}(1) \end{pmatrix}$$

$$T_4 = \begin{pmatrix} \Gamma_{d-1}(d) & 0 & & .. & & .. & 0 \\ \Gamma_{d-2}(d) & \Gamma_{d-1}(d) & 0 & & .. & & 0 \\ \Gamma_{d-3}(d) & \Gamma_{d-2}(d) & \Gamma_{d-1}(d) & 0 & & .. & \\ \vdots & & & & & & \\ \Gamma_{1}(d) & \Gamma_{2}(d) & \Gamma_{3}(d) & & .. & & \Gamma_{d-1}(d) \end{pmatrix}$$

$$D = \begin{pmatrix} \Gamma_{-d+2}(d) & & .. & & \Gamma_{d-1}(d) & & 0 & & .. \\ 0 & & \Gamma_{-d+2}(d) & & .. & & \Gamma_{d-1}(d) & & 0 \\ \vdots & & & & & & & \\ & .. & & 0 & & \Gamma_{-d+2}(d) & & .. & & \Gamma_{d-1}(d) \end{pmatrix}$$

7.5 TWO MAIN APPLICATIONS IN SOLVING PDEs

7.5.1 The Dirichlet Problem

In this section, we will re-develop some prototypical example of application of wavelets in resolving partial differential equations. As we have spoken briefly in the introduction about one-dimensional case to introduce the task, we propose in this section to consider the two-dimensional case. General higher dimensional problems may be treated by analogue ways. We focus in this part on Dirichlet type problems and consider the problem

$$\begin{cases} -\Delta u + \lambda u = f & \text{in } \Omega, \\ u = g & \text{on } \partial\Omega, \end{cases} \tag{7.11}$$

where Ω is an open domain in $\mathbb{R}^2$ with Lipschitz boundary $\partial\Omega$, $f \in L^2(\Omega)$ and $g \in H^{\frac{1}{2}}(\partial\Omega)$.

To approximate the solution u, we will apply the wavelet-Galerkin method. To do this consider a level of decomposition J and consider the approximation of the solution u by means of a wavelet series as

$$u_J(x,y) := \sum_{l,k \in I_J} u_{l,k}^J \varphi_l^J(x) \varphi_k^J(y),$$

where

$$I_J = \{(l,k) \in \mathbb{Z} \times \mathbb{Z} : Support(\varphi_l^J(x)\varphi_k^J(y)) \subset \overline{\Omega}\}.$$

We denote for the next

$$f_J(x,y) = \sum_{l,k} f_{l,k}^J \varphi_l^J(x)\varphi_k^J(y),$$

where

$$f_{l,k}^J = \int_D f(x,y)\varphi_m^j(x)\varphi_n^j(y)dxdy.$$

We also consider the approximation of the function g as

$$g_J(x,y) = \sum_{l,k} g_{l,k}^J \varphi_l^J(x)\varphi_k^J(y),$$

where

$$g_{l,k}^J = \int_D g(x,y)\varphi_m^j(x)\varphi_n^j(y)dxdy.$$

Wavelet-Galerkin method permits reduction of the order of derivative in the original problem (which equals 2 in the present case) by transforming the problem into a variation alone. Indeed, it suffices to multiply the first equation of problem (7.11) by an element of the orthogonal wavelet basis $(\varphi_m^J(x)\varphi_n^J(y))_{m,n}$

$$\int_D u^j(x,y)\varphi_m^j(x)\varphi_n^j(y))dxdy = \sum_{p,q} u_{p,q}^j u_{p,q}^j \delta_m^p \delta_n^q$$

and similarly for the derivatives

$$\frac{\partial u^j}{\partial x}(x,y) = 2^j \sum_{k,q}\left(\sum_p u_{p,q}^j \Gamma_k^p\right)\varphi_{jk}(x)\varphi_q^j(y),$$

$$\frac{\partial u^j}{\partial y}(x,y) = 2^j \sum_{k,q}\left(\sum_p u_{p,q}^j \Gamma_k^p\right)\varphi_k^j(x)\varphi^{jq}(y),$$

$$\frac{\partial \varphi_m^j(x)}{\partial x}\varphi_n^j(y) = 2^j \sum_k \Gamma_k^m \varphi_{jk}(x)\varphi_n^j(y),$$

$$\varphi_m^j(x)\frac{\partial \varphi_n^j(y)}{\partial y} = 2^j \sum_k \Gamma_k^m \varphi_m^j(x)\varphi_{jk}(y).$$

Using next the connection coefficients

$$\Gamma_m^l =: \int_{\mathbb{R}} \varphi_l'(x)\varphi_m(x)dx$$

and the orthonormality of $\{\varphi_{jk}\}$, we obtain

$$\int_D \nabla u^j(x,y).\nabla(\varphi_m^j(x)\varphi_n^j(y))dxdy = \sum_{p,q} u_{p,q}^j(2^{2j}\sum_k \Gamma_k^m \Gamma_k^p \delta_n^q + \Gamma_k^n \Gamma_k^q \delta_m^p),$$

where δ_m^l is equal to 1 when $l = m$, and zero otherwise.
On the other hand, we have

$$\int_D f^j \varphi_m^j \varphi_n^j dxdy = \sum_{p,q} f_{p,q}^j \delta_m^p \delta_n^q,$$

$$\int_{\partial\Omega} u^j \varphi_m^j \varphi_n^j ds = \int_{\mathbb{R}^2}\left(\sum_{p,q} u_{p,q}^j \varphi_p^j(x)\varphi_q^j(y)\right)(\varphi_m^j(x)\varphi_n^j(y))\left(\sum_{k,l} D_{k,l}^j \varphi_{jk}(x)\varphi_l^j(y)dxdy\right)$$

$$= \sum_{p,q} u_{p,q}(2^j \sum_{k,l} D_{k,l}^j \Gamma_{k,m}^p \Gamma_{l,n}^q)$$

and

$$\int_{\partial\Omega} g\varphi_m^j\varphi_n^j ds = \int_{\mathbb{R}^2}\left(\sum_{p,q} g_{p,q}^j\varphi_p^j(x)\varphi_q^j(y)\right)(\varphi_m^j(x)\varphi_n^j(y))\left(\sum_{p,q} D_{k,l}^j\varphi_{jk}(x)\varphi_l^j(y)\right)dxdy$$
$$= 2^j\sum_{p,q,k,l} g_{p,q}^j D_{k,l}^j\Gamma_{k,m}^j\Gamma_{l,n}^q.$$

From these estimations, we get a linear system for unknown $u_{l,k}^J$.

Such a solution is known as the wavelet-Galerkin approximation to the solution u. Moreover, in wavelet theory, we have

$$\|u - u^J\|_{L^2(\Omega)} \le C 2^{-2J}$$

and

$$\|u - u^J\|_{H^1(\Omega)} \le C 2^{-2J}$$

whenever $u \in C^2(\bar{\Omega})$.

Next, we will develop some illustrative experiments to show the feasibility of the wavelet representations as well as their efficient implementations. We consider precisely the problem

$$\begin{cases} -\Delta u - \dfrac{16}{3}u = -16(x^2 + y^2)^2 - 16xy + 4 & \text{in } B(0,2), \\ u = 3xy + 18 & \text{on } \partial B(0,2). \end{cases} \tag{7.12}$$

Here $\Omega = B(0,2)$ is the ball of center 0 and radius 2 in $\mathbb{R}^2$. Note in this example that

$$f(x,y) = -16(x^2 + y^2)^2 - 16xy + 4, \qquad g(x,y) = 3xy + 18$$

and the exact solution is

$$u(x,y) = 3(x^2 + y^2 - 4)(x^2 + y^2 + 1) + 3xy + 18.$$

To continue developing the example, we consider the fictitious domain $D = \{(x,y) \in \mathbb{R}^2 : |x|, |y| < 2\}$.

The relative error is of the form

$$E = \frac{\|u - u_\varepsilon\|_{L^2(\mathbb{R})}}{\|u\|_{L^2(\mathbb{R})}}.$$

7.5.2 The Neumann Problem

In this section, we consider the second typical problem in elliptic differential equation, which consists of the Neumann problem.

We consider here also the elliptic problem considered in the previous section but with Neumann boundary conditions:

$$\begin{cases} -\Delta u + \lambda u = f & \text{in } \Omega, \\ \dfrac{\partial u}{\partial n} = g & \text{on } \partial\Omega, \end{cases} \tag{7.13}$$

where

$$\frac{\partial u}{\partial n} = \nabla. \text{ on } \partial\Omega.$$

n is the outer normal to the oriented boundary of the domain Ω and $\nabla u = \left(\frac{\partial y}{\partial x}, \frac{\partial u}{\partial y}\right)$ is the gradient of u. f and g are assumed to be sufficiently smooth and satisfy further $f \in L^2(\Omega)$ and $g \in H^{-\frac{1}{2}}(\partial\Omega)$.

As in the previous section, we consider the variational formulation of problem (7.13)

$$\int_\Omega (-\Delta uv + \lambda uv)dxdy = \int_\Omega (\nabla u.\nabla v + \lambda uv)dxdy - \int_{\partial\Omega} \nabla u.nvd\sigma = \int_\Omega fvdxdy,$$

where $v \in H^{-1}(\Omega)$ is a test function and $d\sigma$ is the Lebesgue measure on $\partial\Omega$. Otherwise, we may write

$$\int_\Omega (\nabla u.\nabla v + \lambda uv)dxdy = \int_\Omega fvdxdy + \int_{\partial\Omega} \nabla u.nvd\sigma.$$

To solve such an equation, we need, as for the case of Dirichlet problem, a fictitious domain D containing $\overline{\Omega}$. Next, we solve instead the following integral equation for $u \in H^1(D)$

$$\int_\Omega (\nabla u\nabla v + \lambda uv)dx = \int_\Omega fvdx + \int_{\partial\Omega} gvd\sigma, \quad \forall v \in H^1(D).$$

As for the previous section, we consider the wavelet series decompositions of all the functions included at a level J as

$$f = \sum_{i,j\in\Lambda} f_{ij}^J \varphi_i^J(x)\varphi_j^J(y)$$

and

$$g = \sum_{i,j\in\Lambda} g_{ij}^L \varphi_i^J(x)\varphi_j^J(y),$$

where

$$\Lambda = \{(i,j) : supp(\varphi_i^L(x)\varphi_j^L(y)) \subset D\}.$$

Similarly,

$$\chi_\Omega = \sum_{i,j\in\Lambda} \Omega_{ij}^J \varphi_i^J(x)\varphi_j^J(y),$$

and

$$\mu_{\partial\Omega} = \sum_{i,j\in\Lambda} \mu_{ij}^J \varphi_i^J(x)\varphi_j^J(y).$$

Here, we use the calculation of the boundary measure $\mu_{\partial\Omega}$. Then, we have

$$\int_{\partial\Omega} gvds = \int_D \mu_{\partial\Omega} gvdx.$$

Also,

$$\int_\Omega f v dx = \int_D \chi_\Omega f v dx$$

and

$$\int_\Omega (\nabla u \nabla v + uv) dx = \int_D \chi_\Omega (\nabla u \nabla v + uv) dx.$$

Let us suppose the unknown function u admits the following expansion at level L:

$$u = \sum_{i,j \in \Lambda} u_{ij}^L \varphi_i^L(x) \varphi_j^L(y).$$

Then, letting the test function v be of the form

$$v = \varphi_i^L(x) \varphi_j^L(y), (i,j) \in \Lambda,$$

we have

$$
\begin{aligned}
\int_D \nabla u \nabla v &= \int_D \sum_{p,q \in \Lambda} u_{p,q}^L \Big(\big[\tfrac{d}{dx} \varphi_p^L(x) \big] [\varphi_q^L(y)] \big[\tfrac{d}{dx} \varphi_i^L(x) \big] [\varphi_j^L(y)] \\
&\quad + [\varphi_p^L(x)] \big[\tfrac{d}{dy} \varphi_q^L(y) \big] [\varphi_i^L(x)] \big[\tfrac{d}{Dy} \varphi_j^L(y) \big] \Big) dy \\
&= \sum_{p,q \in \Lambda} u_{p,q}^L \Big[\int_\mathbb{R} (\varphi_p^L(x))'(\varphi_i^L(x))' dx \Big] \delta_q^j \\
&\quad + \delta_p^i \int_\mathbb{R} (\varphi_q^L(y))'(\varphi_j^L(y))' dy \\
&= \sum_{p,q \in \Lambda} u_{p,q}^L \Big(\Gamma_{p-i} \delta_q^j + \Gamma_{q-j} \delta_p^i,
\end{aligned}
$$

where

$$\Gamma_k = \int_\mathbb{R} (\varphi_k^L)'(\varphi_0^L)' dx$$

$$\int_D uv dx = u_{ij}^L.$$

Moreover, $\int_\Omega \nabla u \nabla v = \sum_{p,q \in \Lambda} u_{p,q}^L \sum_{m,n \in \Lambda} \Omega_{m,n}^L [\overline{\Gamma}_{p-i,m-i} \overline{\Gamma}_{q-j,n-j} \overline{\Gamma}_{p-i,m-i} \overline{\Gamma}_{q-j,n-j}]$, where

$$\overline{\Gamma}_{k,l} = \int (\varphi_k^L)'(\varphi_l^L)(\varphi_0^L)' dx,$$

$$\overline{\Gamma}_{k,l} = \int \varphi_k^L \varphi_l^L \varphi_0^L dx,$$

$$\int_\Omega uv = \sum_{p,q} u_{p,q}^L \sum_{m,n} \Omega_{m,n}^L \overline{\Gamma}_{p-i,m-i} \overline{\Gamma}_{q-j,n-j}.$$

As a consequence, we get similarly to the Dirichlet case a linear system with unknown $u_{l,k}^J$.

Next, we will develop, as previously, some illustrative experiments to show the feasibility of the wavelet representations as well as their efficient implementations. We consider the problem

$$
\begin{cases}
-\Delta u - 8u = -8(x^2 + y^2)^2 - 32xy + 4 \quad \text{in } B(0,2), \\
\dfrac{\partial u}{\partial n}(x,y) = 4(xy + 6) \quad \text{on } \partial B(0,2).
\end{cases}
\tag{7.14}
$$

Here, $\Omega = B(0, 2)$ is the ball of center 0 and radius 2 in $\mathbb{R}^2$. Note in this example that

$$f(x, y) = -8(x^2 + y^2)^2 - 32xy + 4, \qquad g(x, y) = 4(xy + 6)$$

and the exact solution is

$$u(x, y) = (x^2 + y^2 - 3)(x^2 + y^2 + 1) + 4xy + 4.$$

To continue developing the example, we consider the fictitious domain $D = \{(x, y) \in \mathbb{R}^2 : |x|, |y| < 2\}$. The relative error is of the form

$$E = \frac{\|u - u_\varepsilon\|_{L^2(\mathbb{R})}}{\|u\|_{L^2(\mathbb{R})}}.$$

7.6 APPENDIX

If θ is the primitive of φ vanishing at zero, we obtain the following values

N	=	4	N	=	5
$\theta(0)$	=	0.37993441	$\theta(0)$	=	0
$\theta(2)$	=	1.15338184	$\theta(1)$	=	0, 220614012
$\theta(3)$	=	0, 954427918	$\theta(2)$	=	1, 11667365
$\theta(4)$	=	1, 00636553	$\theta(3)$	=	0, 956269232
$\theta(5)$	=	1, 00050094	$\theta(4)$	=	1.01457715
$\theta(6)$	=	0, 999996157	$\theta(5)$	=	0.998197939
$\theta(7)$	=	1, 0	$\theta(6)$	=	0.999762129
			$\theta(7)$	=	0.999997875
			$\theta(8)$	=	0.999999995
			$\theta(9)$	=	1.

The values of $\varphi * \varphi^{(n)}(k)$, where $1 \leq k \leq 2N - 2$ and n equal 1 or 2

$$N = 3$$

γ_0^1	=	0.00000000000000000	γ_0^2	=	-2.25765306122408704
γ_1^1	=	0.74520547945205251	γ_1^2	=	3.39047619047554738
γ_2^1	=	-0.14520547945205348	γ_2^2	=	-0.87619047619023382
γ_3^1	=	0.01461187214611861	γ_3^2	=	0.11428571428568265
γ_4^1	=	0.00034246575342466	γ_4^2	=	0.00535714285714025

$$N = 5$$

γ_0^1	=	0.00000000000000000	γ_0^2	=	-1.64356899162416981
γ_1^1	=	0.82590601185017376	γ_1^2	=	2.41479035119715890
γ_2^1	=	-0.22882018706694840	γ_2^2	=	-0.64950218998273490
γ_3^1	=	0.05335257193267310	γ_3^2	=	0.11428571428568265
γ_4^1	=	-0.00746139636577562	γ_4^2	=	0.18095355009392441
γ_5^1	=	0.00023923582002393	γ_5^2	=	0.00079462055713373
γ_6^1	=	0.00005404730164476	γ_6^2	=	0.00036714538389740
γ_7^1	=	0.00000025241171136	γ_7^2	=	0.00000165654413596
γ_8^1	=	0.0000000026960480	γ_8^2	=	0.00000000353875999

The values of $\Gamma_k^1(x)$ for $N = 3$ and $N = 4$ are in the following

$$N = 3$$

$\Gamma_{-3}^1(1)$	$=$	-0.00960718235	$\Gamma_{-2}^1(2)$	$=$	0.143760453
$\Gamma_{-2}^1(1)$	$=$	0.246829139	$\Gamma_{-1}^1(3)$	$=$	-0.744882201
$\Gamma_{-1}^1(2)$	$=$	-0.771134002	Γ_{-1}^1	$=$	-1.06420841
$\Gamma_0^1(1)$	$=$	0.827328955	Γ_0^1	$=$	0.0744350803
$\Gamma_0^1(3)$	$=$	0.00453795266	Γ_0^1	$=$	$8.9648414E - 06$
$\Gamma_1^1(2)$	$=$	0.567892796	$\Gamma_1^1(3)$	$=$	0.734376262
$\Gamma_1^1(4)$	$=$	0.745285597	$\Gamma_2^1(4)$	$=$	-0.14539422
$\Gamma_2^1(3)$	$=$	-0.124283154			

$$N = 4$$

$\Gamma_5^1(6)$	$=$	-0.000176038891	$\Gamma_4^1(6)$	$=$	-0.00221729335
$\Gamma_4^1(5)$	$=$	-0.0019240125	$\Gamma_3^1(6)$	$=$	0.0335777806
$\Gamma_3^1(5)$	$=$	0.0329393848	Γ_3^1	$=$	0.0358510492
$\Gamma_2^1(6)$	$=$	-0.191998699	$\Gamma_2^1(5)$	$=$	-0.191800586
$\Gamma_2^1(4)$	$=$	-0.197376814	$\Gamma_2^1(3)$	$=$	-0.179838114
$\Gamma_1^1(6)$	$=$	0.793009587	$\Gamma_1^1(5)$	$=$	0.792979922
$\Gamma_1^1(4)$	$=$	0.79360871	$\Gamma_1^1(3)$	$=$	0.812556806
$\Gamma_1^1(2)$	$=$	0.646060577	$\Gamma_0^1(6)$	$=$	$1.7727329E - 10$
$\Gamma_0^1(5)$	$=$	$7.17551201E - 07$	$\Gamma_0^1(4)$	$=$	$6.92000621E - 0.5$

$\Gamma_0^1(3)$	$=$	0.000784494379	$\Gamma_0^1(2)$	$=$	0.000572469733
$\Gamma_0^1(1)$	$=$	0.507195682	$\Gamma_{-1}^1(1)$	$=$	-0.680140141
$\Gamma_{-1}^1(2)$	$=$	-0.813897103	Γ_{-1}^1	$=$	-0.794074701
$\Gamma_{-1}^1(4)$	$=$	-0.792965829	$\Gamma_{-1}^1(5)$	$=$	-0.793009609
$\Gamma_{-2}^1(1)$	$=$	0.219732583	$\Gamma_{-2}^1(2)$	$=$	0.197774885
$\Gamma_{-2}^1(3)$	$=$	0.191753134	$\Gamma_{-2}^1(4)$	$=$	0.191998477
$\Gamma_{-3}^1(1)$	$=$	-0.0476997576	$\Gamma^1-31(2)$	$=$	-0.0328988496
$\Gamma_{-3}^1(3)$	$=$	-0.0335770348	$\Gamma_{-4}^1(1)$	$=$	0.000717465579
$\Gamma_{-4}^1(2)$	$=$	0.00221665622	$\Gamma_{-5}^1(1)$	$=$	0.000195003311

7.7 EXERCISES

Exercise 1.

Compute the connection coefficients to at most the order 4 of the following scaling function:

a. Haar scaling function.

b. Faber-Schauder scaling function.

c. The 4th derivative of Gaussian wavelet.

Exercise 2.

Consider the following boundary conditions problem,

$$y(0, t) = 0, y(\pi, t) = 0$$

and the initial conditions

$$\frac{\partial y}{\partial t}_{(t=0)} = g(x) \quad and \quad y(x, 0) = 0.$$

By separating variables and proceeding formally, obtain the solution

$$y(x, t) = \sum_{j=1}^{\infty} c_j \sin jx \sin jat,$$

where

$$c_j = \frac{2}{\pi ja} \int_0^\pi g(x) \sin jx dx.$$

Exercise 3.

Solve the following partial differential equation:

$$\frac{\partial y}{\partial x} + \frac{\partial y}{\partial t} = 1, \quad x, t \geq 0,$$

with the initial conditions $y(0, t) = y(x, 0) = 0$ using 2D Haar wavelets.

Exercise 4.

1) Recall the explicit expression of the second kind Chebyshev wavelet.
2) Using the second kind Chebyshev wavelet operational method, solve the following fractional partial differential equation:

$$\frac{\partial^\alpha y}{\partial^\alpha x} + \frac{\partial^\alpha y}{\partial^\alpha t} = 1, \quad x, t \geq 0$$

with zero initial conditions and at $\alpha = \frac{1}{2}$.

Exercise 5.

1) Propose a wavelet solution to problem [7.12] above.
2) Propose a finite difference solution to problem [7.12] above.
3) Develop a Matlab code for the numerical solutions in 1 and 2.
4) Evaluate the error estimates of the two methods.

Exercise 6.

1) Propose a wavelet solution to problem [7.14] above.
2) Propose a finite difference solution to problem [7.14] above.
3) Develop a Matlab code for the numerical solutions in 1 and 2.
4) Evaluate the error estimates of the two methods.

CHAPTER 8

Wavelets for fractal and multifractal functions

8.1 INTRODUCTION

Since their discovery by Mandelbrot, fractals have been proven to be useful tools in modeling many phenomena, such as financial time series and natural views. Next, a whole theory has been developed and has become a proper field in pure mathematics, called the multifractal analysis. Models and concepts that have been firstly described without pure mathematical foundations have been re-developed in such field.

For example, in measure theory, there has been a whole subfield in multifractal analysis concerned with multifractal measures. Besides, a functional analysis subfield has been developed concerning the functional face of multifractal analysis. A large set of functions has been pointed out where the elements are known as multifractal functions.

A first observation of these functions has been pointed out in an experience of measurements of the velocity of a turbulent fluid [222]. It has been proved that the rate of increase of the temperature of a thin wire put in the flow of a turbulent fluid is directly related to the orthogonal component of the velocity of the fluid. Such an experience permitted scientists to know that the regularity measure of a signal may itself be irregular and wide from point to point.

The principal goal of the multifractal analysis of functions is to inform us about the regularity/irregularity of functions and study the measures of their regularity/irregularity pointwise.

Indeed, at a first step the regularity is evaluated by means of specific tools such as Hölder exponent. Next, the support of the function is decomposed into cells containing each of the points with the same regularity computed previously. Such cells are called singularities sets. The final step consists of a combination of measure theory and functional analysis and permits us to evaluate the so-called spectrum of singularities, which is the Hausdorff dimension of the singularities sets by means of some specific functions such as the Legendre transform. Many investigations have been taken to measure the spectrum of singularities for functions. Some are based on modulus of continuity and others are based on wavelets. See [27] and [209].

In this chapter, we aim to develop precisely the link between multifractal analysis of functions and wavelet theory. Indeed, when comparing the two mathematical formulations developed in [27] and [209], it has been proved that the two formulae coincide for the singularity values between 0 and 1. However, in other cases the wavelet method has shown more flexibility.

In [67], [173], [174], [248], [268], [269], studies of special types of functions known as self-similar functions have been developed. It has been noticed that the self-similarity of the analyzed function is inherited by the wavelet transform of the function. The estimations of the transforms permitted us to justify the veracity of the multifractal formalism due to [27] and [209]. This in some parts joins the works of Daubechies and Lagarias in [179, 180], where a result of possible self-similarity for wavelets has been shown.

Cattani in [138] proved that wavelets are effectively a powerful tool in studying self-similar functions. Ben Mabrouk and his collaborators have extended these findings to more general classes of quasi-self-similar functions especially in anisotropic multidimensional cases by showing that anisotropic wavelets permit the wavelet transform of the function to inherit some properties of the analyzed function reminiscent of some Gibbs loss. See [46, 18, 60, 58, 57, 54, 51, 50, 49, 53].

8.2 HAUSDORFF MEASURE AND DIMENSION

Let $N \in \mathbb{N}$, $E \subset \mathbb{R}^N$. We will apply throughout the chapter the notation $|E|$ or sometimes $diam(E)$ to designate the diameter of E:

$$|E| = diam(E) = \sup_{x,y \in E} \|x - y\|,$$

where $\|x - y\|$ is the usual Euclidean distance between x and y in $\mathbb{R}^N$.
For $\eta > 0$, a countable set $\{U_j\}_j$ of subsets of $\mathbb{R}^N$ is said to be an η-covering of E if it satisfies the following simultaneous assumptions:

$$E \subset \bigcup_j U_j \quad \text{and} \quad 0 < |U_j| < \eta, \, \forall j. \tag{8.1}$$

Note that some order relatively to the inclusion property may be immediately noticed for the for η-coverings in the sense that for $\eta_1 < \eta_2$, any ϵ_1-covering of the set E is obviously an η_2-covering.

Denote next, for $s \geq 0$,

$$\mathcal{H}^s_\eta(E) = \inf \sum_j |U_j|^s,$$

where the inf is evaluated on all η-coverings of E. Because of the ordering above, we immediately observe a monotony for $\mathcal{H}^s_\eta(E)$ as a function of η. Indeed, for $\eta_1 < \eta_2$, we observe that

$$\mathcal{H}^s_{\eta_2}(E) \leq \mathcal{H}^s_{\eta_1}(E),$$

which means precisely that $\mathcal{H}^\alpha_\eta(E)$ is non-increasing as a function of η. Consequently, its limit as η goes to 0 exists. Denote therefore

$$\mathcal{H}^s(E) = \lim_{\eta \downarrow 0} \mathcal{H}^s_\eta(E) = \sup_{\eta > 0} \mathcal{H}^s_\eta(E).$$

Lemma 8.1   *1.* $\forall E \subset F \subset \mathbb{R}^N$, $\mathcal{H}^s(E) \leq \mathcal{H}^s(F)$, $\forall s$.

2. $\forall E \subset \mathbb{R}^N$, $\mathcal{H}^s(E) = 0$, $\forall s > N$.

3. $\mathcal{H}^s$ *is an outer metric measure on* $\mathbb{R}^N$.

4. *Borel sets are* $\mathcal{H}^s$ *measurable.*

5. *For all* $\mathcal{H}^s$-*measurable subset* E *of* $\mathbb{R}^N$, *there exists a critical value* $D_0 \in \mathbb{R}_+$ *satisfying*

$$\mathcal{H}^s(E) = 0 \text{ for } s > D_0 \text{ and } \mathcal{H}^s(E) = \infty \text{ for } s < D_0.$$

The proof is a simple exercise. We left it to the reader. We may also refer to [46] for a detailed proof.

Definition 8.2   *1. The restriction of* $\mathcal{H}^s$ *on the set* $\mathcal{B} = \{X \subset \mathbb{R}^N \,; X \text{ is a Borel set}\}$ *is said to be the* α-*dimensional Hausdorff measure.*

2. *The critical value* D_0 *is called the Hausdorff dimension of* E *and will be denoted by* $\dim E$.

We immediately get the following assertions resuming some properties of the Hausdorff dimension of sets.

Corollary 8.3   *1.* $\forall E \subset F \subset \mathbb{R}^N$, $\dim E \leq \dim F$.

2. $\forall E \subset \mathbb{R}^N$ *non-empty,* $0 \leq \dim E \leq N$.

3. *For* $(E_p)_p$, *a countable family of subsets of* $\mathbb{R}^d$,

$$\dim \bigcup_p E_p = \sup_p (\dim E_p).$$

Proof. 1. This is a consequence of Lemma 8.1, Assertion 1. Indeed, let $s > \dim F$. So, $\mathcal{H}^s(F) = 0$. As a result, $\mathcal{H}^s(E) = 0$, which means that $s \geq \dim E$. Letting $s \downarrow \dim F$ in the last inequality, we obtain $\dim F \geq \dim E$.

2. The left-hand inequality $0 \leq \dim E$ is obvious from the definition of $\mathcal{H}^s(E)$. The right-hand inequality is a consequence of Lemma 8.1, Assertion 2, as this assertion yields that $\dim E \leq s$, $\forall s > N$. Letting $s \downarrow N$ in the last inequality, we obtain $\dim F \leq N$.

3. It follows from Lemma 8.1, Assertion 1 that

$$\dim E_n \leq \dim(\bigcup_p E_p), \quad \forall n.$$

As a result,

$$\sup_n \dim E_n \leq \dim(\bigcup_p E_p).$$

Next, it follows from Lemma 8.1, Assertion 3 (the subadditivity of $\mathcal{H}^{\alpha+\epsilon}$) that

$$\mathcal{H}^s\left(\bigcup_n E_n\right) \leq \sum_n \mathcal{H}^s(E_n).$$

Consequently, $\forall s > \sup_n dim E_n$, we obtain

$$\mathcal{H}^s\left(\bigcup_n E_n\right) = 0.$$

Letting $s \downarrow \sup_n dim E_n$, we get the desired result.

8.3   WAVELETS FOR THE REGULARITY OF FUNCTIONS

A very well-known concept in mathematical analysis is the regularity and/or the singularity of functions. We know from elementary functional analysis that being regular or differentiable at a point of its support in some order did not imply the same character in higher orders of differentiability. A well-known example may be the Weierstrass function

$$W(x) = \sum_{n=0}^{\infty} \frac{\sin(2^n x)}{2^n}$$

or more generally

$$W_{a,b}(x) = \sum_{n=0}^{\infty} a^n \sin(b^n x),$$

where a, b are fixed real numbers. In the case, for example, $a = 1/2$ and $b = 2^{-\alpha}$ for some α positive, the maximal Hölder regularity known as the Hölder exponent of such a function is equal to α for all x. As a result, the spectrum of singularities of this function will be equal to 1.

The computation of the Hölder regularity of a function F at a point t in its support consists of finding the 'best' exponent α such that $|F(x) - F(t)| \sim |x - t|^\alpha$ in a neighborhood of t. It looks like the concept of differentiability of F at the point t where $|F(x) - F(t)| \sim |x - t|^1$ whenever F is differentiable at t. The first problem that may appear for the last relation is the fact that it did not allow the computation of the exponent α for many situations such as $F(x) = |x|^\alpha \log|x|$, as the logarithm function is not bounded. We thus adopt some modification to cover such a situation such as

$$C_1|x - t|^{\alpha+\delta} \leq |F(x) - F(t)| \leq C_2|x - t|^{\alpha-\delta}$$

for some $\delta > 0$ small enough.

Finally, for a situation where the function F is higher-order differentiable, we adopt the following definition.

Definition 8.4 (Hölder regularity) *Let F be a real valued function, α a non negative real number and t a point in its support. F is said to be α-Hölder at t if there exists a neighborhood W_t of t, a constant $C > 0$ and a polynomial P of degree less than $[\alpha]$ satisfying*

$$|F(x) - P(x - t)| \leq C|x - t|^\alpha \; ; \forall\, x \in W_t. \tag{8.2}$$

We write $F \in \mathcal{H}^{\alpha}(t)$

In fact, this definition may be re-formulated in other ways to be more easy when applied for computing the Hölder regularity of functions. See [46, 274, 275].

Definition 8.5 (Pointwise Hölder exponent) *The point-wise Hölder exponent of F at the point t is $H_F(t) = \sup\{\alpha \geq 0; \ F \in \mathcal{H}^{\alpha}(t)\}$.*

This definition is somehow critical. It measures indeed the pointwise regularity of the function at a given point, but it did not take into consideration the behavior of F on a whole neighborhood of the point. This may be done by adapting a local exponent as follows.

Definition 8.6 (Local Hölder exponent) *The local Hölder exponent of F at a point t is the greater real number $\tilde{\alpha}$ satisfying*

$$\exists C, \rho_0 > 0; \ \forall \rho < \rho_0, \quad \sup_{x,y \in B(t,\rho)} \frac{|F(x) - F(y)|}{|x - y|^{\tilde{\alpha}}} \leq C.$$

We write as $\tilde{H}_F(t)$ the local Hölder exponent of F at t. When F is differentiable, cut off as usual its Taylor polynomial at the point t.

We now come to the link of the Hölder regularity of functions and wavelet theory. The first fundamental result in this concern is proved in [268] (see also [248, 269, 270, 271, 272, 273]).

Proposition 8.7 *[268]*

- $F \in \mathcal{H}^{\alpha}(\mathbb{R})$ *if and only if $|d_{j,k}(F)| \leq C2^{-(\alpha+1/2)j}$ for all $j, k \in \mathbb{Z}$.*

- *If $F \in \mathcal{H}^{\alpha}(t)$, then for all $j, k \in \mathbb{Z}$ and $|k2^{-j} - x_0| \leq 1/2$,*

$$|d_{j,k}(F)| \leq C2^{-(\alpha+1/2)j} \left(1 + \frac{|k2^{-j} - x_0|}{2^{-j}}\right)^{\alpha}. \tag{8.3}$$

Proof. 1. F is $C^{\alpha}(\mathbb{R})$ means that it is $C^{\alpha}(x)$ for all $x \in \mathbb{R}$. Consequently,

$$
\begin{aligned}
|d_{j,k}(F)| &= 2^{j/2}|\int_{\mathbb{R}} F(x)\psi(2^j x - k)dx| \\
&= 2^{j/2}|\int_{\mathbb{R}} (F(x) - P_{j,k}(x))\psi(2^j x - k)dx| \\
&\leq C2^{j/2} \int_{\mathbb{R}} |x - k2^{-j}|^{\alpha}|\psi(2^j x - k)|dx \\
&\leq C2^{-j/2} \int_{\mathbb{R}} |x2^{-j}|^{\alpha}|\psi(x)|dx \\
&\leq C2^{-(\alpha+1/2)j}.
\end{aligned}
$$

The first equality is the definition of the discrete wavelet transforms. The second is a consequence of the cancelation property of the wavelet. The third one is the fact that

F is $C^\alpha(k2^{-j})$. The next one is an obvious variable change. The last is an immediate rewriting of the previous one. For the converse, denote for j, k fixed and $n \le [\alpha]$ integer

$$U_{j,k}^n = d_{j,k}2^{j/2}2^{nj}\psi^{(n)}(2^j x - k).$$

We have immediately

$$|U_{j,k}^n| \le C2^{-(\alpha+1/2)j}2^{j/2}\frac{2^{[\alpha]j}}{(1 + |2^j x - k|)^{[\alpha]+2}},$$

which means that the series $\sum_{j,k} U_{j,k}^n$ is uniformly convergent around x for any $x \in \mathbb{R}$.
Consider then the polynomial

$$P_{\alpha,x}(h) = \sum_{n=0}^{[\alpha]}\sum_{j,k} U_{j,k}^n \frac{h^n}{n!}.$$

We have next,

$$|F(x + h) - P_{\alpha,x}(h)| \le \sum_{j,k} d_{j,k}2^{j/2}\left|\psi(2^j(x + h) - k) - \sum_{n=0}^{[\alpha]} 2^{nj}\psi^{(n)}(2^j x - k)\frac{h^n}{n!}\right|.$$

Using Taylor's expansion, this is bounded by

$$C\sum_{j,k} d_{j,k}2^{j/2}2^{[\alpha]j}\sup_t |\psi^{([\alpha]+1)}(t)||h|^{[\alpha]+1},$$

which is bounded by

$$C\sum_{j,k} 2^{-(\alpha+1/2)j}2^{j/2}2^{[\alpha]j}|h|^{[\alpha]+1} \le Ch^\alpha.$$

So, F is $C^\alpha(x)$.
2. We have

$$
\begin{aligned}
|d_{j,k}(F)| &= 2^{j/2}|\int_\mathbb{R} F(x)\psi(2^j x - k)dx| \\
&= 2^{j/2}|\int_\mathbb{R} (F(x) - P_{\alpha,x_0}(x))\psi(2^j x - k)dx| \\
&\le C2^{j/2}\int_\mathbb{R} |x - x_0|^\alpha|\psi(2^j x - k)|dx \\
&\le C2^{j/2}\int_\mathbb{R} (|x - k2^{-j}| + |k2^{-j} - x_0|)^\alpha\frac{1}{(1 + |2^j x - k|)^N}dx \\
&\le C2^{-j/2}2^{\alpha j}\int_\mathbb{R} (|t| + \frac{|k2^{-j} - x_0|}{2^{-j}})^\alpha\frac{1}{(1 + |t|)^N}dt \\
&\le C2^{-j/2}2^{\alpha j}(1 + \frac{|x_0 - k2^{-j}|}{2^{-j}})^\alpha.
\end{aligned}
$$

For more backgrounds and examples on the link of wavelets to the Hölder regularity of functions, we may refer to [18], [16], [17], [47], [50], [49], [53], [51], [54], [67], [174].

8.4   THE MULTIFRACTAL FORMALISM

The multifractal formalism for functions is a mathematical formulation stating that the spectrum of singularities of a function may be evaluated by means of a Legendre transform of other quantities extracted from it such as the modulus of continuity and wavelet transform.

There are two variants of the multifractal formalism for functions. The first original one is introduced in [209]. The second is due to [27] and is based on the wavelet transform.

8.4.1   Frisch and Parisi multifractal formalism conjecture

Denote for $\alpha \geq 0$, $d(\alpha)$ the Hausdorff dimension of the singularity set

$$E(\alpha) = \{x \in \mathbb{R}^N ;\ H_F(x) = \alpha\},$$

i.e., the set of all points in the support of the function for which F has the same Hölder exponent equal to α. $d(\alpha)$ is known as the spectrum of singularities of F.

Frisch and Parisi showed that the spectrum may be computed by means of some average quantities extracted from the function. The formula established in [209] used the order of magnitude of the L^p-modules of continuity

$$\zeta(p) = \liminf_{|h| \to 0} \frac{\log \displaystyle\int_{\mathbb{R}^m} |F(x+h) - F(x)|^p dx}{\log |h|}.$$

For $0 < \alpha < 1$, it permits computation of the spectrum $d(\alpha)$ from the Legendre transform of $\zeta(p) - m$ as

$$d(\alpha) = \inf_p (\alpha p - \zeta(p) + m).$$

Indeed, by splitting the integral on the various sets of singularities covering the support of F and observing that the α-level singularity set E^α is covered by a number of $|h|^{-d(\alpha)}$ balls of diameters smaller than $|h|$, we deduce that

$$\int_{\cup B} |F(x+h) - F(x)|^p dx \sim |h|^{\alpha p - d(\alpha) + m}.$$

When $|h| \to 0$, the dominating term is that of smaller power, which means that

$$\zeta(p) = \inf_\alpha (\alpha p - d(\alpha) + m).$$

Next, whenever the function $\alpha \mapsto d(\alpha)$ is concave, we obtain

$$d(\alpha) = \inf_p (\alpha p - \zeta(p) + m).$$

However, this conjecture has been proved next to be critical in the sense that it did not yield the spectrum for all functions, mainly regular functions where the singularities that may appear in higher derivatives do not affect the function $\zeta(p) = p$. This has led researchers to look for general formulations to overcome all cases. This is done by using the wavelet transform.

8.4.2 Arneodo et al wavelet-based multifractal formalism

Arneodo et al proposed in [27] a mathematical formula that relates the spectrum of F to its wavelet transform by setting

$$\eta(p) = \liminf_{a \to 0} \frac{\log \int |C_{a,b}(F)|^p db}{\log a}$$

and conjecturing as usual that

$$d(\alpha) = \inf_p (\alpha p - \eta(p) + m).$$

The idea is based on a wavelet characterization of Besov spaces as

$$B_p^{s/p,\infty}(\mathbb{R}^N) = \left\{ F; \int |C_{a,b}(F)|^p db \le C a^s \text{ for } a \text{ small enough} \right\},$$

which permits the link to the Frisch and Parisi formulation by means of the Nikol'skij spaces

$$\mathcal{N}^{s,p}(\mathbb{R}^N) = \left\{ F \in L^p(\mathbb{R}^N); \int |\partial^\gamma F(x+h) - \partial^\gamma F(x)|^p dx \le C|h|^{\sigma p}, |h| << 1 \right\},$$

where for $s \ge 0$, $s = [s] + \sigma$, $|\gamma| = [s]$ and $p \ge 1$. Indeed, denote the Nikol'skij exponent

$$\xi(p) = \sup\{s; F \in \mathcal{N}^{s/p,p}(\mathbb{R}^N)\}.$$

The authors in [27] established the following result.

Proposition 8.8 .

1. $\eta(p) = \sup\{s; F \in B_p^{s/p,\infty}(\mathbb{R}^N)\}.$

2. *Whenever $\zeta(p) < p$ we have $\eta(p) = \zeta(p)$.*

Proof. 1. Denote by $\widetilde{\eta}(p)$ the right-hand-side quantity and let $s < \delta(p)$. We immediately observe that whenever $F \in B_p^{s/p,\infty}(\mathbb{R}^N)$ we get $\int |C_{a,b}(F)|^p db \le C a^s$. As a result, $s \le \eta(p)$, $\forall s < \widetilde{\eta}(p)$. Consequently, $\widetilde{\eta}(p) \le \eta(p)$. We now prove that $\widetilde{\eta}(p) \ge \eta(p)$. Indeed, for $s < \eta(p)$ there exists $\varepsilon > 0$ such that $\forall a; 0 < a < \varepsilon$ we have

$$s < \frac{\log \int |C_{a,b}(F)|^p db}{\log a}.$$

As a consequence, $F \in B_p^{s/p,\infty}(\mathbb{R}^N)$. Hence, $s \le \widetilde{\eta}(p)$. Letting $s \to \eta(p)$, we get $\eta(p) \le \widetilde{\eta}(p)$.

2. To prove this assertion, we will use the intermediate exponent $\xi(p)$. We will prove firstly that $\zeta(p) = \xi(p)$. Indeed, let $s < \zeta(p)$. There exists $r > 0$ such that $\forall |h| < r$ we have

$$\int |F(x+h) - F(x)|^p dx < |h|^s.$$

Consequently,

$$\int |F(x+h) - F(x)|^p dx < |h|^s = |h|^{\sigma p},$$

where $\sigma = \dfrac{s}{p} - \left[\dfrac{s}{p}\right] = \dfrac{s}{p}$ as $s < p$. Therefore, $s \leq \xi(p)$. Letting $s \to \zeta(p)$, we obtain $\zeta(p) \leq \xi(p)$. We do not check the opposite inequality. Let s be such that $p > s > \xi(p)$. Then $F \in \mathcal{N}^{s/p,p}(\mathbb{R}^m)$. Consequently,

$$\int |\partial^\gamma F(x+h) - \partial^\gamma F(x)|^p dx \leq C|h|^{\sigma p},$$

where σ is as above and $|\gamma| = 0$. As a result, we get

$$\int |F(x+h) - F(x)|^p dx \leq C|h|^s,$$

which yields $s \leq \zeta(p)$ and thus $\xi(p) \leq \zeta(p)$. On the other hand, observe that

$$H^{s+\varepsilon,p}(\mathbb{R}^m) \hookrightarrow B_p^{s,\infty}(\mathbb{R}^m) \hookrightarrow H^{s-\varepsilon,p}(\mathbb{R}^m) \ \forall \ \varepsilon > 0.$$

Consequently, $\xi(p) = \eta(p)$. It holds from all the previous inequalities that

$$\zeta(p) = \xi(p) = \eta(p).$$

Definition 8.1 *A tempered distribution F belongs to the two-microlocal space $D^{\alpha,\beta}(t)$ for some point t if it satisfies*

$$|d_{j,k}(F)| \leq C2^{-(\alpha+1/2)j}\left(1 + 2^{-j}|k2^{-j} - t|\right)^\beta,$$

for all $j, k \in \mathbb{Z}$ with $|k2^{-j} - t| \leq 1$ and where $d_{j,k}$ is the discrete wavelet transform of F at the level j and the position k.

The following theorem yields an a priori upper bound of the spectrum of singularities in Besov spaces based on the wavelet transform.

Theorem 28 *[300]— Let $s > \dfrac{1}{p}$, $F \in B_p^{s,p}(\mathbb{R})$ and $0 < d < 1$. Then the function $F \in \mathcal{H}^{(sp+d-1)/p}(t)$ for all $t \in Support(F) \setminus \mathcal{S}$ for some set $\mathcal{S}$ satisfying $\dim\mathcal{S} = d$.*

Proof. Let $s, p > 0$ and $\alpha = \dfrac{1}{2} - \dfrac{1}{p} + s$. Denote next $W_{j,k} = d_{j,k}2^{\alpha j}$. Whenever $F \in B_p^{s,p}(\mathbb{R})$, we get $\sum_{j,k} |W_{j,k}|^p < \infty$. Let $\eta > 0$ and $I_{j,k}$ be the interval centered on $k2^{-j}$ with length $|W_{j,k}|^\eta$. We immediately observe that $\sum_{j,k} |I_{j,k}|^{p/\eta} < \infty$. Therefore, whenever t belongs to a countable number of intervals $I_{j,k}$, for j large enough, we get

$$|W_{j,k}|^\eta \leq |\dfrac{k}{2^j} - t|, \ \forall \ k.$$

As a consequence,

$$|d_{j,k}| \leq |\frac{k}{2^j} - t|^{1/\eta} 2^{-\alpha j} = 2^{-j(\alpha+1/\eta)} |2^j t - k|^{1/\eta}.$$

The last result permits us to establish some bounds for the multifractal formalism of functions based on wavelets. Indeed, for $p > 0$ we know that

$$\eta(p) = \sup\{s; F \in B_p^{s/p,p}(\mathbb{R})\}.$$

Besides, for $\varepsilon > 0$ we also have $F \in B_p^{(\eta(p)-\varepsilon)/p,p}(\mathbb{R})$. As a result, according to Theorem 28 it holds that

$$d(\alpha) \leq d \ , \ \text{for} \ \alpha = \frac{\eta(p) + d - 1 - \varepsilon}{p}.$$

Consequently,

$$d(\alpha) \leq p\alpha - \eta(p) + \varepsilon + 1, \quad \forall \ p, \varepsilon > 0.$$

Hence,

$$d(\alpha) \leq \inf_{p>0} (p\alpha - \eta(p) + 1).$$

The multifractal formalism affirms precisely that this inequality is in fact an equality. It is not checked for all functions, but a large functional class consisting of self-similar-type functions has been proved to satisfy such a formalism under some suitable assumptions. This is the object of the next section.

8.5 SELF-SIMILAR-TYPE FUNCTIONS

Self-similar-type functions constitute a large class of multifractal functions. These are the most investigated ones by researchers in both pure mathematical fields and applied ones. They also constitute a good example of functions for which the multifractal formalism holds. We propose in this section to review such class of functions and show that the wavelet transform of these functions inherits similar characteristics such as self-similarity, which allows the estimation of such transform.

The most simple examples are the so-called homogeneous ones. Let $F : \mathbb{R}^N \to \mathbb{R}$ be such that $F(x) = \lambda F(rx)$ for all x and for some λ and r. The wavelet transform of F at the scale $a > 0$ and a position $b \in \mathbb{R}^N$ satisfies a homogeneity property such as

$$C_{a,b}(F) = \lambda C_{ra,rb}(F) \ \forall \ a > 0 \text{ and } \forall \ b \in \mathbb{R}^N.$$

A second simple case is due to periodic functions. Let $F : \mathbb{R}^N \to \mathbb{R}$ be such that $F(x + T) = F(x)$ for all x and for some $T \in \mathbb{R}^N$. The wavelet transform of F at the scale $a > 0$ and a position $b \in \mathbb{R}^N$ satisfies

$$C_{a,b}(F) = C_{a,b+T}(F) \ \forall \ a > 0 \text{ and } \forall \ b \in \mathbb{R}^N.$$

The first mathematical definition of self-similar-type functions is due to [269] and called self-affine functions.

Definition 8.9 *A function $F : \mathbb{R}^N \longrightarrow \mathbb{R}$ is said to be self-affine of order $k \geq 0$ if*

- *There exists a bounded domain Ω and affine similitudes $S_i(x) = \rho_i x + b_i$, $i = 1, ..., m$ satisfying*

$$S_i(\Omega) \subset \Omega \quad and \quad S_i(\Omega) \cap S_j(\Omega) = \emptyset$$

- *There exists $\lambda_1, ..., \lambda_m$ such that $0 < \lambda_i < 1$ and a function g $\mathcal{H}^k$ with all derivatives of order less than k having fast decay such that*

$$F(x) = \sum_{i=0}^{m} \lambda_i F(S_i^{-1}(x)) + g(x). \tag{8.4}$$

- *There exists a closed subset $L \subset \Omega$ such that F is not $\mathcal{C}^k$ on it.*

In [269], it has been proved that the multifractal formalism holds for these functions when a uniform minimal regularity occurs. In [67], an extension has been developed to cover some classes of nonlinear self-similar functions associated with nonlinear contractions in one-dimensional case and analytic ones in the complex case. The main idea of extension is based on the fact that the nonlinear similitudes look locally like linear (affine) ones as in the following proposition (see [67]).

Proposition 8.10 *Let $I = [0, 1]$ and S_1 and S_2 be nonlinear contractions on I, and denote*

$$I_i = S_i(I) \quad and \quad H = I \setminus (S_1(]0, 1[) \cup S_2(]0, 1[)).$$

Suppose that the following assertions hold

- *S_1 and S_2 are $\mathcal{C}^1$, strictly increasing or strictly decreasing on I.*

- *Suppose that there exists a constant $C \geq 1$ such that, for all $x \in int(H)$, for all $n \in \mathbb{N}$ and all $i = (i_1, i_2, ..., i_n) \in \{1, 2\}^n$, we have*

$$C^{-1}|I_{i_1}|...|I_{i_n}| \leq |(S_{i_1} \circ ... \circ S_{i_n})'(x)| \leq C|I_{i_1}|...|I_{i_n}|.$$

Then, there exists a $\mathcal{C}^1$-diffeomorphism χ on Ω and some affine contractions $\tilde{S}_i$ such that

$$\tilde{S}_i = \chi \circ S_i \circ \chi^{-1}.$$

Denoting $\tilde{F} = F \circ \chi$, the singularity exponent of F at x is the same as $\tilde{F}$ at $\chi^{-1}(x)$.

Let next Ω be a bounded open subset in $\mathbb{R}$ and S_i nonlinear contractions defined on $\mathbb{R}$ such that

$$S_i(\overline{\Omega}) \subset \Omega \quad and \quad S_i(\Omega) \cap S_j(\Omega) = \emptyset. \tag{8.5}$$

Assume further some constants $\underline{K}$, $\overline{K}$ and C such that

$$0 < \underline{K} \leq |S_i'(x)|| \leq \overline{K} < 1, \quad \forall j = 1, 2 \quad and \quad \forall x \tag{8.6}$$

and

$$|S_i^{(l)}(x)| \leq C, \quad \forall j = 1, 2, \quad l = 2, ..., k+1 \quad and \quad \forall x. \tag{8.7}$$

Due to the separation set condition (8.5), there exists a unique non-empty compact set Γ satisfying $\Gamma = S_0(\Gamma) \cup S_1(\Gamma)$ (see [251]). Next for $i = (i_1, \ldots, i_n) \in \{0, 1\}^n$ we denote

$$S_i = S_{i_1} \circ \cdots \circ S_{i_n}, \quad x_i = S_i(0), \tag{8.8}$$

$$\Omega_i = S_i(\Omega), \quad \rho_i = |\Omega_i| = diam(\Omega_i) \tag{8.9}$$

and

$$\lambda_i = \lambda_{i_1} \ldots \lambda_{i_n}. \tag{8.10}$$

For an infinite sequence i, we put $x_i = \lim_{n \to \infty} x_{(i_1, \ldots, i_n)}$. To each sequence, $i \in \{0, 1\}^{\mathbb{N}}$ corresponds to a unique point x_i in Γ. Let $x \in \mathbb{R}$. We call "D-branch over x" a branch of the tree, of length n, starting from the origin $(0, 1)$, ending at (x, ρ_i) and such that $|x - x_i| \leq D\rho_i$. Let $x \in \Gamma$, $B_j = \{i; \ 2^{-j} \leq \rho_i < 2.2^{-j}\}$ and $B_j(x)$ be the set of D-branches i over x such that $i \in B_j$. Let next

$$\alpha_{\min}(F) = \liminf_{j \to +\infty} \left[\inf_{i \in B_j} \frac{\log |\lambda_i|}{\log \rho_i} \right] \quad \text{and} \quad \alpha_{\max}(F) = \limsup_{j \to +\infty} \left[\sup_{i \in B_j} \frac{\log |\lambda_i|}{\log \rho_i} \right]. \tag{8.11}$$

The principal result resuming the multifractal analysis of self-affine functions and their extension to the nonlinear is resumed in the following theorem.

Proposition 8.11 *(See [269] and [67]) Let F be a self-similar function for which $\sum_{i=1}^{m} |\lambda_i| \rho_i < 1$. Then*

- *Equation (8.4) has a unique solution F in $L^1(\mathbb{R})$, given by the series*

$$F(x) = \sum_{n=0}^{\infty} \sum_{|i|=n} \lambda_i g((S_i)^{-1}(x)). \tag{8.12}$$

- *If furthermore $\alpha_{\min}(F) < k$, then $F \in C^{\alpha_{\min}(F)}(\mathbb{R})$.*

- *F is C^k outside Γ.*

- *For $x \in \Gamma$ and $a_F(x) = \liminf_{j \to \infty} \inf_{i \in B_j(x)} \frac{\log |\lambda_j|}{\log \rho_j} < k$, it holds that $H_F(x) = a_F(x)$.*

- *The spectrum $d_F(\alpha)$ vanishes outside the interval $[\alpha_{\min}(F), \alpha_{\max}(F)]$ and is analytic and concave on this interval.*

- *The maximal value $d_{\max}$ of $d_F(\alpha)$ satisfies $\sum_{i=1}^{m} \rho_i^{d_{\max}} = 1$ and it is attained for some $\alpha = \alpha_0$.*

- *Let p_0 be such that $\eta_F(p_0) = kp_0$ and let $\alpha_1 < \alpha_0$ be the value of the inverse Legendre transform of $\eta_F(q) - d$ at p_0. If $\alpha \leq \alpha_1$, $d_F(\alpha)$ can be obtained by computing the Legendre transform of $\eta_F(q) - d$.*

Next, the multifractal formalism has been extended to more large classes of self-similar-type functions known as quasi-self-similar functions constructed with both linear and nonlinear contractions. See [16], [17], [47], [50].

Definition 8.12 *A quasi-self-similar function F is a series of the form (8.12) in which the λ_j's which are constants become sequences $(\lambda_j^n)_n$ and the multiplicative weights $\lambda_{i_1}\lambda_{i_2}\dots\lambda_{i_n}$ are replaced by non-multiplicative ones $\lambda_{i_1}^1\lambda_{i_2}^2\dots\lambda_{i_n}^n$,*

$$F(x) = \sum_{n=0}^{\infty}\sum_{|i|=n} \lambda_{i_1}^1\lambda_{i_2}^2\dots\lambda_{i_n}^n g(S_i^{-1}(x)), \tag{8.13}$$

where the sequences $(\lambda_j^n)_n$ are uniformly absolutely bounded in $]-1,1[$. i.e., there exist A and B such that

$$For \quad k = 0,1 \quad and \quad \forall n \in \mathbb{N}^*, \quad 0 < \underline{\lambda} \le |\lambda_k^n| \le \overline{\lambda} < 1 \tag{8.14}$$

and satisfy further

$$\sum_i |\lambda_i^n|\rho_i^d < 1. \tag{8.15}$$

Compared to self-similar functions, the techniques for the computation of the spectrum of singularities are different. In the remaining part of this section, we will review some details on the computation of the Hölder regularity for the nonlinear case and left to the reader to develop by following analogue techniques the linear case. See also [269] for original proofs of the linear case.

The first result concerns the minimal global Hölder regularity of self-similar-type functions, which has been taken as an assumption in [269]. Similarly the upper global Hölder regularity is assumed to be finite in [269]. However, it is noticed in [47] and [16] that these assumptions may be proved. We have precisely the following result.

Proposition 8.13 *The following assertions are true.*

1. $0 < \alpha_{\min} \le \alpha_{\max} < \infty$.

2. $\alpha_{min} \le k \implies F \in \mathcal{H}^{\alpha_{min}-\varepsilon}(\mathbb{R}), \forall \varepsilon > 0$ small enough.

Proof. 1. It follows from (8.6) that

$$\underline{K}_1^n \le |\Omega_i| \le \overline{K}_1^n$$

for some constants $\underline{K}_1$ and $\overline{K}_1$ and $i = (i_1, i_2, \dots, i_n) \in B_j$ with $2^{-j} \le |\Omega_i| \le 2^{-j+1}$. Therefore,

$$(-j+1)\frac{\log 2}{\log \underline{K}_1} \le n \le -j\frac{\log 2}{\log \overline{K}_1}.$$

Now, exploiting equation (8.14), we obtain the boundaries

$$\frac{j-1}{j}\frac{\log \overline{\lambda}}{\log \underline{K}_1} < \frac{\log |\lambda_i|}{\log |\Omega_i|} < \frac{j}{j-1}\frac{\log \underline{\lambda}}{\log \overline{K}_1}.$$

Consequently,

$$0 < \alpha_{\min} \leq \alpha_{\max} < \infty.$$

2. According to the well-known Littlewood characterization $F \in \mathcal{H}^{\alpha_{min}-\varepsilon}(\mathbb{R})$ if

$$|F * \psi_l(x)| \leq C2^{-\alpha l}, \forall x \in \mathbb{R},$$

where ψ is an appropriate function in the Schwartz class. In our case, by choosing ψ as the analyzing wavelet and $x = k2^{-l}$, the inequality above reads as

$$|d_{l,k}(F)| \leq C2^{-\alpha l}, \forall l, k.$$

So, consider an analyzing wavelet ψ satisfying

$$\hat{\psi}(\xi) = 0 \text{ on } \{|\xi| \leq 1\} \cup \{|\xi| \geq 8\} \qquad \text{and} \qquad \hat{\psi}(\xi) = 1 \text{ on } \{2 \leq |\xi| \leq 4\}.$$

We next split the function F into a sum

$$F(x) = \sum_{j \geq 0} F_j(x) = \sum_{j \geq 0} \sum_{i \in B_j} \lambda_i g(S_i^{-1}(x)).$$

Denote next $W_{l,j} = F_j * \psi_l$ and $G_{i,l} = (g \circ S_i^{-1}) * \psi_l$. We immediately obtain for all $x \in \Omega_i$,

$$|G_{i,l}(x)| = |\int g(T_i^{-1}(y))\psi_l(x - y)dy| \leq C2^{kj+l} \int |\psi(2^l(x - y))||x - y|^k \, dy,$$

which in turns yields

$$|G_{i,l}(x)| \leq C2^{kj}2^{-kl}, \forall j \leq l, \forall x \in \Omega_i.$$

As a consequence,

$$\sum_{i \in B_j, x \in \Omega_i} |\lambda_i G_{i,l}(x)| \leq C2^{k(j-l)} \sum_{i \in B_j, x \in \Omega_i} |\lambda_i|.$$

Next, denote $B_{j,L}(x) = \{i \in B_j : |x - x_i| \leq L2^{-j}\}$. It is straightforward that $card[B_{j,L}(x)] \leq CL$ for L large enough and some constant C independent of x and j. As a result, for $x \in \Omega_i$, $|x - x_i| \leq |\Omega_i| \leq C2^{-j}$, $x_i = S_i(0)$, one has

$$\sum_{0 \leq j \leq l} \sum_{i \in B_j, x \in \Omega_i} |\lambda_i G_{i,l}(x)| \leq C2^{-kl} \sum_{0 \leq j \leq l} 2^{(-\alpha_{min}+\varepsilon+k)j}$$
$$\leq C2^{(-\alpha_{min}+\varepsilon)l}.$$

Now, for $x \notin \Omega_i$ and $j \leq l$, it holds from the localization and cancellation of ψ that

$$|G_{i,l}(x)| \leq C_N \frac{2^{k(j-l)}}{(1 + 2^j dist(x, \Omega_i))^N}.$$

Similarly, we have

$$\sum_{0 \leq j \leq l} \sum_{i \in B_j, x \notin \Omega_i} |\lambda_i G_{i,l}(x)| \leq C2^{(-\alpha_{min}+\varepsilon)l}.$$

Finally, observing that

$$|W_{l,j}(x)| \leq \sup_t |F_j(t)| \sup_{i \in B_j} |\lambda_i| \leq C2^{-(\alpha_{min}-\varepsilon)j},$$

we obtain for $j > l$ and $x \in \mathbb{R}$,

$$\sum_{j>l} |W_{l,j}(x)| \leq C \sum_{j>l} 2^{(-\alpha_{min}+\varepsilon)j} \leq C2^{-(\alpha_{min}-\varepsilon)l}.$$

It results from the estimations above that

$$|W_{l,x}| \leq C2^{-(\alpha_{min}-\varepsilon)l}, \ \forall \varepsilon > 0, \ \forall x \in \mathbb{R},$$

which means that $F \in \mathcal{H}^{\alpha_{min}-\varepsilon}(\mathbb{R})$.

The next step concerns the computation of the pointwise Hölder regularity of self-similar-type functions. We observe obviously that outside the compact Cantor type Γ, the self-similar-type function F is locally a finite sum of functions generated by g, so it is C^k. Inside the compact Γ, we will show that the wavelet transform inherits a quite self-similar-type functional equation which enables its estimate well. To do this, we need the following result established originally in [67] and which states that locally the composition of the analyzing wavelet with the contractions is asymptotically a wavelet.

Lemma 8.14 ([67]) *Let ψ be an even real compactly supported wavelet with enough vanishing moments. Let $b \in \Omega_i$ and $0 < a < |\Omega_i|$. Then*

$$\psi_{a,b}(S_i(t)) = |(S_i^{-1})'(b)| \, \psi_{a|(S_i^{-1})'(b)|, S_i^{-1}(b)}(t)$$

$$+ \sum_{p=1}^{k-1} \sum_{l=2p}^{k-1+p} a^{l-p} A_i^{(p,l)}(b) \, \psi_{a|(S_i^{-1})'(b)|, S_i^{-1}(b)}^{(p,l)}(t) \qquad (8.16)$$

$$+ (R_{a,b}^{i,k})(t),$$

where $\psi^{(p,l)}(t) = t^l \psi^{(p)}(t)$ is a compactly supported wavelet,

$$A_i^{(p,l)}(b) = \frac{1}{p!} \, |(S_i^{-1})'(b)|^{l+1} \sum_{\substack{2 \leq q_1,\ldots,q_p \leq k \\ q_1+\cdots+q_p=l}} \prod_{m=1}^{p} \frac{S_i^{(q_m)}(S_i^{-1}(b))}{q_m!} \, ,$$

$$|A_i^{(p,l)}(b)| \leq C|\Omega_i|^{p-l-1}$$

and $(R_{a,b}^{i,k})$ is a function supported in $|t - S_i^{-1}(b)| \leq Ca|\Omega_i|^{-1}$ such that

$$|(R_{a,b}^{i,k})(t)| \leq Ca^{k-1}|\Omega_i|^{-k} \quad \forall t \qquad (8.17)$$

and

$$\|(R_{a,b}^{i,k})(t)\|_{L^1(\mathbb{R})} \leq Ca^k|\Omega_i|^{-k-1} \, . \qquad (8.18)$$

The following theorem yields the Hölder regularity of self-similar-type functions.

Theorem 29 *For $x \in K$, denote*

$$\widetilde{H}_F(x) = \liminf_{n \to \infty} \frac{\log |\lambda^n_{i_1(x),\dots,i_n(x)}|}{\log |\Omega_{i_1(x),\dots,i_n(x)}|}$$

where $(i_1(x),\dots,i_n(x),\dots)$ is the unique coding sequence associated with x. Then, whenever $\widetilde{H}_F(x) < k$, we have $H_F(x) = \widetilde{H}_F(x)$.

Proof. Let for j fixed

$$\Lambda_j(x) = \sup_{i \in B_j(x)} |\lambda_i| \quad \text{and} \quad \Gamma_j(x) = \sum_{l=1}^{j} \Lambda_l(x) 2^{-A(j-l)},$$

where $A > \alpha_{max}$ fixed. Standard calculus yields that

$$A(x) = \liminf_{j \to \infty} \frac{\log \Lambda_j(x)}{-j \log 2} = \liminf_{j \to \infty} \frac{\log \Gamma_j(x)}{-j \log 2}.$$

Next, we split, as previously, F to be

$$F(x) = \sum_{j=0}^{J-1} \sum_{i \in B_j} \lambda_i g(S_i^{-1}(x)) + \sum_{i \in B_J} \lambda_i F_J(S_i^{-1}(x)),$$

where the copy F_J is

$$F_J(x) = \sum_{n=0}^{\infty} \sum_{i \in \bigcup_{p \in \Delta_n} B_p} \lambda_{i_1}^{J+1} \dots \lambda_{i_n}^{J+n} g(S_i^{-1}(x)), \tag{8.19}$$

where $card(\Delta_n)$ is finite independent of n. Consider next for $x \in K$ and $J \in \mathbb{N}$ large enough such that $\Lambda_J(x) \geq \frac{1}{2}\Gamma_J(x)$, $j^0 = (j_1^0, \dots, j_n^0) \in B_J(x)$, $b \sim k2^{-j} \in \Omega_{j^0}$ and $a \sim 2^{-J}$ be such that

$$|x - k2^{-J}| \leq C2^{-J} \quad \text{and} \quad |d_{J,k}(F)| \geq C2^{-J/2}\Lambda_J(x). \tag{8.20}$$

Without loss of the generality, we may assume that $\Lambda_J(x)$ is reached at j^0. Since $F \notin \mathcal{H}^k(x_0)$ for some point x_0 in Γ, then for every N, if $|i| = N$ and $x_0 \in \Omega_i$ we have also $F_J \notin \mathcal{C}^k(S_i^{-1}(x_0))$. As a result, there exists sequences $a_n \to 0$, b_n and $C_n \to +\infty$ satisfying

$$|b_n - x_0| \leq a_n \quad \text{and} \quad \forall N, \quad |C_{a_n,b_n}(F_J)| \geq C_n a_n^k \tag{8.21}$$

uniformly. Choosing $b = T_{j^0}(b_n)$ for n large enough and $a = a_n|T'_{j^0}(b_n)|$, we obtain

$$|d_{J,k}(F)| \geq C\Lambda_J(x) \geq C2^{-J\widetilde{H}_F(x)}.$$

As a result, $H_F(x) \leq \widetilde{H}_F(x)$. The opposite inequality is easier. Indeed, let for $x \in K$ and $\beta < \widetilde{H}_F(x)$,

$$P_{[\beta]}F_x(h) = \sum_{j=0}^{\infty} \sum_{i \in B_j} \lambda_i P_{[\beta]}(g \circ S_i^{-1})_x(h)$$

where $P_{[\beta]}(g \circ S_i^{-1})_x(h)$ is the Taylor polynomial of $g \circ S_i^{-1}$ of order $[\beta]$ at the point x. We immediately obtain by standard calculus

$$|F(x+h) - P_{[\beta]}F_x(h)| \leq C|h|^{\beta-\varepsilon}$$

for $|h|$ small enough and all $0 < \varepsilon < \beta$. As a consequence $\beta - \varepsilon \leq H_F(x)$. Letting $\varepsilon \to 0$ and next $\beta \to \tilde{H}_F(x)$, we obtain $\tilde{H}_F(x) \leq H_F(x)$.

The last step in multifractal analysis of functions using wavelet techniques is the computation of spectrum of irregularities which constitutes the multifractal formalism. In the case of exact self-similar functions, consider the measure

$$\mu(\Omega_i) = |\lambda_i| = |\lambda_{i_1}\lambda_{i_2}\ldots\lambda_{i_n}| \,, \forall\, i = (i_1, i_2 \ldots, i_n). \tag{8.22}$$

It yields a Gibbs measure ν_p concentrated on $E(\alpha)$ as

$$\nu_p(I_i) = \mu(I_i))^p |I_i|^{-\varphi(p)}, \tag{8.23}$$

where the parameter p and the function φ are related to the function analyzed. We immediately observe that

$$\mu(\Omega_{ij}) \simeq \mu(\Omega_i)\mu(\Omega_j) \,; \forall\, i = (i_1, i_2 \ldots, i_n),\ j = (j_1, j_2 \ldots, j_m). \tag{8.24}$$

Similarly, we have

$$|\Omega_{ij}| \simeq |\Omega_i|.|\Omega_j| \,; \forall\, i = (i_1, i_2 \ldots, i_n),\ j = (j_1, j_2 \ldots, j_m) \tag{8.25}$$

and

$$\limsup_{n\to+\infty} \frac{1}{n} \log\left(\sup_{|i|=n} |\Omega_i|\right) < 0. \tag{8.26}$$

Denote next

$$C_n(x,y) = \frac{1}{n} \log\left(\sum_{|i|=n} |\lambda_i|^x |\Omega_i|^{-y}\right)$$

and $C(x,y)$ its limit as $n \to \infty$. Hence, $\mathcal{C}^2$ is convex on $\mathbb{R}^2$, non-increasing on x and non-decreasing on y. Denote next $\Delta = \{(x,y);\ C(x,y) < 0\}$. There exists $\varphi : \mathbb{R} \to \overline{\mathbb{R}}$ non-decreasing and concave such that $int(\Delta) = \{(x,y) \in \mathbb{R}^2;\ y < \varphi(x-0)\}$, where int stands for the usual topological interior. Besides, $C(t, \varphi(t)) = 0$ for all $t \in \mathbb{R}$ and φ is $\mathcal{C}^1$ for all t such that $D_2C(t, \varphi(t)) \neq 0$. Let next $x \in K$ be such that $H_F(x) = \alpha$, and $r, s > 0$, $p \in \mathbb{R}$ and j be such that $2^{-j} \leq r < 22^{-j}$. It holds that

$$\frac{\nu_p(B(x,r))}{r^s} \sim \sum_{i \in B_j(x)} \frac{\nu_p(\Omega_i)}{|\Omega_i|^s} \sim \sup_{i \in B_j(x)} \frac{\nu_p(\Omega_i)}{|\Omega_i|^s} \sim \sup_{i \in B_j(x)} |\lambda_i|^p |\Omega_i|^{-\varphi(p)}.$$

So, for $s > \alpha\, p - \varphi(p)$, we obtain

$$\limsup_{r\to 0} \frac{\nu_p(B(x,r))}{r^s} = +\infty.$$

As a result, $\mathcal{H}^s(E(\alpha)) = 0$ and, consequently, $dim(E(\alpha)) \le s$ for all $s > \alpha\, p - \varphi(p)$ and for all $p \in \mathbb{R}$. Therefore,

$$d(\alpha) \le \inf_p (\alpha\, p - \varphi(p)).$$

We now investigate the lower bound of $d(\alpha)$. Let for $x \in \Gamma$, $i(x) = (i_1(x), i_2(x), ..., i_n(x), ...) \in \{0,1\}^{\mathbb{N}}$ its unique coding sequence and for $n \in \mathbb{N}$, $\Omega_n(x)$ the unique cell of order n containing x. Let also

$$\widehat{E(\alpha)} = \{x \; ; \; \lim_{n \to +\infty} \frac{\log |\lambda_{(i_1(x), i_2(x), ..., i_n(x))}|}{|\Omega_{(i_1(x), i_2(x), ..., i_n(x))}|} = \alpha\}$$

and

$$\widehat{E(\alpha)} = \{x \; ; \; \liminf_{n \to +\infty} \frac{\log \nu_p(\Omega_n(x))|}{|\Omega_n(x)|} \ge \inf_p (\alpha\, p - \varphi(p))\}.$$

It holds that $\nu_p(\widehat{E(\alpha)}) > 0$. Therefore, $\forall x \in \widehat{E(\alpha)}$ and $\forall s < \inf_p(\alpha\, p - \varphi(p))$, we get

$$\limsup_{r \to 0} \frac{\nu_p(B(x,r))}{r^s} < \infty,$$

which yields that $dim(\widehat{E(\alpha)}) \ge s$. Consequently,

$$d(\alpha) \ge \inf_p (\alpha\, p - \varphi(p)).$$

We now investigate the case of quasi-self-similar functions. We will apply the so-called Forstmann's method (see Appendix 8.7). The measure μ is defined in this case by

$$\mu(\Omega_i) = |\lambda_i| = |\lambda_{i_1}^1 \lambda_{i_2}^2 \dots \lambda_{i_n}^n|, \; \forall\, i = (i_1, i_2 \dots, i_n) \tag{8.27}$$

and 0 elsewhere. Denote next for p, q real numbers,

$$\mathcal{L}(p,q) = \liminf_{\varepsilon \downarrow 0} \left\{ \sum_s^* \mu(U_s)^p |U_s|^{-q} \; ; \; U_s \in \mathcal{F}_n \text{ and } |U_s| \le \varepsilon \right\}$$

and

$$\Theta(p,q) = \frac{1}{n} \log \sum_{U \in \mathcal{F}_n}^* \mu(U)^p |U|^{-q}.$$

Let also

$$\varphi(p) = \sup\{\, y \; ; \; \Theta(p,q) = \limsup_{n \to +\infty} \Theta_n(p,q) < 0\}.$$

Assume also that φ is of finite values on an open interval in $\mathbb{R}$ and consider the sequence of partition $\mathcal{F}_n = \{\Omega_i \,; \, |i| = n\}$ and $\zeta(U) = |\lambda_i|^p |\Omega_i|^{-\varphi(p)}$ and also 0 elsewhere. Denoting next $\alpha = \varphi'(p)$ and considering the sets

$$U_\alpha = \left\{ x \,; \; \lim_{n \to +\infty} \frac{\log |\lambda_i|}{\log |\Omega_i|} = \alpha \right\}, \text{ and } V_\alpha = \left\{ x \,; \; \liminf_{n \to +\infty} \frac{\log \widetilde{\zeta}(\Omega_i)}{\log |\Omega_i|} \ge \alpha\, p - \varphi(p) \right\},$$

we deduce from Lemma 8.17 that $U_\alpha \subset V_\alpha$. On the other hand, we know from classical Billingsley theory that $dim(U_\alpha) \geq \alpha\,p - \varphi(p)$. As a result we get

$$d(\alpha) \geq \alpha\,p - \varphi(p).$$

8.6 APPLICATION TO FINANCIAL INDEX MODELING

In this part, we reproduce with few details an application already developed by one of the authors and collaborators in [56] and originally by [175] on financial time series modelling. We will show precisely that multifractal functions, especially self-similar-type ones, may be good candidates to describe well financial markets. However, we will apply different data relatively to existing works.

Let $X(t)$, $t > 0$, be a time series issued from a financial index such as the well-known SP500, the most known US market index. We seek self-similar-type function F_X that minimizes the error $\|X - F_X\|$ relatively to an adapted norm on the sample support. In other words, the problem consists of finding a function F_X satisfying Definition 8.9 or Definition 8.12, allowing the best approximation of $X(t)$.

Consider the wavelet decomposition series of $X(t)$ relatively to an orthonormal wavelet basis $\{\psi_{j,k}\}$ in $L^2([0,1])$,

$$X(t) = \sum_k a_{J_0,k}\Phi_{J_0,k}(t) + \sum_{j=J_0}^{+\infty}\sum_k d_{j,k}\psi_{j,k}(t). \tag{8.28}$$

Denote as usual

$$AX_\varphi(t) = \sum_k a_{J_0,k}\Phi_{J_0,k}(t)$$

the approximation or the trend part of the series $X(t)$ and

$$DX_\psi(t) = \sum_{j=0}^{\infty}\sum_{k=0}^{2^j-1} d_{j,k}\psi(2^j t - k)$$

its detail component. To avoid the problem of vanishing wavelet coefficients and also the possibility to obtain abnormal values, we proceed differently from [175] by evaluating the scaling law between $\widehat{d}_{j,k}$ and $\widehat{d}_{j-1,[k/2]}$, where $[.]$ stands for the integer part and where the hat $\,\widehat{}\,$ is one of the estimators of the wavelet coefficients already introduced in Chapter 6 by setting

$$\widehat{d}_{j,k} = \nu_j^k \widehat{d}_{j-1,[k/2]}.$$

This permits us to obtain a sequence $(\nu_j^k)_{j,k}$ as in the following lemma.

Lemma 8.15 *The following assertions hold.*

1. $\widehat{d}_{j+1,k} = \widehat{d}_{0,0}\nu_1^{i_1(k)}\nu_2^{i_2(k)}\nu_3^{i_3(k)}...\nu_{j-2}^{i_{j-2}(k)}\nu_{j-1}^{i_{j-1}(k)}\nu_j^{i_j(k)}\nu_{j+1}^{i_{j+1}(k)}.$

2. $i_0(k) = k$ *and* $i_p(k) = \left[\dfrac{i_{p-1}}{2}\right]$, $\forall p \geq 1$.

3. $\dfrac{k}{2^j} = \sum_{p=1}^{j} \dfrac{i_p(k)}{2^p}.$

4. $\nu_j^k = \dfrac{\widehat{d}_{j,k}}{\widehat{d}_{j-1,[k/2]}}.$

Consequently,

$$DX_\psi(t) = \widehat{d}_{0,0} \sum_{j=0}^{\infty} \sum_{k=0}^{2^j-1} \nu_1^{i_1(k)} \nu_2^{i_2(k)} \nu_3^{i_3(k)} ... \nu_{j-2}^{i_{j-2}(k)} \nu_{j-1}^{i_{j-1}(k)} \nu_j^{i_j(k)} \psi(2^j t - k), \qquad (8.29)$$

which is a self-similar-type representation. However, to obtain an exact self-similar-type function, we should define a number of sequences independent of the decomposition levels $j \in \mathbb{N}$. The authors in [174, 175] provided an optimal construction by resolving an optimization problem leading to

$$\lambda_i^1 = \nu_i^1 \quad \text{and} \quad \lambda_i^n = \sum_{0 \le k < 2^{n-1}} P_k^n \nu_{2k+i}^n,$$

where for $k = 0, ..., 2^{n-1} - 1$,

$$P_k^n = \dfrac{\widehat{d}_{n-1,k} \prod_{j=1}^{n-1} (\lambda_{i_j}^j)^{n-j}}{\prod_{j=1}^{n-1} ((\lambda_1^j)^2 + (\lambda_2^j)^2)}.$$

$DX_\psi(t)$ becomes

$$DX_\psi(t) = \widehat{d}_{0,0} \sum_{j=0}^{\infty} \sum_{|i|=j} \prod_{s=1}^{j} \lambda_{i_s}^s \psi(2^j t - k). \qquad (8.30)$$

This means that DX_ψ is a quasi-self-similar process with

$$\Omega = [0,1], \quad S_i(x) = \dfrac{x+i}{2}; \ i = 1, 2 \quad \text{and} \quad G = \psi.$$

In [56], the authors have developed a more general model leading to a quasi-self-similar approximation. The readers may refer to such a reference for comparison and more details.

The model is next applied on the well-known SP500 index closing values during the period from September 20, 2004, to October 10, 2008 (sample size $N = 7260$). A wavelet decomposition is conducted in order to show eventual fluctuations in the series. This is illustrated by Figure 8.1.

Next, we plotted the spectrum of singularities of such series, which being nonlinear (concave) confirms the multifractal nature of the signal and which in turns motivates the self-similar-type modeling illustrated by Figure 8.3 below.

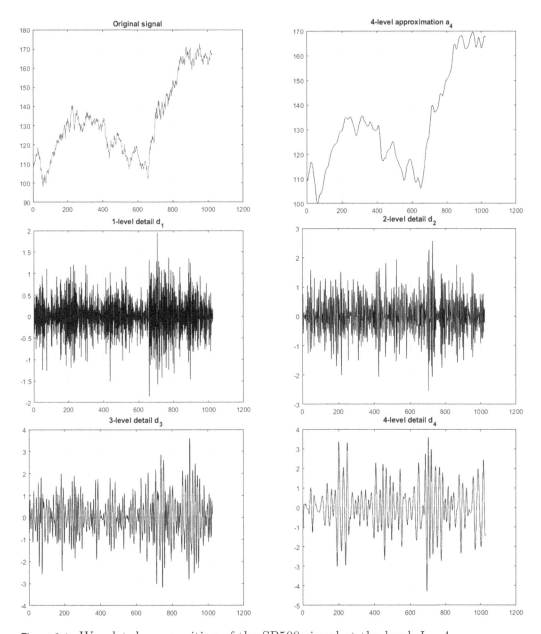

Figure 8.1 Wavelet decomposition of the SP500 signal at the level $J = 4$.

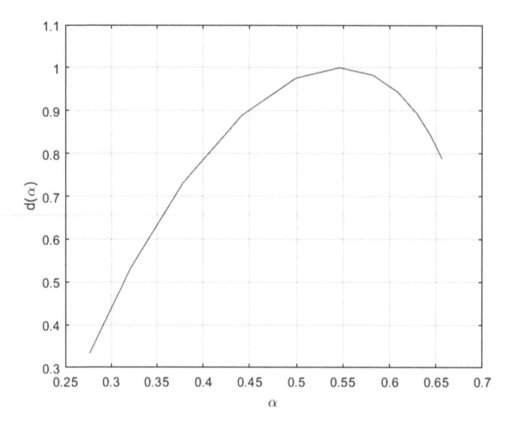

Figure 8.2 SP500 spectrum of singularities.

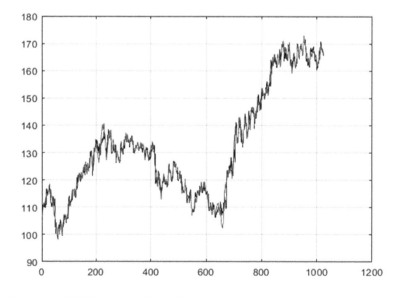

Figure 8.3 Original SP500 and self-similar-type model.

8.7   APPENDIX

Consider a Borel probability measure μ on Ω and a sequence of partitions $(\mathcal{F}_n)_n$ of Ω such that

- $\mathcal{F}_{n+1}$ is a refinement of $\mathcal{F}_n$.

- For $x \in \Omega$, let $\Omega_n(x) \in \mathcal{F}_n$ that contains x. Then $|\Omega_n(x)| \to 0$ as $n \to \infty$.

- $\sup\limits_{U \in \mathcal{F}_n} |U| \to 0$ as $n \to \infty$.

Denote $\mathcal{F} = \bigcup\limits_n \mathcal{F}_n$.

Definition 8.16 $\zeta : \mathcal{F} \to \mathbb{R}_+$ *is said to be a Forstmann's function on Ω if 0 is adherent to the sequence* $\left(\sup\limits_{U \in \mathcal{F}} \zeta(U) \right)_n$ *and if $\mathcal{H}(\zeta) > 0$ where*

$$\mathcal{H}(\zeta) = \liminf_{\varepsilon \downarrow 0} \left\{ \sum_s \zeta(U_s); \ \{U_s\} \ \varepsilon - covering \ of \ \Omega \right\}.$$

Forstmann's lemma reads as follows.

Lemma 8.17 *Let ζ be a Forstmann's function. There exists a probability measure $\tilde{\zeta}$ on Ω and constants $C, \varepsilon > 0$ such that*

$$\tilde{\zeta}(U) \leq C\zeta(U), \quad \forall U \in \mathcal{F}, |U| \leq \varepsilon.$$

8.8   EXERCISES

Exercise 1.

Let $F : \mathbb{R} \to \mathbb{R}$, $\alpha \geq 0$, be non-integer and t in the support of F. Show that F is α-Hölder at t if it satisfies

(i) For all $\gamma < \alpha$,
$$\lim_{h \to 0} \frac{|F(t+h) - P(h)|}{|h|^\gamma} = 0.$$

(ii) For all $\gamma > \alpha$,
$$\limsup_{h \to 0} \frac{|F(t+h) - P(h)|}{|h|^\gamma} = +\infty,$$

where P is a polynomial of degree less than $[\alpha]$.

Exercise 2.

Let $F : \mathbb{R} \to \mathbb{R}$, $0 < \alpha < 1$, and t in the support of F. Show that F is α-Hölder at t if it satisfies

$$\exists C, \rho_0 > 0; \ \forall \rho < \rho_0, \quad \sup_{x,y \in B(t,\rho)} |F(x) - F(y)| \leq C\rho^\alpha.$$

Exercise 3.

Consider the function

$$F(x) = \begin{cases} |x|^\alpha \sin(\dfrac{1}{|x|^\beta}) & \text{for } x \neq 0; \\ 0 & \text{for } x = 0. \end{cases}$$

with $0 < \alpha < \min(1, \beta)$. Show that

a. $H_F(0) = \alpha$.

b. $\tilde{H}_F(0) = \dfrac{\alpha}{1 + \beta}$.

Exercise 4.

Let f be an m-times continuously differentiable function and let $f_{j,k} = 2^{j/2} \langle f, \Phi_{j,k} \rangle$ its approximation coefficient relatively to a scaling function Φ with L vanishing moments. Prove that

$$|f_{j,k} - f(k2^{-j})| \leq C 2^{-mj}$$

with some constant C depending only on L.

Exercise 5.

1) Show that the polynomial P in the Definition 8.4 is unique.
2) Let f be α-Hölder at a point x and F a primitive of f near x. Does it hold that F $(\alpha + 1)$-Hölder at x?

Exercise 6.

For every $\varepsilon > 0$, let f_ε be the fractional primitive of a function f of order ε defined by

$$f_\varepsilon = (-\Delta)^{\varepsilon/2} f$$

and $H_{f_\varepsilon}(x)$ the Hölder exponent of f_ε at the point x.
1) Prove that $H_{f_\varepsilon}(x) \geq H_f(x) + \varepsilon$, $\forall \varepsilon > 0$.
2) Deduce that the following limit exists

$$\lim_{\varepsilon \to 0} \frac{\partial H_{f_\varepsilon}(x)}{\partial \varepsilon}.$$

Exercise 7.

For $s \in {]}0, 1[$, denote

$$F(x) = \sum_{n \geq 0} \frac{\sin(2^n x)}{2^{sn}}, \quad x \in \mathbb{R}.$$

1) Prove that $F \in \mathcal{H}^s(\mathbb{R})$.
2) Prove that the wavelet coefficient $d_{(j,k)}(F)$ satisfies

$$|d_{j,k}(F)| \leq C2^{-(\alpha+1/2)j}, \ \forall j, k \in \mathbb{Z}.$$

3) Deduce that $H_F(x) = s, \ \forall x \in \mathbb{R}$.

Exercise 8.

Let μ be a positive Borel probability measure on $[0,1]$ satisfying

$$\mu(B(t,r)) \leq Cr^s, \ \forall t \in [0,1], \ \forall 0 < r < 1.$$

For $t \in [0,1]$, denote

$$H_\mu(t) = \liminf_{r \to 0^+} \frac{\log(\mu(B(t,r)))}{\log r},$$

where $B(t,r)$ denotes the open ball with center t and radius r. Consider next the function F defined by its wavelet series

$$F = \sum_{j,k} \mu(I_{j,k})\psi_{j,k},$$

where $I_{j,k} = [k2^{-j}; (k+1)2^{-j})$.
1) Prove that $F \in \mathcal{H}^s([0,1])$.
2) Prove that $\mathcal{H}_F = H_\mu$ on $[0,1]$.
3) Deduce the equality $d_F \equiv d_\mu$, where d_F and d_μ are, respectively, the spectrum of singularities of F and μ.
4) Study the case where

$$\tilde{F} = \sum_{j,k} 2^{-\alpha j} (\mu(I_{j,k}))^\beta \psi_{j,k}$$

for some constants $\alpha, \beta > 0$.
5) Let $(\eta_{j,k})_{j,k}$ be a sequence of random variable i.i.d. with common law $\mathcal{N}(0,1)$ and consider the function

$$G = \sum_{j,k} \mu(I_{j,k})\eta_{j,k}\psi_{j,k}.$$

Prove that almost surely $H_G = H_F$ on $[0,1]$.

Exercise 9.

For $j \in \mathbb{N}$, consider the partition $\mathcal{F}_j$ of $[0,1)$ into dyadic intervals $I_{j,k} = [k2^{-j}; (k+1)2^{-j})$. Let also μ be a positive Borel probability measure on $[0,1]$ and denote for $q \in \mathbb{R}$,

$$\tau_\mu(q) = \liminf_{j \to \infty} \frac{1}{\log 2^{-j}} \log\left(\sum_k \mu(I_{j,k})^q\right).$$

Prove that

$$d_\mu(\alpha) \leq \tau_\mu^*(\alpha) = \inf_{q \in \mathbb{R}}(\alpha q - \tau_\mu(q)).$$

Exercise 10.

Denote for $x \in \mathbb{R}$, δ_x the well-known Dirac weight at x. Consider next the measure

$$\mu = \sum_{j \geq 1} \frac{1}{j^2 2^j} \sum_{k \geq 1} \delta_{2k+1}.$$

Denote next

$$\Gamma_x = \{\eta \geq 0\,;\, |x - (2k+1)2^{-j}| \leq 2^{-\eta j} \text{ for a finite number of } j, k\}$$

and $R_x = \sup \Gamma_x$.
1) Prove that $R_x \in [1, +\infty]$ for all $x \in [0, 1]$.
2) For every $s \in [1, +\infty]$ find $x \in [0, 1]$ such that $R_x = s$.
3) Prove that $H_\mu(x) = \dfrac{1}{R_x}$ for all x.
4) Deduce that $Support(d_\mu) = [0, 1]$.
5) Prove that $d_\mu(\alpha) = \alpha$ for all α.

Bibliography

[1] A. Abouelaz, R. Daher and L. El Mehdi, Harmonic analysis associated with the generalized q-Bessel operator, International Journal of Analysis and Applications, 10 (1) (2016), 17-23.

[2] R. Abreu-Blaya, J. Bory-Reyes and P. Bosch, Extension theorem for complex clifford algebras-valued functions on fractal domains, Boundary Value Problems, 2010, Article ID 513186, 9 pages.

[3] P. Abry, P. Goncalves and J. Levy Vehel. Scaling, Fractals and Wavelets. ISBN 978-1-84821-072-1. Wiley, 2009.

[4] R. A. Adams, Sobolev spaces. Academic press. 1978.

[5] G. D. Akrivis, Finite difference discretization of the Kuramoto-Sivashinsky equation, Numerische Mathematik, 63 (1992), 1-11.

[6] B. Aktan, M. Ozturk, N. Rhaeim and A. Ben Mabrouk, Wavelet-based systematic risk estimation: an application on Istanbul Stock Exchange, International Research Journal of Finance and Economics, 1 (23) (2009), 34-45.

[7] S. T. Ali, J. P. Antoine and J. P. Gazeau, Coherent states, wavelets and their generalizations, Graduate texts in contemporary physics, Springer, 2000.

[8] R. F. Al Subaie and M. A. Morou, The continuous wavelet transform for a Bessel type operator on the half line, Mathematics and Statistics, 1(4) (2013), 196-203.

[9] M. Alvarez and G. Sansigre, On polynomials with interlacing zeros, in: C. Brezinski, et al. (Eds.), Polynomes Orthogonaux et Applications. Proceedings, Bar-le-Duc 1984, Springer, Berlin, 1985, pp. 255-258.

[10] M. H. Annaby and Z. S. Mansour, q-Taylor and interpolation series for Jackson q-difference operators, Journal of Mathematical Analysis and Applications, 344 (2008), 472-483.

[11] J.-P. Antoine, R. Murenzi and P. Vandergheynst, Directional wavelets revisited: Cauchy wavelets and symmetry detection in patterns, Applied and Computational Harmonic Analysis 6 (1999), 314-345.

[12] J.-P. Antoine, R. Murenzi and P. Vandergheynst, Two-dimensional directional wavelets in image processing, International Journal of Imaging Systems and Technology, 7(3) (1996), 152-165.

[13] J.-P. Antoine and P. Vandergheynst, Wavelets on the n-sphere and related manifolds, Journal of Mathematical Physics 39 (1998), 3987-4008.

[14] A. Antoniadis, G. Gregoire and I. W. McKeague, Wavelet methods for curve estimation. Journal of the American Statistical Association, 89 (1994), 1340-1353.

[15] A. Antoniadis and G. Oppenheim, Wavelets and statistics. Lecture notes in statistics, 103, Berlin/New York, Springer-Verlag, 1995.

[16] J. Aouidi and M. Ben Slimane, Multifractal formalism for quasi self similar functions, Journal of Statistical Physics, 108 (2002), 541-590.

[17] J. Aouidi and M. Ben Slimane, Multifractal formalism for non self similar functions, Integral Transforms and Special Functions, 15(3) (2004), 189-207.

[18] J. Aouidi and A. Ben Mabrouk, Multifractal analysis of some weighted quasi self similar functions, International Journal of Wavelets, Multiresolution and Information Processing, 9(6) (2011), 965-987.

[19] A. Aral, V. Gupta and R. P. Agarwal, Applications of q-calculs in operator theory, Springer, New York, 2013, pp. 1-262.

[20] S. Arfaoui, I. Rezgui and A. Ben Mabrouk, Wavelet Analysis on the Sphere: Spheroidal Wavelets. Walter de Gruyter (March 20, 2017), ISBN-10: 311048109X, ISBN-13: 978-3110481099.

[21] S. Arfaoui and A. Ben Mabrouk, Some ultraspheroidal monogenic Clifford Gegenbauer Jacobi polynomials and associated wavelets. Advances in Applied Clifford Algebras (2017) 27: 2287.

[22] S. Arfaoui and A. Ben Mabrouk, Some old orthogonal polynomials revisited and associated wavelets: two-parameters Clifford-Jacobi polynomials and associated spheroidal wavelets. Acta Applicandae Mathematicae 155(1) (2018), 177–195.

[23] S. Arfaoui and A. Ben Mabrouk, Some generalized Clifford-Jacobi polynomials and associated spheroidal wavelets, Analysis in Theory and Applications, Accepted June 2020. http://arxiv.org/abs/1704.03513

[24] S. Arfaoui, A. Ben Mabrouk and C. Cattani, New type of Gegenbauer-Hermite monogenic polynomials and associated Clifford wavelets. Journal of Mathematical Imaging and Vision, 62(1) (2020), 73-97.

[25] S. Arfaoui, A. Ben Mabrouk and C. Cattani, New type of Gegenbauer-Jacobi-Hermite monogenic polynomials and associated continuous Clifford wavelet transform, Acta Applicandae Mathematicae, (2020), 1-35.

[26] A. Arneodo, E. Bacry, S. Jaffard and J. F. Muzy, Oscillating singularities on cantor sets: a grand-canonical multifractal formalism, Journal of Statistical Physics, 87(1/2) (1997), 179-209.

[27] A. Arneodo, E. Bacry and J. F. Muzy, Random cascades on wavelet dyadic trees, Journal of Mathematical Physics, 39(8) (1998), 4142-4164.

[28] A. Arneodo, E. Bacry and J. F. Muzy, Singularity spectrum of fractal signals from wavelet analysis, Journal of Statistical Physics 70(314) (1993), 635-674.

[29] A. Arneodo, E. Bacry and J. F. Muzy, Wavelet analysis of fractal signals. Direct determination of the singularity spectrum of fully developed turbulence data, Springer-Berlin, 1991.

[30] M. J. Atia, An example of nonsymmetric semi-classical form of class $s = 1$, Generalization of a case of Jacobi sequence, International Journal of Mathematics and Mathematical Sciences, 24(10) (2000), 673-689.

[31] M. J. Atia, Explicit representations of some orthogonal polynomials, Journal of Integral Transforms and Special Functions, 18(10) (2007), 731-742.

[32] M. J. Atia and J. Alaya, Some generalized Jacobi polynomials, Computers & Mathematics with Applications, 45(4-5) (2003), 843-850.

[33] M. J. Atia, M. Benabdallah and R. S. Costas-Santos, Zeros of polynomials orthogonal with respect to a signed weight, Indagationes Mathematicae, 23(1-2) (2012), 26-31.

[34] M. J. Atia and S. Chneiguir, The exceptional Bessel polynomials, Journal of Integral Transforms and Special Functions, 25(6) (2014), 470-480.

[35] M. J. Atia and J. Zeng, An explicit formula for the linearization coefficients of Bessel polynomials, arXiv:1109.4660v1 [math.CA], 21 September 2011, pp. 1-10.

[36] M. J. Atia and S. Leffet, On orthogonal polynomials: semi classical and of second category, Journal of Integral Transforms and Special Functions, 23(1) (2012), 35-47.

[37] M. J. Atia, F. Marcellan and I. A. Rocha, On semi classical orthogonal polynomials: a quasi-definite functional of class1, Facta Universitatis, Series: Mathematics and Informatics, 17 (2002), 13-34.

[38] M. Audin, Un cours sur les fonctions speciales, Universite de Strasbourg, Novembre 2012.

[39] A. Z. Averbuch and V. A. Zheludev, Wavelet transforms generated by splines. International Journal of Wavelets, Multiresolution and Information Processing, 5(2) (2007), 257-291.

[40] A. Z. Averbuch, P. Neittaanmaki and V. A. Zheludev, Spline and Spline Wavelet Methods with Applications to Signal and Image Processing Volume I (2014). Springer. ISBN 978-94-017-8925-7.

[41] Z. Avazzadeh, M. H. Heydari and C. Cattani, C. Legendre wavelets for fractional partial integro–differential viscoelastic equations with weakly singular kernels, The European Physical Journal Plus, 134(7) 368 (2019), 13 pages. https://doi.org/10.1140/epjp/i2019-12743-6

[42] C. Azizieh, Modélisation des series financieres par un modele multifractal. Memoire d'actuaire, Universite Libre de Bruxelle, 2002.

[43] D. Barlet and J. L. Clerc, Le comportement a l'infini des fonctions de Bessel generalisees, Advances in Mathematics 61 (1986), 165-183.

[44] E. Baspinar, G. Citti and A. Sarti, A geometric model of multi-scale orientation preference maps via Gabor functions, Journal of Mathematical Imaging and Vision 60 (2018), 900–912 (2018). https://doi.org/10.1007/s10851-018-0803-3.

[45] F. Bowman, Introduction to Bessel Functions. Dover Publications, 2010.

[46] A. Ben Mabrouk and J. Aouidi, Lecture note on wavelet multifractal analysis and self similarities, LAP 2012, ISBN:978-3-8465-9646-3.

[47] A. Ben Mabrouk, Multifractal analysis of some non isotropic quasi-self-similar functions, Far East Journal of Dynamical Systems 7(1) (2005), 23-63.

[48] A. Ben Mabrouk and N. Ben Abdallah, Study of some multinomial cascades, Advances and Applications in Statistics 6(3) (2006), 95-303.

[49] A. Ben Mabrouk, Study of some nonlinear self-similar distributions, International Journal of Wavelets, Multiresolution and Information Processing, 5(6) (2007), 907-916.

[50] A. Ben Mabrouk, On some nonlinear non isotropic quasi self similar functions, Nonlinear Dynamics 51 (2008), 379-398.

[51] A. Ben Mabrouk, An adapted group dilation anisotropic multifractal formalism for functions, Journal of Nonlinear Mathematical Physics 15(1) (2008), 1-23.

[52] A. Ben Mabrouk, Higher order multifractal measures, Statistics and Probability Letters 78 (2008), 1412-1421.

[53] A. Ben Mabrouk, Wavelet analysis of nonlinear self-similar distributions with oscillating singularity, International Journal of Wavelets, Multiresolution and Information Processing 6(3) (2008), 1-11.

[54] A. Ben Mabrouk, Directionlets and some nonlineare self similarities, International Journal of Mathematical Analysis, 5(26) (2011), 1273-1285.

[55] A. Ben Mabrouk, N. Ben Abdallah and Z. Dhifaoui, Wavelet decomposition and autoregressive model for the prevision of time series, Applied Mathematics and Computation 199(1) (2008), 334-340.

[56] A. Ben Mabrouk, N. Ben Abdallah and M. E. Hamrita, A wavelet method coupled with quasi self similar stochastic processes for time series approximation, International Journal of Wavelets, Multiresolution and Information Processing, 9(5) (2011), 685-711.

[57] A. Ben Mabrouk and J. Aouidi, Multifractal analysis of weighted quasi-self-similar functions, International Journal of Wavelets, Multiresolution and Information Processing, 9(6) (2011), 965-987.

[58] A. Ben Mabrouk, M. Ben Slimane and J. Aouidi, A wavelet multifractal fromalism for simultaneous singularities of functions, International Journal of Wavelets, Multiresolution and Information Processing, 12(1) (2014).

[59] M. M. Ibrahim Mahmoud, A. Ben Mabrouk and M H. A. Hashim, Wavelet multifractal models for transmembrane proteins- series, International Journal of Wavelets, Multiresolution and Information Processing, 14(6) (2016) 1650044 (36 pages).

[60] A. Ben Mabrouk, M. Ben Slimane and J. Aouidi, Mixed multifractal analysis for functions: general upper bound and optimal results for vectors of self-similar or quasi-self-similar of functions and their superpositions, Fractals, 24(4) (2016) 1650039 (12 pages).

[61] A. Ben Mabrouk, M. L. Ben Mohamed and K. Omrani, Numerical solutions for PDEs modeling binary alloy-solidification dynamics. Proceedings of 2007 International Symposium on Nonlinear Dynamics. Journal of Physics, conference series 96 (2008) 012067.

[62] A. Ben Mabrouk, H. Kortass and S. Ben Ammou, Wavelet estimators for long memory in stock markets, International Journal of Theoretical and Applied Finance 12(3) (2009), 297-317.

[63] A. Ben Mabrouk, I. Kahloul and S.-E. Hallara, Wavelet-based prediction for governance, diversification and value creation variables, International Research Journal of Finance and Economics, 60 (2010), 15-28.

[64] A. Ben Mabrouk and O. Zaafrane, Wavelet fuzzy hybrid model for physico financial signals, Journal of Applied Statistics, 40(7) (2013), 1453-1463.

[65] F. Ben Nasr, Analyse multifractale de mesures, CRAS. Paris, Mathematiqus 319(I) (1994), 807-810.

[66] M. Ben Said and J. El kamel, Product formula for the generalized q-Bessel function, Journal of Difference Equation and Application (2016), DOI:10.1080.

[67] M. Ben Slimane, Etude du Formalisme Multifractal pour les Fonctions, These de Mathematiques Appliquees, Ecole Nationale des ponts et Chaussees, 1996.

[68] S. Bertoluzza, Adaptive wavelet collocation method for the solution of Burgers equation, Transport Theory and Statistical Physics, 25 (1996).

[69] S. Bertoluzza, C. Canuto and K. Urban, On the adaptive computation of integrals of wavelets, Istituo di analisi numirica del C.N.R, Pavia (1999).

[70] S. Bertolozza and S. Falletta, Wavelets on $]0,1[$ at large scales, Journal of Fourier Analysis and Applications, 9 (2003), 261-288.

[71] N. Bettaibi, F. Bouzeffour, H. Elmonser and W. Binous, Elements of harmonic analysis related to the third basic zeros order Bessel function, Journal of Mathematical Analysis and Applications, 342 (2008), 1203-1219.

[72] G. Beylkin, On the representation of operators in bases of compactly supported wavelets, SIAM. Journal on Numerical Analysis, 6(6) (1992), 1716-1740.

[73] S. Bouaziz, The q-Bessel wavelet packets, Advances in Analysis, 1(1) (2016), 27-39.

[74] F. Bouzeffour and H. Ben Mansour, On the zeros of the big q-Bessel functions and applications, arXiv: 1311.1165v1, 2013.

[75] F. Brackx, R. Delanghe and F. Sommen, Clifford analysis, Pitman Publication, 1982.

[76] F. Brackx, R. Delanghe and F. Sommen, Cauchy-Kowalewski theorems in Clifford analysis: a survey. Proceedings of the 11th Winter School on Abstract Analysis, (1984), pp. 55–70.

[77] F. Brackx and F. Sommen, Clifford-Hermite wavelets in Euclidean space, Journal of Fourier Analysis and Applications 6(3) (2000), 299–310.

[78] F. Brackx and F. Sommen, The generalized Clifford-Hermite continuous wavelet transform, Advances in Applied Clifford Algebras 11(1) (2001), 219–231.

[79] F. Brackx, N. De Schepper and F. Sommen, The Clifford-Laguerre continuous wavelet transform, Bulletin of the Belgian Mathematical Society, 10 (2003), 201-215.

[80] F. Brackx, N. De Schepper and F. Sommen, The Clifford-Gegenbauer polynomials and the associated continuous wavelet transform, Integral Transforms and Special Functions, 15(5) (2004), 387-404.

[81] F. Brackx, N. De Schepper and F. Sommen, The Clifford-Fourier transform, Journal of Fourier Analysis and Applications 11(6) (2005), 669–681.

[82] F. Brackx, N. De Schepper and F. Sommen, The two-dimensional Clifford-Fourier transform, Journal of Mathematical Imaging and Vision, 26 (2006), 5-18.

[83] F. Brackx, N. De Schepper and F. Sommen, Clifford-Jacobi polynomials and the associated continuous wavelet transform in euclidean space. Tao Qian, Mang I. Vai and Xu Yuesheng, Eds, Applied and Numerical Harmonic Analysis, (2006), pp. 185-198.

[84] F. Brackx, N. De Schepper and F. Sommen, Clifford-Hermite and Two-dimensional Clifford-Gabor Filters for Early Vision. Proceedings of the 17th International Conference on the Application of Computer Science and Mathematics in Architecture and Civil Engineering, 2006, Editors, K. Gurlebeck and C. Konke, Bauhaus-Universitat Weimar.

[85] F. Brackx, N. De Schepper and F. Sommen, Clifford-Hermite-monogenic operators, Czechoslovak Mathematical Journal, 56(4) (2006), 1301–1322.

[86] F. Brackx, N. De Schepper and F. Sommen, The Fourier transform in Clifford analysis, Advances in Imaging and Electron Physics, 156 (2009), 55-201.

[87] F. Brackx and F. Sommen, Clifford Bessel wavelets in Euclidean space, Mathematical Methods in Applied Sciences, 25(16-18) (2002), 1479-1491.

[88] F. Brackx, E. Hitzer and S. J. Sangwine, Quaternion and Clifford Fourier Transforms and Wavelets. In History of Quaternion and Clifford-Fourier Transforms and Wavelets, (2013), pp. 12–28; Springer Basel.

[89] G. Brown, G. Michon and J. Peyrierre, On the multifractal analysis of measures, Journal of Statistical Physics, 66 (1992), 775-779.

[90] R. Bujack, G. Scheuermann and E. Hitzer, A General Geometric Fourier Transform. In Quaternion and Clifford Fourier transforms and wavelets, Trends in Mathematics 27 (2013), Editors, Hitzer, E. and Sangwine, S. J., Birkhauser, Basel, pp. 155–176.

[91] R. Bujack, G. Scheuermann and E. Hitzer, A general geometric Fourier transform convolution theorem, Advances in Applied Clifford Algebras, 23(1) (2013), 15–38.

[92] R. Bujack, H. De Bie, N. De Schepper and G. Scheuermann, Convolution products for hypercomplex Fourier transforms, Journal of Mathematical Imaging and Vision, 48(2014), 606–624.

[93] R. Bujack, G. Scheuermann and E. Hitzer, Demystification of the geometric Fourier transforms and resulting convolution theorems, Mathematical Methods in the Applied Sciences, 39(7) (2015), 1877–1890.

[94] A. Bultheel, Wavelets with applications in signal and image processing, Publisher: Dept Computer Science, KU Leuven, October 25, 2002, 181 pages.

[95] M. Carlini, S. Castellucci, G. Sun, J. Leng, C. Cattani and A. Cardarelli, A wavelet-based optimization method for biofuel production, Energies 11(2) (2018), 17 pages, doi:10.3390/en11020377.

[96] P. Carre and M. Berthier, Color Representation and Processes with Clifford Algebra, Chapter 6, In Advanced Color Image Processing and Analysis, Editors: C. Fernandez-Maloigne, Springer, New York, (2013), pp. 147–179.

[97] C. Cattani, Wavelet based approach to fractals and fractal signal denoising, Transactions on Computational Science, 6 (2009), 143-162.

[98] C. Cattani, Fractals based on harmonic wavelets. ICCSA'09: Proceedings of the International Conference on Computational Science and Its Applications: Part I, July 2009, Pages 729–744https://doi.org/10.1007/978-3-642-02454-2-56

[99] C. Cattani and A. Kudreyko, Application of wavelet-basis for solution of the fredholm type integral equations. In: Taniar D., Gervasi O., Murgante B., Pardede E., Apduhan B.O. (eds) Computational Science and Its Applications – ICCSA 2010. ICCSA 2010. Lecture Notes in Computer Science, vol 6017. Springer, Berlin, Heidelberg. https://doi.org/10.1007/978-3-642-12165-4-13.

[100] C. Cattani and A Kudreyko, Mutiscale analysis of the Fisher equation. In: Gervasi O., Murgante B., Laganà A., Taniar D., Mun Y., Gavrilova M.L. (eds) Computational Science and Its Applications – ICCSA 2008. ICCSA 2008. Lecture Notes in Computer Science, vol 5072. Springer, Berlin, Heidelberg. https://doi.org/10.1007/978-3-540-69839-5-89.

[101] C. Cattani and A Kudreyko, On the discrete harmonic wavelet transform, Mathematical Problems in Engineering, Article ID 687318, 7 pages, 2008. https://doi.org/10.1155/2008/687318.

[102] C. Cattani and A. Kudreyko, Harmonic wavelet method towards solution of the Fredholm type integral equations of the second kind, Applied Mathematics and Computation, 215(12, 15 February)(2010), 4164-4171.

[103] C. Cattani, Harmonic wavelet approximation of random, fractal and high frequency signals, Telecommunication Systems 43(2010), 207–217. https://doi.org/10.1007/s11235-009-9208-3.

[104] C. Cattani, Haar wavelet splines, Journal of Interdisciplinary Mathematics 4(1) (2001), 35-47.

[105] C. Cattani, Shannon wavelets for the solution of integro-differential equations, Mathematical Problems in Engineering (2010), Article ID 408418 - 22 pages, https://doi.org/10.1155/2010/408418.

[106] C. Cattani, Fractional calculus and Shannon wavelet, Mathematical Problems in Engineering (2012), 2012 —Article ID 502812, 26 pages - https://doi.org/10.1155/2012/502812.

[107] C. Cattani, Connection coefficients of Shannon wavelets, Mathematical Modelling and Analysis 11(2) (2006), 117-132.

[108] C. Cattani, Harmonic wavelet solutions of the Schrodinger equation, International Journal of Fluid Mechanics Research 30(5) (2003), 10 pages.

[109] C. Cattani, Haar wavelet-based technique for sharp jumps classification, Mathematical and Computer Modelling: An International Journal 39 (2-3) (2004), 255-278.

[110] C. Cattani, On the existence of wavelet symmetries in archaea DNA, Computational and Mathematical Methods in Medicine (2012), Article ID 673934, 21 pages, https://doi.org/10.1155/2012/673934.

[111] C. Cattani, Multiscale analysis of wave propagation in composite materials, Mathematical Modelling and Analysis 8(4) (2003), 267-282.

[112] C. Cattani, The wavelet-based technique in dispersive wave propagation, International Applied Mechanics 39(4) (2003), 493-501.

[113] C. Cattani and M Pecoraro, Nonlinear differential equations in wavelet bases, Acoustic Bulletin, 3(4) (2000), 1-10.

[114] C. Cattani, Sinc-fractional operator on Shannon wavelet space, Frontiers in Physics 6(2018), Article 118, 16 pages. https://doi.org/10.3389/fphy.2018.00118

[115] C. Cattani and Y. Y. Rushchitskii, Solitary elastic waves and elastic wavelets, International Applied Mechanics 39(6) (2003), 741-752.

[116] C. Cattani, O. Doubrovina, S. Rogosin, S. L. Voskresensky and E. Zelianko, On the creation of a new diagnostic model for fetal well-being on the base of wavelet analysis of cardiotocograms, Journal of Medical Systems 30(6) (2006), 489-494.

[117] C. Cattani, Wavelet algorithms for DNA analysis. Chapter 35 In Algorithms in Computational Molecular Biology: Techniques, Approaches and Applications, Book Editor(s): Mourad Elloumi Albert Y. Zomaya, First published: 23 December 2010, https://doi.org/10.1002/9780470892107.ch35.

[118] C. Cattani, Harmonic wavelet solution of Poisson's problem, Balkan Journal of Geometry & Its Applications 13(1) (2008), 27-37.

[119] C. Cattani, A review on harmonic wavelets and their fractional extension, Journal of Advanced Engineering and Computation 2(4) (2018), 224-238.

[120] C. Cattani, Wavelet based approach to fractals and fractal signal denoising, Transactions on Computational Science 6 (2009), 143-162.

[121] C. Cattani, Wave propagation of Shannon wavelets. In: Gavrilova M. et al. (eds) Computational Science and Its Applications - ICCSA 2006. ICCSA 2006. Lecture Notes in Computer Science, vol 3980. Springer, Berlin, Heidelberg. https://doi.org/10.1007/11751540-85.

[122] C. Cattani, Second order Shannon wavelet approximation of C^2-functions, University Politehnica of Bucharest Scientific Bulletin-Series A-Applied Mathematics and Physics 73(3) (2011), 73-84.

[123] C. Cattani and L. M. Sanchz Ruiz, Discrete differential operators in multidimensional Haar wavelet spaces, International Journal of Mathematics and Mathematical Sciences 44 (2004), 2347-2355, Article ID 480617, https://doi.org/10.1155/S0161171204307234.

[124] C. Cattani, Wavelet solutions of evolution problems, Accademia Peloritana dei Pericolanti-Classe di Scienze FF. MM. NN. 80 (1) (2002), 21-34.

[125] C. Cattani and A. Ciancio, Energy wavelet analysis of time series, Accademia Peloritana dei Pericolanti - Classe di Scienze FF. MM. NN., LXXX (1) (2002), 105-115.

[126] C. Cattani and A. Ciancio, Wavelet estimate of time series, Accademia Peloritana dei Pericolanti - Classe di Scienze FF.MM.NN., 78-79 (1) (2001), 159-170.

[127] C. Cattani and A. Ciancio, Wavelet clustering in time series analysis, Balkan Journal of Geometry and Its Applications 10(2) (2005), 33-44.

[128] C. Cattani, A. Ciancio, Wavelet analysis of linear transverse acoustic waves, Accademia Peloritana dei Pericolanti - Classe di Scienze FF. MM. NN., LXXX (1) (2002), 83-103.

[129] C. Cattani, Local Fractional Calculus on Shannon Wavelet Basis. In: Fractional Dynamics, Publisher: De Gruyter, Editors: C. Cattani, H. Srivastava, Xiao-Jun Yang, January 2016, DOI: 10.1515/9783110472097-002.

[130] C. Cattani, Wavelet solutions in elastic nonlinear oscillations. 11th Conference on Waves and Stability in Continuous Media, Porto Ercole (Grosseto), Italy, 3-9 June 2001: WASCOM 2001: proceedings, pp. 144–149.

[131] C. Cattani and I Bochicchio, Wavelet analysis of chaotic systems, Journal of Interdisciplinary Mathematics 9(3) (2006), 445–458.

[132] C. Cattani, Wavelet, approach to stability-of-orbits analysis. International Applied Mechanics 42(6) (2006), 721–727.

[133] C. Cattani, P Mercorelli, F Villecco, K Harbusch, A theoretical multiscale analysis of electrical field for fuel cells stack structures. International Conference on Computational Science and Its Applications, (2006), 857-864.

[134] C. Cattani, Harmonic wavelet solution of Poisson's problem with a localized source. Atti dell'Accademia Peloritana dei Pericolanti Classe di Scienze Fisiche, Matematiche e Naturali 86(2) (2008), 14 pages.

[135] C. Cattani, Fractals based on harmonic wavelets. ICCSA'09: Proceedings of the International Conference on Computational Science and Its Applications: Part IJuly 2009 Pages 729–744. https://doi.org/10.1007/978-3-642-02454-2-56.

[136] C. Cattani, Harmonic wavelet analysis of nonlinear waves, Technische Mechanik 28(2) (2008)

[137] C. Cattani, M. Ehler, M. Li, Z. Liao and M. Hooshmandasl, Scaling, self-similarity, and systems of fractional order. Abstract and Applied Analysis 201412014

[138] C. Cattani, Wavelet analysis of self similar functions. Journal of Dynamical Systems and Geometric Theories 9 (1) (2013), 75-97

[139] C. Cattani, Sparse representation with harmonic wavelets. 2009 Sixth International Conference on Fuzzy Systems and Knowledge Discovery -12009

[140] C. Cattani and M Scalia, Wavelet analysis of spike train in the Fitzhugh model. Transactions on Computational Science VI (1) (2009), 163-179.

[141] C. Cattani and M Scalia, Wavelet analysis of pulses in the Fitzhugh model. International Conference on Computational Science and Its Applications, 2008

[142] C. Cattani, Wavelet solution for the momentless state equations of an hyperboloid shell with localized stress. International Conference on Computational Science and Its Applications, 490-49912007

[143] C. Cattani and LMS Ruiz, On the differentiable structure of Meyer wavelets. International Conference on Computational Science, 1004-101112007

[144] C. Cattani and A. Ciancio, On the Haar wavelet analysis of jumps. Proceedings of The 3-rd International Colloquium, Mathematics in Engineering and Numerical Physics, October 7-9 , 2004, Bucharest, Romania, p. 72-81.

[145] C. Cattani, Wavelets in the propagation of waves in materials with microstructure. 2003 IEEE International Workshop on Workload Characterization (IEEE Cat. No -12003

[146] C. Cattani, Shannon wavelets theory Mathematical Problems in Engineering (2008), Article ID 164808, 24 pages.

[147] C. Cattani, Wavelet analysis of solitary wave evolution. International Conference on Computational Science and Its Applications, 604-613, 2005

[148] C. Cattani, Haar wavelet splines. Journal of Interdisciplinary Mathematics 4 (1) (2001), 35-47.

[149] C. Cattani and RJ Jaroslavich, Wavelet and Wave Analysis as applied to Materials with Micro or Nanostructure, World Scientific, 2007

[150] C. Cattani, Harmonic wavelets towards the solution of nonlinear PDE. Computers & Mathematics with Applications 50 (8-9) (2005), 1191-1210.

[151] C. Cattani, Shannon wavelets theory. Mathematical Problems in Engineering 20081022008

[152] C. Cattani, Signorini cylindrical waves and Shannon wavelets. Advances in Numerical Analysis, 2012, Article ID 731591, 24 pages.

[153] C. Cattani, Cantor waves for Signorini hyperelastic materials with cylindrical symmetry. Axioms 9 (1), 222020

[154] N. N. Čentsov, Evaluation of an unknown distribution density from observations. Soviet Mathematics. Doklady, 3 (1962), 1559–1562.

[155] C. Chatfied, The Analysis of Time Series, An Introduction, Chapman / Hall 1989 Fourth Edition.

[156] Y. Chen, M. E. H. Ismail and K. A. Muttalib, Asymptotics of basic Bessel functions and q-Laguerre polynomials, Journal of Computational and Applied Mathematics 54(1994), 263-272.

[157] M. Q. Chen, C. Hwang and P. Shih, A wavelet Galerkin method for solving population balance equations. Computers & Chemical Engineering, 20(2) (1996), 131-145.

[158] C. Chevalley, The Algebraic Theory of Spinors and Clifford Algebras: Collected Works, Volume 2 (Collected Works of Claude Chevalley) (v. 2), Springer, 1996.

[159] R. Chteoui, S. Arfaoui and A. Ben Mabrouk, Study of a generalized nonlinear Euler-Poisson-Darboux system - numerical and Bessel based solutions. Journal of Partial Differential Equations, 33 (2020), 313-340.

[160] S. M. Choo, S. K. Chung and Y. J. Lee, Convergence of finite difference scheme for generalized Cahn-Hillard and Kuramoto-Sivashinsky equations. To appear in Comput. Math. Appl.

[161] C. K. Chui and Jian-zhong Wang (1991). A cardinal spline approach to wavelets. Proceedings of the American Mathematical Society. 113 (3): 785-793. doi:10.2307/2048616. JSTOR 2048616. Retrieved 22 January 2015.

[162] C. K. Chui and Jian-Zhong Wang (April 1992). On Compactly Supported Spline Wavelets and a Duality Principle. Transactions of the American Mathematical Society. 330 (2): 903-915. doi:10.1090/s0002-9947-1992-1076613-3. Retrieved 21 December 2014.

[163] C. K Chui (1992). An Introduction to Wavelets. Academic Press.

[164] A. Ciancio and C. Cattani, Analysis of singularities by short haar wavelet transform. International Conference on Computational Science and Its Applications, 828-83832006

[165] V Ciancio and C. Cattani, On the propagation of transverse acoustic waves in the form of harmonic wavelets in isotropic media. Conference on Applied Geometry-General Relativity and the Workshop of Global -42003

[166] Cifter A and Ozun A (2007) Multiscale systematic risk: An application on ISE 30, MPRA Paper 2484, University Library of Munich: Germany

[167] W. K. Clifford, On the classification of geometric algebras, Mathematical Papers, (1882), 397–401.

[168] W. K. Clifford, Applications of Grassmann's extensive algebra. American Journal of Mathematics, 1(4) (1878), 350.

[169] T. Conlon, M. Crane and H.J. Ruskin (2008) Wavelet multiscale analysis for hedge funds: Scaling and strategies. Physica A: Statistical Mechanics and Its Applications, 387 (21) (2008), 5197-5204.

[170] M. J. Craddock and J. A. Hogan, The fractional Clifford-Fourier kernel. The Erwin Schrödinger Internationa Institute for Mathematical Physics ESI, Vienna, Preprint ESI 2411, 2013.

[171] A. K. Common and F. Sommen, Axial monogenic functions from holomorphic functions. Journal of Mathematical Analysis and Applications 179(2) (1993), 610–629.

[172] T. Cooklev, G.I. Berbecel and A.N. Venetsanopoulos. Wavelets and differential–dilation equations. IEEE Transactions on Signal Processing 48 (8) (2000), 2258—2268.

[173] K. Daoudi, J. L. Véhel and Y. Meyer, Construction of continuous functions with prescribed local regularity. Constructive Approximation, 14(3) (1998), 349-385.

[174] K. Daoudi, Généralisations des systémes de fonctions itérées: applications au traitement du signal. Thèse de mathématiques appliquées, Paris 9 Dauphine, 1996.

[175] K. Daoudi and J. Lévy Véhel, Signal representation and segmentation based on multifractal stationarity. Signal Processing 82 (2002), 2015-2024.

[176] G. Dattoli and A. Torre, Symmetric q-Bessel functions. Le Matematiche Vol. LI(1996), 153-167.

[177] I. Daubechies, Ten Lectures on Wavelets, Society for Industrial and Applied Mathematics, Philadelphia, PA, USA (1992)

[178] Daubechies, I. Orthonormal bases of compactly supported wavelets. Communications on Pure and Applied Mathematics 1988, 41, 909-996.

[179] I. Daubechies and J. C. Lagarias, Two-scale difference equations, 1. existence and global regularity of solutions. SIAM Journal on Mathematical Analysis, 22(5) (1991), 1388-1410.

[180] I. Daubechies and J. C. Lagarias, Two-scale difference equations, 2. local regularity, infinite products of matrices and fractals of solutions. SIAM Journal on Mathematical Analysis, 24 (1992), 1031-1079.

[181] H. De Bie, Clifford algebras, Fourier transforms and quantum mechanics, arXiv: 1209.6434v1, September 2012, 39 pages.

[182] H. De Bie and N. De Schepper, The fractional Clifford-Fourier transform. Complex Analysis and Operator Theory 6(5) (2012), 1047–1067.

[183] H. De Bie and Y. Xu, On the Clifford-Fourier transform. ArXiv:1003.0689, December 2010, 30 pages.

[184] H. De Bie and Y. Xu, On the Clifford-Fourier transform. International Mathematics Research Notices 22 (2011), 5123–5163.

[185] R. Delanghe, On regular-analytic functions with values in a Clifford algebra. Mathematische Annalen 185 (1970), 91–111.

[186] R. Delanghe, F. Sommen and V. Souček, Clifford Algebra and Spinor-valued Functions. Springer Netherlands, 1992.

[187] R. Delanghe, Clifford analysis: history and perspective. Computational Methods and Function Theory, 1 (1) (2001), 107-153.

[188] B. Delyon and A. Juditsky, On the computation of wavelet coefficients. Journal of Approximation Theory, 88(1), (1997).

[189] B. Delyon and A. Juditsky, On minimax wavelet estimators. Applied Computational Harmonic Analysis, 3 (1996), 215-228.

[190] N. De Schepper, Multi-dimensional continuous wavelet transforms and generalized Fourier transforms in Clifford analysis. PhD Thesis,Ghent University, 2006.

[191] E. Deriaz, Ondelettes pour la Simulation des Ecoulements Fluides Incompressibles en Turbulence. Thèse de Doctorat de l'Institut National Polytechnique de Grenoble, 2006.

[192] E. Deriaz and V. Perrier, Divergence-free Wavelets in 2D and 3D, application to the Navier-Stokes equations, Journal of Turbulence, 7(3) (2006), 1-37.

[193] L. Dhaouadi, On the q-Bessel Fourier transform. Bulletin of Mathematical Analysis and Applications, 5(2) (2013), 42-60.

[194] L. Dhaouadi and M. J. Atia, Jacobi operators, q-dfference equations and orthogonal polynomials, arXiv:1211.0359v1, 2 Nov 2012, 22 pages.

[195] L. Dhaouadi and M. Hleili, Generalized q-Bessel operator. Bulletin of Mathematical Analysis and Applications, 7(1) (2015), 20-37.

[196] L. Dhaouadi, A. Fitouhi and J. El Kamel, Inequalities in q-Fourier analysis. Journal of Inequalities in Pure and Applied Mathematics, 7(5) (2006), Article 171, 14 pages.

[197] L. Dhaouadi, W. Binous and A. Fitouhi, Paley-Wiener theorem for the q-Bessel transform and associated q-sampling formula. Expositiones Mathematicae 27 (2009), 55-72.

[198] R. DiSario, H. Saroglu, J. McCarthy and H. Li, Long memory in the volatility of an emerging equity market: The case of Turkey. International Markets Institutions and Money 18 (2008), 305-312.

[199] D. Donoho and I. Johnstone, Minimax estimation via wavelet shrinkage, Tech. Report, Stanford University, 1991.

[200] D. Donoho and I. Johnstone, Ideal spatial adaptation by wavelet shrinkage, Biometrika 81 (1994), 425-455.

[201] D. Donoho and I. Johnstone, Adapting to unknown smoothness via wavelet shrinkage, Journal of the American Statistical Association 90 (1995), 1200-1224.

[202] D. Donoho, I. Johnstone, G. Kerkyacharian and D. Picard, Wavelet shrinkage: Asymptopia. Journal of the Royal Statistical Society, Series B 57 (1995), 301-369.

[203] D. Donoho, I. Johnstone, G. Kerkyacharian and D. Picard, Density estimation by wavelet thresholding. Annals of Statistics 24 (1996), 508-539.

[204] H. Exton, q-Hypergeometric functions and applications, Ellis Horwood Series in Mathematics and Application (E. Horwood, Chichester, West Sussex, 1983).

[205] Y. A. Farkov, B-spline wavelets on the sphere. Proc. Intern. Workshop Self-Similar Systems (July 30 - August 7, 1998). Editors V.B.Priezzhev, V.P.Spiridonov. - Dubna: Joint Institute for Nuclear Research, 1999, pp. 79-82.

[206] J. M. Fedou, Note sur les fonctions de Bessel. Discete Mathematics 139 (1995), 473-480.

[207] V. Fernandez, The CAPM and value at risk at different time-scales. International Review of Financial Analysis 15 (2006), 203-219.

[208] G. Ferrarese and C. Cattani, Generalized frames of references and intrinsic Cauchy problem in General Relativity. Physics on Manifolds, 93-10931994

[209] U. Frisch and G. Parisi, Fully developped turbulence and intermittency. Proc. Int. Summer school Phys., Enrico Fermi, North Holland (1985), pp. 84-88.

[210] R. Fueter, Zur Theorie der regularen funktionen einer quaternionenvariablen. Monatshefte für Mathematik und Physik 43 (1935), 69–74.

[211] A. Fitouhi and L. Dhaouadi, Positivity of the generalized translation associated with the q-Hankel transform. Constructive Approximation 34 (2011), 453-472.

[212] A. Fitouhi and N. Bettaibi, Wavelet transforms in quantum calculus. Journal of Nonlinear Mathematical Physics 13(3) (2006), 492-506.

[213] A. Fitouhi, N. Bettaibi and W. Binous, Inversion formulas for the q-Riemann-Liouville and q-Weyl transforms using wavelets. Fractional Calculus and Applied Analysis 10(4) (2007), 327-342.

[214] A. Fitouhi, M. M. Hamza and F. Bouzeffour, The q-j_α Bessel function. Journal of Approximation Theory 115(2002), 144-166.

[215] A. Fitouhi and A. Safraoui, Paley-Wiener theorem for the q^2-Fourier-Rubin transform. Tamsui Oxford Journal of Mathematical Sciences 26(3) (2010), 287-304.

[216] A. Fitouhi, K. Trimeche and J. L. Lions, Transmutation operators and generalized continuous wavelets, Preprint, Faculty of Science of Tunis (1995).

[217] M. Frazier and Sh. Zhang, Bessel wavelets and the Galerkin analysis of the Bessel operator. Journal of Mathematical Analysis and Applications 261(2) (2001), 665-691.

[218] W. Freeden and V. Michel, Multiscale Potential Theory (With Applications to Geoscience), Birkhäuser Verlag, Boston, 2004.

[219] W. Freeden and M. Schreiner, Orthogonal and non-orthogonal multiresolution analysis, scale discrete and exact fully discrete wavelet transform on the sphere. Constructive Approximation, 14(4) (1998), 493-515.

[220] W. Freeden and U. Windheuser, Spherical wavelet transform and its discretization. Advances in Computational Mathematics 5 (1996), 51-94.

[221] N. Fukuda, T. Kinoshita and K. Yoshino, Wavelet transforms on Gelfand-Shilov spaces and concrete examples. Journal of Inequalities and Applications 119 (2017), 1-24.

[222] Y. Gagne, Etude expérimentale de l'intermittence et des singularités dans le plan complexe en turbulence développée. Thèse de l'Université de Grenoble, 1987.

[223] T. Gao, Z.-G. Liu, S.-H. Yue, J.-Q. Mei and J. Zhang, Traffic video-based moving vehicle detection and tracking in the complex environment. Cybernetics and Systems: An International Journal 40(7) (2009), 569–588.

[224] G. Gasper and M. Rahman, Basic Hypergeometric Series, Encyclopedia of Mathematics and its Application, Combridge university Press, Cambridge, UK, vol.35(1990).

[225] R. Gencay, F. Seluk and B. Whitcher, An Introduction to Wavelets and Other Filtering Methods in Finance and Economics. Academic Press, 2001.

[226] R. GenÃğay, B. Whitcher and F. SelÃğuk, Systematic risk and time scales. Quantitative Finance 3 (2003), 108-116.

[227] R. GenÃğay, B. Whitcher and F. SelÃğuk, Multiscale systematic risk. Journal of International Money and Finance 24 (2005), 55-70.

[228] J. C. Goswami and A. K. Chan, Fundamentals of Wavelets Theory, Algorithms and Applications, John Wiley & Sons, Inc. Royaume-Uni, 1999.

[229] A. Grossman and J. Morlet, Decomposition of Hardy functions into square integrable wavelets of constant shape. SIAM Journal on Mathematical Analysis 15(4) (1984), 723-736.

[230] B.-Y. Guo, J. Shen and L.-L. Wang, Generalized Jacobi polynomials/functions and their applications. Applied Numerical Mathematics 59 (2009), 1011-1028.

[231] D. T. Haimo, Integral equations associated with Hankel convolution, Transactions of the American Mathematical Society 116 (1965), 330-375.

[232] P. Hall, G. Kerkyacharian and D. Picard, Adaptive minimax optimality of block thresholded wavelet estimators. Statistica Sinica 9(1) (1996), 33-49.

[233] P. Hall and P. Patill, Formelae for mean integrated squared error of nonlinear wavelet-based density estimators. Annals of Statistics 23(3) (1995), 905-928.

[234] P. Hall and P. Patill, On wavelet methods for estimating smooth functions. Bernouilli 1 (1995), 41-58.

[235] A. E. Hamza and M. H. Al-Ashwal, Leibniz's rule and Fubini's theorem, associated with power quantum difference operators. International Journal of Mathematical Analysis 9 (55) (2015), 2733-2744.

[236] W. Hardle, G. Kerkyacharian, D. Picard, and A. Tsybakov, Wavelets, approximation and statistical applications. Seminar Berlin-Paris (1997)

[237] C. Heil, A Basis Theory Primer. Birkhauser. pp. 177-188.

[238] M. H. Heydari, M. R. Hooshmandasl, C. Cattani and Ming Li, Legendre Wavelets Method for Solving Fractional Population Growth Model in a Closed System, Mathematical Problems in Engineering (2013), Article ID 161030, 8 pages.

[239] M. H. Heydari, M. R. Hooshmandasl, A. Shakiba and C. Cattani, Legendre wavelets Galerkin method for solving nonlinear stochastic integral equations. Nonlinear Dynamics 85 (2016), 1185-1202.

[240] M. H. Heydari, M. R. Hooshmandasl and C. Cattani, Wavelets method for solving nonlinear stochastic It-Volterra integral equations. Georgian Mathematical Journal 27(1) (2020), 81-95. doi 10.1515/gmj-2018-0009

[241] M. H. Heydari, M. R. Hooshmandasl, F. Mohammadi and C. Cattani, Wavelets method for solving systems of nonlinear singular fractional Volterra integro-differential equations. Communications in Nonlinear Science and Numerical Simulation 19 (1) (2014), 37-48

[242] M. H. Heydari, M. R. Hooshmandasl, F. M. M. Ghaini and C. Cattani, Wavelets method for the time fractional diffusion-wave equation. Physics Letters A 379 (3) (2015), 71-76

[243] M. H. Heydari, M. R. Hooshmandasl, F. M. M Ghaini and C. Cattani, Wavelets method for solving fractional optimal control problems. Applied Mathematics and Computation 286 (2016), 139-154

[244] M. H. Heydari, M. R. Hooshmandasl, G. Barid Loghmani and C. Cattani, Wavelets Galerkin method for solving stochastic heat equation. International Journal of Computer Mathematics 93 (9) (2016), 1579-1596

[245] M. H. Heydari, M. R. Hooshmandasl and C. Cattani, A new operational matrix of fractional order integration for the Chebyshev wavelets and its application for nonlinear fractional Van der Pol oscillator equation. Proceedings-Mathematical Sciences 128 (2) (2018), 2642018

[246] I. I. Hirschman, Variation diminishing Hankel transform, Journal d'Analyse Mathématique 8(1) (1860), 307-336.

[247] MR Hooshmandasl, MH Heydari and C. Cattani, Numerical solution of fractional sub-diffusion and time-fractional diffusion-wave equations via fractional-order Legendre functions. The European Physical Journal Plus 131 (8) (2016), 268252016.

[248] M. Holschneider, Wavelets An Analysis Tool, Mathematical Monographs. Clarendon Press. Oxford. 1995, pp. 1-438.

[249] H. C. Hsin, T. Y. Sung, Y. S. Shieh and C. Cattani, A new texture synthesis algorithm based on wavelet packet tree. Mathematical Problems in Engineering, 2012, Article ID 305384, 12 pages.

[250] Z. J. Huang, G. H. Huang and L. Cheng, Medical image segmentation of blood vessels based on Clifford algebra and Voronoi diagram. Journal of Software 13(6) (2018), 361–373

[251] J. Hutchinson, Fractals and self-similarity. Indiana University Mathematics Journal 30 (1981), 713-747.

[252] B. R. Hunt and V. Y. Kaloshin, How projections affect the dimension spectrum of fractal measures. Nonlinearity, 10 (1997), 1031-1046.

[253] M. M. Ibrahim Mahmoud, A. Ben Mabrouk and M. H. A. Hashim, Wavelet multifractal models for transmembrane proteins- series. International Journal of Wavelets, , Multiresolution and Information Processing 14(6) (2016) 1650044 (36 pages).

[254] M. F. I. M. Idris, Z. A. Dahlan and H. K. Jusoff, The performance of two mothers wavelets in function approximation. Journal of Mathematics Research September, 1(2) (2009), 135-143.

[255] F. In and S. Kim, The hedge ratio and the empirical relationship between the stock and futures markets: A new approach using wavelet analysis. Journal of Business 79 (2006), 799-820.

[256] F. In and S. Kim, A note on the relationship between Fama-French risk factors and innovations of ICAPM state variables. Finance Research Letters 4 (2007), 165-171.

[257] F. In, S. Kim, V. Marisetty and R. Faff, Analysing the performance of managed funds using the wavelet multiscaling method. Review of Quantitative Finance and Accounting 31 (2008), 55-70.

[258] M. E. H. Ismail, The zeros of basic Bessel functions, the functions $J_{v+ax}(X)$ and associated orthogonal polynomials, Journal of Mathematical Analysis and Applications 86 (1982), 1-19.

[259] M. E. H. Ismail, The basic Bessel functions and polynomials. SIAM Journal on Mathematical Analysis, 12(3) (1981).

[260] M. E. H. Ismail and R. Zhang, q-Bessel functions and Rogers -Ramanujan type identities. 1508.06861, 2015-arXiv.

[261] F. H. Jackson, The application of basic numbers to Bessel's and Legendre's functions. Proceedings of the London Mathematical Society 2 (2) (1903-1904), 192-220.

[262] F. H. Jackson, On a q-definite integrals. Quarterly Journal of Pure and Applied Mathematics, 41 (1910), 193-203.

[263] F. H. Jackson, On basic double hypergeometric functions. Quarterly Journal of Mathematics, 13 (1951), 69-82.

[264] F. H. Jackson, Transformations of q-series. Messenger of Mathematics 39 (1910), 145-153.

[265] F. H. Jackson, On basic hypergeometric functions, Quarterly Journal of Mathematics (Oxford) 13 (1942), 69-82.

[266] F. H. Jackson, On q-functions and a certain difference operator. Transactions of the Royal Society of Edinburgh 46 (1909), 253-281.

[267] S. Jaffard, Construction et proprietes des bases d'ondelettes. Remarques sur la controlabilite exacte, These de doctorat de l'Ecole polytechnique, Palaiseau, 1989.

[268] S. Jaffard, Multifractal formalism for functions, Part 1: Results valid for all functions, SIAM Journal on Mathematical Analysis 28(4) (1997), 944-970.

[269] S. Jaffard, Multifractal formalism for functions, Part 2: Selfsimilar functions. SIAM Journal on Mathematical Analysis 28(4) (1997), 971-998.

[270] S. Jaffard, Pointwise smoothness, two-microlocalization and wavelet coefficients. Publications Mathématiques, 35 (1991), 155-168.

[271] S. Jaffard, Exposant de Hölder en des point donnés et coefficients d'ondelettes, CRAS. Paris, 308(I) (1989), 79-81.

[272] S. Jaffard, Construction of functions with prescribed Hölder and chirps exponents. Revista Matematica Iberoamericana 16(2) (2000), 331-349.

[273] S. Jaffard and B. Mandelbrot, Local regularity of non smooth wavelet expansions and application to the Polya function. Advances in Mathematics 120 (1996), 265-282.

[274] S. Jaffard and Y. Meyer, Pointwise behavior of functions. Memoirs of the American Mathematical Society, Vol. 123 N. 587 (1996).

[275] S. Jaffard and Y. Meyer, On the pointwise regularity of functions in critical Besov spaces. Journal of Functional Analysis 175 (2000), 415-434.

[276] V. K. Jain, Some expansions involving basic hypergeometric functions of two variables. Pacific Journal of Mathematics 91 (2) (1980), 349-361.

[277] YS Juang, HC Hsin, TY Sung and C. Cattani, Fast texture synthesis in adaptive wavelet packet trees. Mathematical Problems in Engineering 2013

[278] V. G. Kac and P. Cheung, Quantum calculus, Universitext, Springer-Verlag, New York, (2002).

[279] J. P. Kahane and P. G. Lemarie-Rieusset, Fourier Series and Wavelets. Gordon and Breach, 1996.

[280] G. Kerkyacharian and D. Picard, Density estimation in Besov spaces. Statistics and Probability Letters 13 (1992), 15–24.

[281] G. Kerkyacharian, D. Picard and K. Tribouley, Adaptive density estimation. Bernoulli 2 (1996), 229–247.

[282] J. Koekoek and R. Koekoek, A note on the q-derivative operator. arXiv:math/9908140v1, 27 Aug 1999.

[283] H. T. Koelink and R. F. Swarttouw, On the zeros of the Hahn-Exton q-Bessel function and associated q-Lommel polynomials. Journal of Mathematical Analysis and Applications 186(1994), 690-740.

[284] H. T. Koelink and W. Van Assche, Orthogonal polynomials and Laurent polynomials related to the Hahn-Exton q-Bessel function. Constructive Approximation 11(1995), 477-512.

[285] O. Le Cadet, Méthodes d'ondelettes pour la segmentation d'images: applications à l'imagerie mèdicale et au tatouage d'images, Thèse de doctorat en Mathématiques appliquées , Université Joseph Fourier Grenoble, 2004.

[286] S. Lehar, Clifford algebra: A visual introduction. A topnotch WordPress.com site, March 18, 2014.

[287] P. G. Lemarié-Rieusset, Analyses multi-résolutions non orthogonales, commutation entre projecteurs et dérivation et ondelettes vecteurs à divergence nulle, Revista Matematica Iberoamericana, 8(2) (1992), 221-236.

[288] Clifford Algebras and Spinors, Cambridge University Press, 2001.

[289] M. Lounsbery, Multiresolution Analysis for Surfaces of Arbitrary Topological Type. PhD thesis, University of Washington, 1994.

[290] M. Lounsbery, T. O. Derose and J. Warren, Multiresolution Surfaces of Arbitrary Topological Type. Department of Computer Science and Engineering 93-10-05, University of Washington, October 1993. Updated version available as 93-10-05b, January, 1994.

[291] M. Lounsbery, T. D. DeRose and J. Warren. Multiresolution analysis for vsurfaces of arbitrary topological type. ACM Transactions on Graphics 16(1) (1997), 34-73.

[292] N. G. Makarov, On the distortion of boundary sets under conformal mappings. Proceedings of the London Mathematical Society V 51 (1985), 369-384.

[293] S. G. Mallat, Multifrequency channel decompositions of images and wavelet models. IEEE Transactions on Acoustics, Speech and Signal Processing [see also IEEE Transactions on Signal Processing] 37 (12) (1989), 2091-2110.

[294] S. Mallat, A theory of multiresolution signal decomposition: The wavelet representation. IEEE Transactions on Pattern Analysis and Machine Intelligence 11 (1989), 674-693.

[295] S. Mallat, A Wavelet Tour of Signal Processing, 3rd ed. Academic Press, December 2008.

[296] M. Mansour, Generalized q-Bessel functions and its properties. Advances in Difference Equations, 121 (2013), 11 pages. https://doi.org/10.1186/1687-1847-2013-121.

[297] M. Mansour and M. M. Al-shomarani, New q-analogy of modified Bessel function and the quantum algebra. Journal of Computational Analysis and Applications 15(4) (2013), 655-664.

[298] R. Masson, Méthodes d'ondelettes en simulation numériques pour les problèmes elliptiques et de point selle. Thèse de Doctorat de l'université Paris 6. 1999.

[299] B. Mauroy, Hydrodynamique dans le poumon. Relations entre flux et geometrie. (In french). Thesis in Mathematics. Ecole normale supérieure de Cachan (2004).

[300] Y. Meyer and S. Jaffard, Méthodes d'ondelettes pour l'analyse des fonctions farctales et multifractales, Cours de DEA, ENS CACHAN, 1997.

[301] S Micula and C. Cattani, On a numerical method based on wavelets for Fredholm-Hammerstein integral equations of the second kind. Mathematical Methods in the Applied Sciences 41 (18) (2018), 9103-9115.

[302] P. Mignot, Une méthode d'estimation du spectre multifractal. CRAS. Paris, 327(I) (1998), 689-692.

[303] F. Mohammadi and C. Cattani, A generalized fractional-order Legendre wavelet Tau method for solving fractional differential equations. Journal of Computational and Applied Mathematics 339 (2018), 306-316.

[304] P. Monasse and V. Perrier, Construction d'ondelette sur l'intervalle pour la prise en compte des conditions aux limites. CRAS de Paris. T321. I. (1995), 1163-1169.

[305] A. F. Moreno, F. Marcellan and B. P. Osilenker, Estimates for polynomials orthogonal with respect to some Gegenbauer-Sobolev type inner product. Journal of Inequalities and Applications 3 (1999), 401-419.

[306] U. A. Müller, M. M. Dacorogna, R. B. Olsen, O. V. Pictet, M. Schwartz and C. Morgenegg, Statistical study of foreign exchange rates, empirical evidence of a price change scaling law, and intrady analysis. Journal of Banking and Finance 14 (1990), 1189-1208.

[307] E. Neuman, Inequalities involving modified Bessel functions of the first kind. Journal of Mathematical Analysis and Applications 171 (1991), 532-536.

[308] I. Y. Novikov and S. B. Stechkin, Basic wavelet theory. Russian Mathematical Surveys, 53(6) (1998), 1159-1231.

[309] E. Olivier, Analyse multifractale de fonctions continues. CRAS. Paris, 326(I) (1998), 1171-1174.

[310] M. A. Olshanetsky and V. B. K. Rogov, The modified q-Bessel functions and the q-Bessel Macdonald Functions, arXiv:q-alg/9509013v1, 11 Sep 1995.

[311] S. Oney, The Jackson Integral, International Journal of Mathematical Analysis and Applications, May 2007, pp. 1-10.

[312] R. S. Pathak, The wavelet transform of distributions. Tohoku Mathematical Journal 56 (2004), 411-421.

[313] R. S. Pathak, Continuity and inversion of the wavelet transform, Downloaded by [UOV University of Oviedo], 24 October 2014, pp. 85-93.

[314] R. S. Pathak and M. M. Dixit, Bessel wavelet transform on certain function and distribution spaces. Journal of Analysis and Applications 1 (2003), 65-83.

[315] R. S. Pathak and M. M. Dixit, Continuous and discrete Bessel wavelet transforms. Journal of Computational and Applied Mathematics, 160 (2003), 240-250.

[316] A. Pathak and R. K. Singh, Solution of system of generalized Abel's integral equation using Gegenbauer wavelets. International Journal of Computer Applications, 95(1) (2014), 1-4.

[317] R. S. Pathak, S. K. Upadhyay and R. S. Pandey, The Bessel wavelet convolution product, Rendiconti del Seminario Matematico Università e Politecnico di Torino 96(3) (2011), 267-279.

[318] D. P. Pena, Cauchy-Kowalevski extensions, Fueter's theorems and boundary values of special systems in Clifford analysis, A PhD thesis in Mathematics, Ghent University, 2008.

[319] D. B. Percival and A. T. Walden, Wavelet methods for time series analysis, Camridge University Press, NY, 2000.

[320] V. Perrier and C. Basdevant, Characterization of Besov spaces by means of the continuous wavelet transform. Reprint 1995.

[321] Y. B. Pesin, Dimension theory in dynamical systems, Contemporary Views and Applications, Chicago Lactures in Mathematics, 1998.

[322] A. Prasad, A. Mahato and M. M. Dixit, Continuity of the Bessel wavelet transform on certain Beurling-type function spaces. Journal of Inequalities and Applications 29 (2013), 1-9.

[323] A. Prasad, A. Mahato, V. Singh and M. M. Dixit, The continuous fractional Bessel wavelet transformation. Boundary Value Problems, 40 (2013), 1-16.

[324] M. Rahman, An addition theorem and some product formulas for q-Bessel functions. Canadian Journal of Mathematics 40(1988), 1203-1221.

[325] M. Rahman and U. Saeed, Gegenbauer wavelets operational matrix method for fractional differential equations. Journal of Korean Mathematical Society 52(5) (2015), 1069-1096.

[326] D. A. Rand, The singularity spectrum $f(\alpha)$ for Cookie-Cutters. Ergodic Theory and Dynamical Systems 9 (1989), 527-541.

[327] J. M. Redondo, Fractal models of density interfaces. IMA. Conf. Ser. 13, (ed. M. Frage, JCR, Hunt, JC Vassilicos) p. 353-370. Elsevier.

[328] R, Rhaiem, S. Ben Ammou and A. Ben Mabrouk, Wavelet estimation of systematic risk at different time scales, Application to French stock markets. International Journal of Applied Economics and Finance 1(2) (2007), 113-119.

[329] R. Riedi, An improuved multifractal formalism and self-similar measures. Journal of Mathematical Analysis and Applications 189 (2) (1995), 462-490.

[330] I. Rezgui and A. Ben Mabrouk, Some generalized q-Bessel type wavelets and associated transforms. Analysis in Theory and Applications (2017), 1-15.

[331] M. Riesz, Marcel, Clifford Numbers and Spinors. Editors, Bolinder, E. Folke and Lounesto, Pertti, Springer Netherlands, 1993.

[332] V. B. K. Rogov, q-Bessel-Macdonald functions, arXiv: math.QA/0010170.

[333] U. Saeed and M. Rehman, Hermite wavelet method for fractional delay differential equations. Journal of Difference Equations, (2014), Article ID 359093, 8 pages.

[334] R. K. Sakena and R. Kumar, Certain transformations of basic hypergeometric functions of two variables. Le Mathematiche, XLIV (1989), 333-344.

[335] A. Salem, Existance of the neutrix limit of the q-analogue of the incomplete gamma function and its derivatives. Applied Mathematics Letters 25 (2012), 363-368.

[336] A. Salem, The neutrix limit of the q-Gamma function and its derivatives. Applied Mathematics Letters, 23 (2010), 1262-1268.

[337] K. Sau, R. K. Basaka and A. Chanda, Image compression based on block truncation coding using Clifford algebra. Procedia Technology 10 (2013), 699–706.

[338] D. W. Scott, Multivariate density estimatimation, Theory, practice and visualisation. New York: Wiley, 1992.

[339] J. Shen, Refinement differential equations and wavelets. Methods and Applications of Analysis 5 (3) (1998), 283—316.

[340] J. V. B. Soares, J. J. G. Leandro, R. M. Cesar-Jr, H. F. Jelinek and M. J. Cree, Retinal vessel segmentation using the 2-D Morlet wavelet and supervised classification. IEEE 2006. ArXiv:cs.Cv/0510001 v2. 04/11/2006.

[341] F. Sommen and H. De Schepper, Introductory Clifford analysis. Operator Theory (2015), 1–27.

[342] S. Stankovic, L. Stankovic and I. Orovic, Compressive Sensing Approach in the Hermite Transform Domain. Mathematical Problems in Engineering, (2015), 9 pages, Article ID 286590, https://doi.org/10.1155/2015/286590.

[343] E. M. Stein and G. Weiss, Introductions to Fourier analysis on educlidean spaces, New Jersey, Princeton University Press, 1971, pp. 1-154.

[344] R. F. Swarttouw, The Hahn-Exton q-Besel functions, Ph. D. Thesis, Deft University (1992).

[345] M. Thuillard, Wavelets in soft computing, world scientific series in Robotics and intelligent systems, vol(25), pp. 1-246.

[346] H. Triebel, Theory of function spaces II. Birkhauser Verlag. Basel. Boston. Berlin, 1992.

[347] K. Trimeche, Generalized Harmonic Analysis and wavelet Packets. Gordon and Breach. Science Publishers, Amesterdam etc. (2001).

[348] M. Unser (1997). Ten good reasons for using spline wavelets (PDF). Proc. SPIE Vol. 3169, Wavelets Applications in Signal and Image Processing V: 422-431. Retrieved 21 December 2014.

[349] M. Upadhyay, q-fractional differentiation and basic hypergeometric transformation. Anals Polinici Math, 25 (1971), 113-128.

[350] S. K Upadhyaya, R. N. Yadavb and L. Debnathc, On continuous Bessel wavelet transformation associated with the Hankel-Hausdorff operator. Integral Transforms and Special Functions, 23(5) (2012), 315-323.

[351] C. Valens, A really friendly guide to wavelets, UNM computer science, 1999, 1-19.

[352] Vannucci M. (1995), Nonparametric density estimation using wavelets. Technical report, 95-26, ISDS, Duke university.

[353] Vannucci M. (1996), On the application of wavelets in statistics. PhD thesis. Dipartimento di statistica. Universita di Florence. Italia.

[354] S. G. Venkatesh, S. K. Ayyaswamy, S. Raja Balachandar and K. Kannan, Wavelet solution for class of nonlinear integro-differential equations. Indian Journal of Science and Technology, 6(6) (2013), 4670-4677.

[355] Vidakovic B. (1999), Statistical modeling by wavelets. Wiley, New York.

[356] N. Vieira, Cauchy-Kovalevskaya extension theorem in fractional Clifford analysis. Complex Analysis and Operator Theory, 9(5) (2015), 1089-1109.

[357] M. Volker, Lectures On Constructive Approximation: Fourier, Spline, and wavelet methods on the real line, the sphere, and the Ball, Applied and Numerical Harmonic Analysis, Birkhäuser, 2013.

[358] D. F. Walnut, An Introduction to Wavelet analysis, Applied and Numerical Harmonic Analysis, Birkhauser, Boston, Basel, Berlin, 2002.

[359] Walter G. G. (1994), Wavelets and other orthogonal systems with application. CRC Press, Boca, Raton, FL.

[360] L. Wang, R. Hu, J. Zhang, and Y. Ma, On the Vortex Detection Method Using Continuous Wavelet Transform with Application to Propeller Wake Analysis. Mathematical Problems in Engineering (2015), Article ID 242917, 9 pages.

[361] J. Wang and H. Peng, Constructing fuzzy wavelet network modeling. International Journal of Information Technology, 11(6) (2005) , 68-74.

[362] G. N. Watson, A Treatise on the theory of Bessel functions, Combridge University Press. Combridge, Second edition (1966).

[363] I. Weinreich, A construction of C^1-wavelets on the two-dimensional sphere. Applied and Computational Harmonic Analysis, 10 (2001), 1-26.

[364] J. Winkler, A uniqueness theorem for monogenic functions. Annales Academi Scientiarum Fennic, Series A. I. Mathematica, 18 (1993), 105-116.

[365] P. Wojtaszczyk, *A mathematical introduction to wavelets*, Cambridge University Press, 1997.

[366] D. W. Wu, Asymptotic normality of the multiscale wavelets density estimate. Communications in Statistics, Theory and Methods, 25(9) (1996), 1957-1970.

[367] Y. Xiaojun, Y. Zhao, D. Baleanu, C. Cattani and D. F. Cheng, Local fractional discrete wavelet transform for solving signals on Cantor sets. Mathematical Problems in Engineering 2013. DOI: 10.1155/2013/560932.

[368] Y. Xiaojun, D. Baleanu, H. M. Srivastava and J. A. T. Machado, On Local Fractional Continuous Wavelet Transform 2013

[369] Y. Xie and Y. Zhang, A wavelet network model for short-term traffic volume forecasting. Journal of Intelligent Transportation Systems: Technology, Planning, and Operations 10(3) (2006), 141-150.

[370] X. Xiong, X. Zhang, W. Zhang and C. Li, Wavelet-based beta estimation of China stock market, Proceedings of 4th. International Conference on Machine Learning and Cybernetic, Guangzhou. IEEE: 0-7803-9091-1, 18-21 August, 2005, pp. 3501-3505.

[371] H. Yamada, Wavelet-based beta estimation and Japanese industrial stock prices. Applied Economics Letters 12 (2005), 85-88.

[372] M. Zemni, M. Jallouli, A. Ben Mabrouk and M. A. Mahjoub, Explicit Haar-Schauder multiwavelet filters and algorithms. Part II: Relative entropy-based estimation for optimal modeling of biomedical signals. International Journal of Wavelets, Multiresolution and Information Processing 17(5) (2019), 1950038.

[373] M. Zemni, M. Jallouli, A. Ben Mabrouk and M. A. Mahjoub, ECG Signal Processing with Haar-Schauder Multiwavelet. Proceedings of the 9th International Conference on Information Systems and Technologies - ICIST 2019. doi:10.1145/3361570.3361611

[374] Ch. Zhang, Transformations de q-Borel-Laplace au moyen de la fonction Theta de Jacobi. Comptes rendus de l'Académie des Sciences 331 (2000), 31-34.

[375] Ch. Zhang, Sur les fonctions q-Bessel de Jackson. Journal of Approximations Theory 122 (2003), 208-223.

[376] Ch. Zhang, Sur la fonction q-Gamma de Jackson. Aequationes Mathematicae 62 (2001), 60-78.

Index